DESIGNING LANDSCAPE ARCHITECTURAL EDUCATION

No single project or endeavour is immune to the issues that the climate crisis brings. The climate crisis encompasses a broad register of "symptoms" – increased global temperatures and sea-level rise, droughts and extreme bushfire events, salinification and desertification of fertile land, and the list goes on. It reveals and amplifies complex causal relationships that are inherently present and traverse scales, sectors and communities divulging a range of impacts and inequalities. This publication asks designers and academic practitioners to describe their own work through an ecological lens, and then to articulate design approaches for developing new practices in landscape architecture teaching.

Designing Landscape Architectural Education: Studio Ecologies for Unpredictable Futures, the Landscape Architecture Design Studio Companion, serves as a resource for academic practitioners in the preparation and delivery of "design-research studios" and students seeking guidance for design methodologies as a part of their landscape architectural education. It draws on the manifold issues of the climate crisis as a set of drivers to examine the utilisation of a range of innovative design approaches to address the current and future priorities of the discipline.

The landscape architecture discipline is evolving rapidly to respond to both a broadening and intensification of changes in the environmental, social and political conditions. These changing conditions require innovation that extend the core competencies of landscape architects. This book addresses two fundamental questions – what are the design competencies required of landscape architects to equip them to deal with the complexities brought forth by contemporary society, and as a result, how could we design the future design studio?

Rosalea Monacella is a faculty member of the Landscape Architecture Program at the Harvard University Graduate School of Design. Her expertise is in the careful indexing and shifting of dynamic resource flows that inform the landscape of the city. Her design research practice explores the notion of the "thickened ground" through a careful and rigorous investigation of an expanded ecology of economic, ecological and social systems that shape the metabolic and material flows of the city. Speculating on alternative near-future cities and how they might respond to climate change, changing resource flows and ecologies of energy.

Bridget Keane is a lecturer in the Landscape Architecture programme at RMIT University. As a landscape architect/academic, her research and practice are concerned with landscape as a dynamic material system and considers ways in which global systems impact local landscapes through speculating on alternate futures of how we might live in response to issues of climate change, ecologies of waste and the effects of extractive industries.

DESIGNING LANDSCAPE ARCHITECTURAL EDUCATION

Studio Ecologies for Unpredictable Futures

Edited by Rosalea Monacella and Bridget Keane

Routledge
Taylor & Francis Group

LONDON AND NEW YORK

Cover image: Siena Scarff Design.

First published 2023
by Routledge
4 Park Square, Milton Park, Abingdon, Oxon OX14 4RN

and by Routledge
605 Third Avenue, New York, NY 10158

Routledge is an imprint of the Taylor & Francis Group, an informa business

© 2023 selection and editorial matter, Rosalea Monacella and Bridget Keane; individual chapters, the contributors

The right of Rosalea Monacella and Bridget Keane to be identified as the authors of the editorial material, and of the authors for their individual chapters, has been asserted in accordance with sections 77 and 78 of the Copyright, Designs and Patents Act 1988.

British Library Cataloguing-in-Publication Data
A catalogue record for this book is available from the British Library

Library of Congress Cataloging-in-Publication Data
A catalog record has been requested for this book

ISBN: 9780367703660 (hbk)
ISBN: 9780367703653 (pbk)
ISBN: 9781003145905 (ebk)

DOI: 10.4324/9781003145905

Typeset in Bembo
by Deanta Global Publishing Services, Chennai, India

Every effort has been made to contact copyright-holders. Please advise the publisher of any errors or omissions, and these will be corrected in subsequent editions.

CONTENTS

CONTRIBUTORS

Ana Abram is a licenced landscape architect, and lecturer (teaching) at the Bartlett, UCL. Her current and recent appointments include MA/MLA co-directorship, teaching Studio 1 and RC11 on B-Pro MArch UD programmes, and leading technology and environment modules. She is a co-founder of the award-winning Amphibious Lab practice, which deals with innovative approaches like Landscape 3D printing. She lectures and exhibits her work at internationally acclaimed venues, like Venice Biennale, Harvard GSD, CMU, AIA, AA, UVA. Ana also has a rich track record of built projects.

Sean Burkholder is the Andrew Gordon Assistant Professor of Landscape Architecture at the University of Pennsylvania's Weitzman School of Design. His work bridges the seemingly incommensurable subjects of environmental sensing and modelling that includes working with ecologists and engineers on large coastal infrastructure projects and the exploration of new methods of knowledge creation that call into question Western science as our primary referent for progress and agency. Burkholder is co-founder of the research and design practice Proof Projects and a member of the Dredge Research Collaborative.

Bradley Cantrell is a landscape architect and scholar whose work focuses on the role of computation and media in environmental and ecological design. He is Chair and Professor of Landscape Architecture at the University of Virginia, School of Architecture. His work in Louisiana over the past decade points to a series of methodologies that develop modes of modelling, simulation and embedded computation that express and engage the complexity of overlapping physical, cultural and economic systems.

Leena Cho is an Assistant Professor in Landscape Architecture, and Co-Director of the Arctic Design Group, at the University of Virginia. Her research focuses on material agencies of Arctic landscapes and field-based scientific practices that produce landscapes in the Arctic. Integrating wide-ranging disciplines, her work spans from Alaska to Siberia and has been funded by numerous organisations, including the US Embassy in Iceland, the World Bank, the National Science Foundation and the Anchorage Museum.

Elisa Cristiana Cattaneo researches experimental ecological design and its theoretical implications, considering the design as a weak field, evolvable and renewable through new transdisciplinary approaches between art and science. Holding a Doctorate in Architectural and Urban Design, she was a visiting scholar at MIT and at the Graduate School of Design, Harvard. She is founder of independent research agency Weakcircus, a research platform on contemporary design. She has been an active designer and has been teaching at Politecnico of Turin.

Dana Cupkova holds an Associate Professorship at Carnegie Mellon University's School of Architecture. She directs EPIPHYTE Lab, an interdisciplinary design and research collaborative engaging issues of environmental ethics. Cuokova's design work engages the built environment at the intersection of design-space and ecology, engaging computational methods, thermodynamic processes and geometric logics of material formations relative to embodied energy and landscape behaviour.

Kate Davies and **Liam Young** (Unknown Fields, UK/AU) is a nomadic design research studio directed by Kate Davies and Liam Young. They venture out on expeditions into the shadows cast by the contemporary city, to uncover the industrial ecologies and precarious wilderness its technology and culture set in motion. These distant landscapes – the iconic and the ignored, the excavated, irradiated and the pristine, are embedded in global systems that connect them in surprising and complicated ways to our everyday lives. Unknown Fields makes provocative objects and films from this expedition work, exploring the dispersed narratives that coalesce to form a contemporary city. They chronicle their expeditions in a book series titled *Unknown Fields: Tales from the Dark Side of the City* and their work has received a BAFTA nomination and has been published extensively including platforms such as the Guardian, BBC, Wired and New Scientist. Their projects have been collected by institutions such as The New York Metropolitan Museum, Victoria and Albert Museum, M+ in Hong Kong and the Museum of Applied Arts and Sciences in Sydney.

Jennifer Deger is a Professor in Digital Humanities at Charles Darwin University. She is a founding member of the north Australian arts collective, Miyarrka Media (Phone & Spear: a Yuta Anthropology, Goldsmiths Press 2019, winner of the Gregory Bateson Book Prize) and co-editor and co-curator of *Feral Atlas: The More-Than-Human Anthropocene* (Stanford University Press, 2020). Her work pursues connections between digital and environmental ecologies through transdisciplinary collaboration and intermedial experiment.

Anya Domlesky is an urban designer currently working at SWA Group as Director of Research. She runs XL Lab, the firm's innovation lab undertaking practice-based research. As a designer, she has worked largely on issues around urbanisation, coasts and water. She holds an MLA from Harvard University Graduate School of Design and an MArch II from McGill University.

Craig Douglas is a landscape architect and scholar whose work focuses on innovative techniques and methodologies that explore the agency of representation in landscape architectural design. He is Assistant Professor at Harvard University, Graduate School of Design. His work explores the landscape as a dynamic material process in a constant state of flux through analytical and conceptual approaches integrating drawing, modelling, simulation and sensing to make visible and reconstitute the landscape as a complex temporal and material manifold of differential space.

Rosetta S Elkin is a landscape architect, educator and practitioner known for her close reading of plant life through a range of media from site-specific installations, international exhibits and field-based research. Among her awards, Elkin is the recipient of the American Academy of Rome Prize and the Harvard University Climate Change Award. She is author and co-author of articles, book chapters and monographs, including *Tiny Taxonomy* (Actar, 2017) and *Dryland: Afforestation and the Politics of Plant Life* (Minnesota, 2022).

Michael Ezban is an architect, landscape designer and educator. His work explores multispecies urbanism, and the design of landscapes and buildings that mediate relations between humans and other animals. His first book, *Aquaculture Landscapes: Fish Farms and the Public Realm* (Routledge, 2019), is focused on the landscape architecture of farms, reefs, parks and cities that are designed to entwine the lives of fish and humans. Ezban is a Clinical Assistant Professor at the School of Architecture, Planning & Preservation, University of Maryland.

Billy Fleming is the Wilks Family Director of the Ian L McHarg Center in the Weitzman School of Design, Co-director of the "climate + community project", and a former senior fellow with Data for Progress. Fleming is co-editor of *A Blueprint for Coastal Adaptation* (Island Press, 2021) and a co-editor and co-curator of the book, and now internationally travelling exhibit, *Design with Nature Now* (Lincoln, 2019). He is also the lead author of *The 2100 Project: An Atlas for the Green New Deal*, *Field Notes toward an Internationalist Green New Deal* and *The Indivisible Guide*.

Zach Fluker is the Co-founder of the architectural practice ao-ft. Prior to becoming an architect, he studied industrial design and worked as a cabinet maker. He is a graduate of both Emily Carr University of Art and Design and the Architectural Association. His research into interfacing digital and physical environments and computational fabrication has led him to collaborate with several practices in the UK and Canada. In parallel to practice, Fluker is a Lecturer (Teaching) at the Bartlett, UCL, in both the postgraduate and undergraduate programmes.

Formafantasma (Andrea Trimarchi and Simone Farresin) is a research-based design studio investigating the ecological, historical, political and social forces shaping the discipline of design today. Whether designing for a client or developing self-initiated projects, the studio applies the same rigorous attention to context, processes and details. Formafantasma's analytical nature translates into meticulous visual outcomes, products and strategies.

Pia Fricker, Vice Head of the Department of Architecture, holds the Professorship for Computational Methodologies in Landscape Architecture and Urbanism at Aalto University. Her interdisciplinary research is focusing on defining novel ways technologies are fusing boundaries and transforming traditional design methods and workflows. Currently, she is exploring new types of sustainable urban development, developing future-oriented strategies for global challenges like sea-level rise, densification and loss of biodiversity, supported by cutting-edge computational design thinking.

Teresa Galí-Izard is a Professor of Landscape Architecture at the ETH Zürich and Director of the Master of Science in Landscape Architecture programme. She is a principal of ARQUITECTURA AGRONOMIA, a landscape architecture firm based in Barcelona. Galí-Izard is the author of *The Same Landscapes: Ideas and Interpretations* and is trained as an agronomist at Polytechnic University of Catalonia.

Rania Ghosn is Founding Partner of Design Earth with El Hadi Jazairy and Associate Professor of architecture and urbanism at the Massachusetts Institute of Technology. Her practice engages design as a speculative medium for making visible and public the geographies of the climate crisis. She is founding editor of the journal *New Geographies*, editor of *Landscapes of Energy* (2009) and co-author of *Geographies of Trash* (2015), *Geostories: Another Architecture for the Environment* (2nd ed., 2020) and *The Planet After Geoengineering* (2021).

Luke Harris, **Cara Turett** and **Bonnie Kate Walker** are Research Associates in the Institute for Landscape and Urban Studies at ETH Zürich. They hold MLAs from the University of Virginia and worked as landscape designers in New York City prior to joining ETH. They co-founded the landscape design and research collective Office of Living Things.

Zaneta Hong is an Assistant Professor at Cornell University. Prior to Cornell, she worked at Harvard GSD, University of Virginia and UT Austin. Her work has been recognised by the Graham Foundation and Environmental Design Research Association and she was recently awarded the 2018–19 Garden Club of America Rome Prize, MacDowell Fellowship and Certificate of Teaching Excellence from the Harvard Derek Bok Center. In addition to teaching, Zaneta is Co-Director of Alterior Office and a Research Design Consultant for GA Collaborative.

Justine Holzman is a landscape researcher and designer, and member of the Dredge Research Collaborative. Currently training as a historian of science at Princeton University, her doctoral studies focus on the history of ecology and environmental history in the 20th-century United States. Holzman was previously an assistant professor of landscape architecture at the Daniels Faculty of Architecture, Landscape, and Design at the University of Toronto, where she taught studio and seminar courses in the graduate and undergraduate programmes.

Sheng-Yang (William) Huang is a PhD Candidate and a B-Pro Tutor at the Bartlett, UCL. He teaches design theory and computing skills for RC18 and taught for Cheng Shiu University as an Adjunct Lecturer. Huang holds a BArch (THU), a MArch II (NCTU) and a MRes Architecture and Digital Theory with Distinction (UCL). He is currently dedicated to the studies of generative architectural design with deep neural networks as the creative instruments of the collective mind.

Jane Mah Hutton is a landscape architect, teaching at the University of Waterloo School of Architecture. Her research looks at the extended material flows of common construction materials. Recent books include *Reciprocal Landscapes: Stories of Material Movements, Landscript 5: Material Culture – Assembling and Disassembling Landscapes*, and *Wood Urbanism: From the Molecular to the Territorial*, co-edited with Daniel Ibañez and Kiel Moe. Her research has been supported by the Robert and Stephanie Olmsted Fellowship at Macdowell, a Research Fellowship at the Canadian Centre for Architecture, and an EDRA Great Places Book Award.

Bridget Keane is a lecturer in the Landscape Architecture programme at RMIT University. As a landscape architect/academic, her research and practice are concerned with landscape as a dynamic material system and considers ways in which global systems impact local landscapes through speculating on alternate futures of how we might live in response to issues of climate change, ecologies of waste and the effects of extractive industries.

Jesse M Keenan is an Associate Professor of Real Estate within the faculty of the School of Architecture at Tulane University. As a globally recognised thought leader, Keenan's research focuses on the intersection of climate change adaptation and the built environment, including aspects of design, engineering, regulation, planning and financing. Keenan holds degrees in the law (JD, LLM) and science of the built environment (MSc), including a PhD from the Delft University of Technology.

Nina-Marie Lister is Professor and Graduate Director of Urban & Regional Planning at Ryerson University where she founded and directs the Ecological

Design Lab. She is also Principal of PLANDFORM, a creative studio practice in ecology, planning and urbanism. Lister is co-editor of *The Ecosystem Approach* (Columbia, 2008), *Projective Ecologies* (Actar, 2014, 2020) and author of more than 60 scholarly and professional publications. Her work as a planner and landscape ecologist connects people to nature in cities, through green infrastructure design for climate resilience, biodiversity and human wellbeing.

Alexis Liu is a designer, researcher and lecturer working across the fields of landscape architecture, urbanism and narrative environments. She graduated with an MLA from RMIT University in 2016. She worked as a landscape architect in Aspect Studios and a lecturer in narrative environment design at SIVA. She is currently a lecturer at the University of Greenwich where she is also a Landscape Urbanism PhD researcher exploring the urbanisation of rural-urban fringe areas.

Enriqueta Llabres-Valls is a Lecturer in Architecture and Urbanism at the Bartlett School of Architecture, UCL, and Mittelsten Scheid Guest Professor in the Faculty of Architecture and Engineer at Wuppertal University. Her research interest expands from the studies of the built environment to local development concepts such as environmental policy and regulation, globalisation and inequalities. She leads, with Zach Fluker, Research Cluster 18 in the MArch Urban Design B-Pro at the Bartlett. She co-founded, with Eduardo Rico, the awarded practice Relational Urbanism. Relational Urbanism continues its mission as Relational Urbanism Lab under the umbrella of Llabres Tabony Architects co-founded with Aiman Tabony in 2017.

Fadi Masoud is an Assistant Professor at the University of Toronto's Daniels Faculty of Architecture, Landscape and Design, and Director of the Centre for Landscape Research. He is the author of *Terra-Sorta-Firma: Reclaiming the Littoral Gradient* and sits on Waterfront Toronto's design review panel.

James Melsom is Senior Lecturer in Landscape Architecture at UTS, Sydney, Australia and co-founder of LANDSKIP, and educator, researcher and practitioner in landscape architecture and spatial technologies. As a registered Swiss Landscape Architect (BSLA), he specialises in landscape design, sensing, scanning and the analysis of dynamic territorial systems. He is the author of many publications including book chapters, peer-reviewed conference and journal articles on the integration of modelling, sensing and representation within the spatial design fields.

Marc Miller is interested in relationships between environments, speculation and public perception. His research examines landscape imagery and worldbuilding, focusing on how television was used by landscape architects and other designers in the past and exploring how it might be used as a medium for fabulations of future landscapes.

Rosalea Monacella is a faculty member of the Department of Landscape Architecture at the Harvard University Graduate School of Design. Her expertise is in the careful indexing and shifting of dynamic resource flows that inform the landscape of the city. Her design research practice explores the notion of the "thickened ground" through a careful and rigorous investigation of an expanded ecology of economic, ecological and social systems that shape the metabolic and material flows of the city. Speculating on alternative near-future cities and how they might respond to climate change, changing resource flows and ecologies of energy.

Kate Orff, FASLA, is Professor and Director of the Urban Design Program (MSAUD) at Columbia University, and founder of SCAPE landscape architecture.

Roberto Pasini, PhD, is a Professor of Landscape Architecture at the Alma Mater Studiorum Università di Bologna. He has previously taught at the Universidad de Monterrey and at Harvard University and served as a juror for the EU Prize for Cultural Heritage. He is author of *Landscape Paradigms* (Springer, 2019) and *Symbiotic Matorral* (Libria, 2019). Pasini is the recipient of the Premio di Architettura HC Andersen from the Accademia Nazionale di San Luca.

Maj Plamenitas is an inventor, researcher and designer. His recent appointments include Co-directorship of the Landscape Architecture programme at the Bartlett UCL, directorship of Linkscale and Co-founding of the Amphibious Lab. Plamenitas is best known for his pioneering work in the field of Cross Scale Design (2011– ongoing) and Growing Islands and Resilient Shorelines with In-situ Forces and Materials (2010–ongoing) for which he received a Holcim Foundation Award. He lectures and exhibits internationally in venues including the Venice Biennale, Royal Academy of Arts, AA, Harvard GSD and CMU, among others.

Jose Alfredo Ramírez is an architect, Co-director of AA Groundlab and Co-director of the Landscape Urbanism MArch/MSc Postgraduate Program at the AA. At Groundlab, he leads research design projects on how the built environment impacts climate change, landscape design and urbanisation in the Global South, and has led the development of projects at the junction of architecture, landscape and urbanism in a variety of contexts. Ramírez has published and lectured worldwide on landscape urbanism as well as on the work of Groundlab.

Chris Reed is Founding Director of Stoss. He is recognised internationally as a leading voice in the transformation of landscapes and cities, and he works alternately as a researcher, strategist, teacher, designer and adviser. Reed is particularly interested in the relationships between ecology and landscape and infrastructure, social spaces and cities. His work collectively includes urban revitalisation initiatives, climate resilience efforts, adaptations of former industrial sites, vibrant public spaces that cultivate a diversity of social uses and numerous landscape installations. He has overseen riverfront work in Green Bay, Little Rock, Milwaukee, Louisville, Cape

Cod, Dallas and Grand Rapids. Reed is a recipient of the 2012 Cooper-Hewitt National Design Award in Landscape Architecture, a Fellow of the American Society of Landscape Architects, and the 2017 Mercedes T. Bass Landscape Architect in Residence at the American Academy in Rome.

Clara Olóriz Sanjuán is Co-director of AA Groundlab and studio master at AA MArch/MSc Landscape Urbanism Program. She has taught design and theory courses at ETSAUN, Leeds Beckett, Tec de Monterrey and other universities, and lectured in professional and academic venues worldwide. Her PhD, developed at the AA and ETSAUN, studies the relationship between technology and architecture. Most recently she received a Graham Foundation award for her book *Landscape as Territory* (2019).

Elise Shelley is an Assistant Professor at the University of Toronto, where she teaches design studio and site technology courses in both the Architecture and Landscape Architecture programmes. As a practitioner in architecture and landscape architecture in Toronto, Shelley's projects and research focus on the interdisciplinary nature of urban public space.

Peter Del Tredici is a botanist specialising in the growth and development of trees. He worked at the Arnold Arboretum of Harvard University for 35 years as a propagator, a curator, an editor of *Arnoldia*, a research scientist and director of Living Collections. He also taught in the Department of Landscape Architecture at the Harvard Graduate School of Design for 24 years. His botanical interests include the ecology of hemlocks and the natural and cultural history of the Ginkgo tree. His widely acclaimed book, *Wild Urban Plants of the Northeast: A Field Guide*, catalogues the spontaneous vegetation that flourishes in cities and makes the case that it can improve the quality of urban life for all their inhabitants.

Paola Viganò is Professor in Urban Theory and Urban Design at the EPFL (Lausanne) and at IUAV (Venice). Viganò founded the architectural study studio with Bernardo Secchi in 1990. Together, they participated in the elaboration of the urban master plans of many European cities. She leads the laboratory of Urbanism (Lab-U) at the École Polytechnique Fédérale de Lausanne (EPFL). Her research focuses on the study of new forms of urbanisation and explores the concept of cities as renewable resources. She analyses the concept of the horizontal metropolis as a vision for planetary urbanisation.

Charles Waldheim is John E Irving Professor of Landscape Architecture at the Harvard University Graduate School of Design, where he directs the Office for Urbanization and hosts the *Future of the American City* podcast. He also serves as the Ruettgers Curator of Landscape at the Isabella Stewart Gardner Museum in Boston, where he convenes *The Larger (Landscape) Conversation* series. Waldheim is recipient of the Rome Prize Fellowship from the American Academy in Rome, the

Visiting Scholar Research Fellowship at the Canadian Centre for Architecture, the Sanders Fellowship at the University of Michigan and the Cullinan Chair at Rice University. He has been a visiting scholar at the Architectural Association School of Architecture in London and the Bauhaus in Dessau, Germany.

Ed Wall is Head of Landscape Architecture and Urbanism at the University of Greenwich, Visiting Professor at Politecnico di Milano, and, in 2017, was City of Vienna Visiting Professor of Urban Culture, Public Space and the Future – Urban Equity and the Global Agenda (TU Vienna). His research explores the intersection of environmental justice, urban public space and design experimentation. Wall has written widely, recently guest editing *Architectural Design: The Landscapists* (2020) and co-editing *Landscape and Agency: Critical Essays* (2017) and *Landscape Citizenships* (2022). He has a PhD from the Cities Programme of the London School of Economics.

Jane Wolff is an Associate Professor at the University of Toronto's Daniels Faculty of Architecture, Landscape and Design, and the author *of Bay Lexicon and Delta Primer: A Field Guide to the California Delta.*

FOREWORD: THE SPACE OF THE STUDIO

It has become fashionable to deride studio teaching as an anachronism at best, or a perverse form of ritualised abuse at worst. The current preoccupations with societal conditions, climate change and economic inequity which are inflecting studio briefs across the design disciplines represent the most recent evidence of studio teaching's engagement with the *instrumental externalities* of the wider world. Rather than a threat to the purported autonomy of the design disciplines, these conditions and contexts inform studio teaching, injecting it with specificity, context and resistance. However, it would be a misreading to understand the design studio as engaging with these conditions as a kind of simulacra of professional practice. Quite the contrary, rather than emulating professional practice, studio pedagogy is best understood as the unique form and format of knowledge production in the design disciplines through design research. It is in this understanding of the space of the studio that Rosalea Monacella and Bridget Keane and their three dozen contributors gathered here have offered a new reading of the studio as a venue for the revision of landscape architecture pedagogy, and thereby, the field itself.

The founding of landscape architecture as an academic discipline came only a few decades after the consolidation of architecture itself as a professional field of study in univeristies. The first coursework and academic programmes in architecture emerged in the second half of the 19th century, as American universities moved towards specialisation in the sciences. Many of these schools referred to themselves as schools of architecture, applied arts or fine arts. A small number developed identities composed of laundry lists of their constituent professional fields including architecture, landscape architecture and planning. By the mid-1930s, a third alternative identity for the modern American school of architecture was formulated around the European conception of *design*. Design was first formulated in the academic context of new schools of design in London and the

US in the second half of the 20th century. These schools were explicitly intended to introduce women, immigrants and the working classes to the applied arts adjacent to architecture. In this sense, they were offered an alternative to architecture schools based in the fine arts and in universities which largely refused to admit women or persons of colour to their degree programmes. Design was intended as a progressive social and professional métier which by transcending the traditional professional identities, was both more accepting of a diverse student body while simultaneously training students in the ability to work across various professions. The *design studio* in this context was a much more open and pliable space in contrast to the longstanding Beaux-Arts tradition of atelier affiliations found in the earliest architecture programmes. Rather than an entrée to, or a mirror of, the space of practice, the design studio was from the outset an altogether different space.

Over the past century, the design studio has proven to be an effective and adaptable venue for peer review, knowledge production and design pedagogy. While it is true that its exceptional character can often make the studio seem an ill-fitted format to the modern research university, on the contrary, we see the resurgence of interest in design research across campuses. The design studio, in this sense, can be seen to anticipate the current interests on collaborative and hands-on learning, flipped classrooms and applied research engaging with the wider world around. Monacella and Keane's proposition of *Studio Ecologies* offers a compelling account of how we might rethink the foundations of the landscape architecture discipline through the very venue of the design studio itself. The title of this volume, *Studio Ecologies*, is at once ambitious and aspirational, yet it is ultimately modest in relation to the scope of the book's claims. Monacella and Keane's introduction, orchestration and ordering of the contributions reveal the much larger and more urgent challenge of reconstructing a studio praxis to substantially revise the assumptions upon which landscape architecture education rests. This is ultimately a manifesto, albeit a collective one, in which Monacella and Keane and their co-conspirators propose nothing less than an agenda for reforming landscape studio teaching and therefore the field itself. This is timely and urgent work, and we would do well to reflect on its implications for both teaching and research as our field reconstructs its commitments to the wider world outside the studio.

Charles Waldheim

PREFACE

Designing Landscape Architectural Education: Studio Ecologies for Unpredictable Futures is intended to accomplish two primary purposes. The first is to serve as a resource for academic practitioners in preparing and delivering design studio courses. The second is to serve students seeking guidance and insight into design methodologies as a part of their landscape architectural education. The book draws on the manifold issues of the climate crisis. These are framed as a set of compelling drivers for examining a range of ecological design methods and techniques from around the world.

As editors located in distinctly different landscapes and institutions in the world, we embraced a continuous and productive dialogue in crafting this work. This dialogue continues to be expressed through the multiplicity of positions and approaches captured in the following pages. This wealth of perspectives offers a productive difference of approaches, rather than a singular mediated canon of references.

Landscape architecture as a discipline is evolving rapidly as it responds to both broadening and intensifying changes in environmental, social and political conditions. These changing conditions require development and innovation in the core competencies of landscape architects. This book addresses two fundamental questions. First, what are the skills required of landscape architects to equip them to deal with the complexities brought forth by the climate crisis? Then comes a critical consequential question: how can we design the education of future practitioners?

The work of this publication focuses on the symptoms and systems of the climate crisis. It asks designers and academic practitioners to describe their own work through an ecological lens, and then to articulate design approaches for developing new practices in landscape architecture teaching and research. A focus on design studio pedagogies is foregrounded here, rather than giving prominence to scientific or engineering-based approaches. That conscious choice recognises that design is understood to offer ways to intertwine projects, stakeholders, fieldwork, material processes and ever-changing future scenarios.

For the most part, this book was written during the first year of the global coronavirus pandemic. A multitude of stories, insights, analyses and concerns have attached themselves to our collective experience of the pandemic. One striking and persistent effect of the pandemic has been to evidence drastic systemic inequities across the globe – not least some of the very inequities that are frontline effects of the climate crisis. We now have a clear view of those who are the most vulnerable, and that view brings into more direct light the global, collective, cross-disciplinary pursuit required to address the climate crisis.

Our aim is to demonstrate a re-evaluation of professional competencies that are modified to adapt, flexibly and practically, to the complexities of the climate crisis and its unpredictable futures. The works are organised within the sections with the understanding that no holistic view is possible. That being so, the intention is not to construct a unifying position or to integrate disparate approaches. Rather our determination has been to establish threads of inquiry and to stimulate generative and multiple gathering. These guiding objectives mean the selection is thematic rather than comprehensive. Overall, the contributions reflect situations in the Global North: an acknowledgement that these places and institutions have a different type of work to do given shared legacies of imperial/colonial action that are collectively responsible for the climate crisis.

As a guide to reading, each section is introduced by an allegorical interlude. These seek to locate the editors and authors in place, but also orientate each in relation to the conceptual framework of each section. The allegories attempt to emphasise that the climate crisis, practices of landscape architecture and pedagogical approaches in the Global North belong to intertwined legacies. The narrative provides orientation towards unpicking, other-worlding and rewiring as actions for the future.

This publication includes 33 contributions organised across the five curated threads: Material Ecologies, Generative Lineages, Generative Processes of Fieldwork, Sensing Landscapes and Urban Ecologies. The threads of enquiry bind the contributions of a range of landscape architectural practitioners and academics who evidence an array of ideas of ecology and design approaches linked to the global climate crisis. Each thread describes, through the lens of ecological thinking, design methods and techniques that engage innovatively with key professional skills of site analysis, fieldwork, material investigations and design processes.

Part 1, "Material Ecologies," examines ecological thinking from a material perspective with contributions from authors who have explored ecology as a responsive and evolving set of systems.

Part 2, "Generative Lineages," explores pedagogical practices that critique imperial knowledge systems and extend upon – or intertwine – diverse lineages within the extents of the discipline that project into multiple variations of practice.

Part 3, "Generative Processes of Fieldwork," focuses on techniques of observation. Contributors offer approaches in which the relation between observer and the subject is a form of co-production.

Part 4, "Sensing Landscapes," offers contributors a platform for posing modes of observation and simulation, beyond the ideal or "natural" flow, that shift the perceived inert view of the landscape.

Part 5, "Expanded Ecologies," considers the relationships between drivers of the climate crisis and economic valuation models, and social and political systems. Contributors explore ways of embedding new value systems that declare a shift from current paradigms of production.

We would also like to extend our gratitude to Rob Sheehan for copyediting this volume.

STUDIO ECOLOGIES

The first Intergovernmental Panel on Climate Change (IPCC) assessment (Houghton, Jenkins and Ephraums, 1990) outlined the consequences of climate change and the need for a global response. Over the three decades that followed, the impact of the report's findings (and subsequent IPCC reports) slowly informed landscape architecture programmes, predominantly by altering delivery aspects of landscape architecture design studios. For the most part, the responses manifested in the subject matter of the design studio course through incorporating, and oscillating between, a range of distinct scientific problem-solving foci.

The cumulative influence of this slow infiltration of climate change consciousness into landscape architecture programmes means that design studio teaching has moved, slowly but progressively, beyond pondering the potential impact of scientific facts. It has begun the redesign of design studios, purposefully taking responsibility for the urgent need to engage students in ways that ensure they have a developed understanding of the climate crisis. This early phase in reconstructing landscape architectural design studio teaching is alive to the urgency of shifting how global climate challenges are approached through the design studio in academic institutions. It underlines the imperative to adopt into the design curriculum innovation which empowers landscape architecture, from a disciplinary perspective, to prioritise alternative approaches to future landscapes. Those alternatives must generate possibilities that meaningfully respond to climate change and climate defence.

The reconstruction of landscape architectural design studio teaching is a project in the making. There is a long way to go. We must move quickly. The climate crisis must be met by an ethical momentum that impels academics to reconsider how we educate landscape architects, specifically through the pedagogical framework of the design studio and its delivery. This transformation must involve profound reconfiguration of, and innovation in, discrete knowledge

DOI: 10.4324/9781003145905-1

systems within the design studio curriculum, including the design project course's associated techniques, approach and nature. This pursuit challenges the design studio instructor to consider how the "designer" and the "design project" generate ideas for alternative and indeterminable futures through applying diversified design literacy. This proposed design studio orientation makes it possible to discard the saviour complex which accords primacy to problem-solving goals that often purport solutions grounded in reinforcing historical power structures, social inequities and exploitation of the environment and ecosystem. The proposed transformation comprises new modes of action. The scope of our brief is to examine and unfold approaches to landscape architecture design studio teaching. This expanded lens broadens our vision which in turn obliges us to deal openly with ecological and ethical understandings.

Framing design studio pedagogy through an ethical and ecological lens suggests a different toolset for the design studio, "the instructor" and the "learner". It is an exacting toolset that positions and considers connections, expressions and repercussions of what we would regard as learning activities, learning processes and their interactions. The creative potential of ecological thinking has been emphasised in the landscape architecture discipline in works such as *Ecology and Landscape as Agents of Creativity* (Corner, 1997). This was followed by *Projective Ecologies* (Reed and Lister, 2014) that further emphasised ecology as having multiple frames of reference. Ecological thinking enables the inclusion of economic, social, political and other forces into a set of relations that affect the landscape. Ecological thinking is an interconnected approach that facilitates design capacity building for the student whom it enables to generate an approach to complexity. Including ethics in the framing of pedagogical thinking – and therefore reflected in the design studio's phasing and feedback loops – acknowledges the expressions and repercussions the ecological framework sets in motion.

Design Thinking

A pertinent, and brief, characterisation of the landscape architecture design studio holds that the general pedagogical experience is of an intense "space" of learning and exploration. This space incorporates one or more instructors, students and a defined design approach that describes specific learning objectives and outcomes delivered through a project. The project is progressed within a learning environment enmeshed with different forms of active learning and instruction. The transformative pedagogical shift proposed for the curriculum suggests the instructor and students approach learning through an ecological framework in which processes and activities in the landscape architecture design studio are considered as interrelated non-scalable systems of production. This framework includes a vertical learning structure that involves sequential steps and phases, and a horizontal structure which is akin to a cyclical decision-making approach. These cycles have feedback loops internal to each phase. They are influenced by the material the design studio is engaged with, and the feedback loops occurring across phases within a course (Rowe, 1987), inherently classifying the design studio ecologically in its performance and logic.

This learning and teaching process represents design thinking and communicating that is continually forming as a cognitive unfolding between, and within, the phases and through the underlying studio positioning. The design studio pedagogy is an ecological act of design research; it is defined as "in" design, "through" design and "for" design (Deming and Swaffield, 2011). A pedagogy with these attributes aims to reinforce a shift from problem-solving as a single process in which the end goal is prescribed and apparent. It is a pedagogy that affirms an alternative approach in which the production of knowledge – which is an act of thinking, planning, creating and producing designs – is structured through ethical entanglements and an understanding of their causality in the world.

Ecology and ethics in teaching and learning are considered separately for the purpose of understanding their distinctive roles:

- "Ecology" in landscape architecture design studios must be framed as a multifaceted term. Its use oscillates across a suite of objectives in the design studio, including standing as a metaphorical device for conceptualising relationships between living and non-living things. As a process for formulating relationships and processes between different systems, ecology refers to a multivalent dialogue between different conversations, stories and narratives. Ecology embraces connections, interconnections, relationships, and material and immaterial formations and processes. With these definitional features in mind, reframing the design studio argues for a shift towards a broader, more "whole subjectivity positioning" (Guattari, 2014). This approach to reframing enables us to tackle the immense challenge of climate change through design that requires new connections, collaborations, assemblies of knowledge systems and the potential to influence – and ideally transform – current doctrines of capital and power. The pedagogical model of "Thinking Ecologically" expands beyond its previous delineations to encompass ethics and a reworking of the model's agency. Through this reworking, interactions are considered entangled events that cannot be framed as cause and effect resulting from discrete, singular moments. Conceptually, it moves beyond emergence thinking and non-linear processes to interactivity where ethics are enmeshed in a multiplicity of engagements. As Barad eloquently states, "a delicate tissue of ethicality runs through the marrow of being. There is no getting away from ethics – mattering is an integral part of the ontology of the world in its dynamic presencing. Not even a moment exists on its own" (Barad, 2007).

  Ecological thinking as a pedagogical approach acknowledges human thought processes through neural pathways that enable thinking, feeling and acting within the design studio structure and studio environment. This learning process consciously respects creative thinking and communication as processes that occur simultaneously. Ecology also entails the potential for a reciprocal shaping that considers the brain itself as matter that can be shaped through interactions with the environment (Sitskoorn, 2010). The positioning "within" enabled by ecological thinking is crucial to the urgency presented by the climate crisis. It highlights the importance of the student understanding the ongoing nature of change and the particular perturbations brought about by change. Ecological thinking makes space for responding rather than reacting, and for attuning to the various amplifications the climate crisis brings. Considering the design studio in this manner acknowledges the urgency of the climate crisis and resists the inadequacies of "solutions that can occur quickly, maintain the current state of affairs, lack any sense of realism, and further entrench power" (Whyte, 2020).

- A declaration of ethics of the landscape architecture design studio, and the design projects that follow such a declaration, would see the learning process imbued with the direct and implied effects of design on society and the biosphere. It would see a conscious integration into design studio pedagogy of the agency of the designer in managing those effects. The aim is to ingrain design ethics and ecological thinking into how design studios are pedagogically framed and delivered, and to contribute to the academic positioning of each design project. This intentional declaration of the design studio has two objectives:

  1. To break the historical lineages that separate knowledge streams within the discipline, and to acknowledge the influence and seepage between streams, and the reconfigured relationships of knowledge streams, systems, and design production.
  2. To instil an understanding of how to engage in human–world relations actively, and how encounters connect to and intersect with the broader world, and how they invoke or implicate processes of "worlding" (Haraway, 2016) that have an effect beyond a set of physical situations.

Models for the Landscape Architecture Design Studio

In the landscape architecture design studio, students are taught to think critically, taught to model design approaches and techniques and are challenged to contribute constructively through the positioning of the design project. The design studio is recognised as a learning space that demands "multiple intelligences" (Gardner, 2006). It traverses the visual-spatial, the interpersonal, the logical, all of which draw on different knowledge streams contained within a programme's multi-course structure and delivery. Here, the design studio is posited as the primary avenue for learning, and the moment for coalescing and synthesising knowledge derived from other courses in a programme's curriculum (such as history and theory, communication/representation, site design and engineering, natural and cultural systems, construction documentation and administration). This synthesis enables the amalgam of incongruent streams of knowledge to produce unique tools and approaches expressed together through the design project. The role of synthesising knowledge is to be conceived as an engagement with multiple forms of intelligence and knowledge interactions.

It should be emphasised that this form of teaching and learning establishes a distinction between professional practice and academic environments. Professional practice is concerned with training future practitioners, while academic environments are chiefly concerned with knowledge production – the know-how, know-what and know-who. The design studio model has moved beyond a 19th-century ambition of design education as an exercise of skilling for applied professional practice. It is now clearly seen as a design research model that prioritises ongoing development and contribution of different design approaches.

Three overarching models of pedagogical thinking and knowledge production set the stage for how to configure the landscape architecture design studio: Model for Ecological and Ethical Entanglements, Model for Inter-activity and Model for Reconfigurations. Each model is taken in turn below.

Model for Ecological and Ethical Entanglements

Systems thinking is preferenced as a generative process in the design studio. It is framed as an approach to seeing and knowing the world through an ecological and material premise that argues for a different set of relationships for understanding. In this approach, knowledge is situated and understood according to the conditions within which it is produced. Barad (2007: 377) describes it like this:

> Embodiment is a matter not for being specifically situated in the world but rather of being of the world in its dynamic specificity … Ethos is therefore not about the right response to a radically exteriorized other, but about responsibility and accountability for the lively relationalities of becoming of which we are a part.

Through this focus, the territorial domains of ecological systems are not finite or singular; they are instead composed of ensembles of the multiple events, entities and processes of their environments. These are porous material processes that relate and embrace social and political agency, and which include the often non-scalable "general rights of humanity" (Guattari, 2014). These ecological systems attempt to respond to the "mounting pile of ruins" – the residue of scalability (Tsing, 2012) – generated from paradigms of endless growth and production that for decades have preoccupied processes of economic growth, trade and development. This design studio approach is not an either/or argument for scalable or non-scalable processes. Instead, it is a description of how design approaches encapsulating carefully framed tools and techniques might enable reciprocity between these systems, such that diverse forms of production are proportional and reciprocal.

The diagram in this design studio model is an apparatus for viewing and projecting entanglements between the nature of the ecological, social and political agents that are speculated upon in the design project. These entanglements are non-additive to the design but attuned to the interactivity of material forces, and relationships between material phenomena, in the design process and the landscape.

Model for Inter-activity

Change is needed to the way planning codes and legal frameworks are associated with territorial domains, land ownership, the stewardship of fragile ecosystems and natural resources, the development of the public realm, our homes and the broader urban environment. Policy and planning frameworks in countries such as the United States and Australia have been based on antiquated systems of colonisation that prioritise land claims and settlement agendas, and which conform to a constrained set of homogenised regulatory and land use planning frameworks. These frameworks exacerbate inequality, oppressive cultural indifference and the ease of continued diversions of accountability for land exploitation.

"Land and site" in the landscape architecture design studio tends to be regarded as a bounded entity distinct from its surroundings – a designated area or property allotment which is determined through property ownership, designated overlays and historical subdivision that occurs through transforming land use and development agendas. Historically, pedagogical agendas have reinforced this approach, consequently limiting the designer's role in determining site and its extents, the design brief and who should be involved in its deployment – all of which reinforce historical regulatory frameworks.

The design approach of material interactivity of different landscapes and "site" relationships prioritises a shift in the way site is defined. This approach argues for "site" to be considered as a "relational construct" (Kahn and Burns, 2021) that reveals the ecological processes of known and unknown conditions. It describes transformations and registers novel combinations of unseen amalgamation between and within in which matter and time are inferred as scales of measure and order.

This design model incorporates a range of tools and techniques of observation that purposefully operationalise intrinsic parameters and constraints in this design approach. This diverse array of tools and techniques generates multiple translations of diverse, interacting ecological systems, techniques and technologies for examining and generating forms of analysis, discovery and the formation of design briefs. It is open for alternative, unpredictable assemblies of heterogenous relations in which the landscape and human interaction are considered inseparable. It uses various tools and techniques that inscribe information between the physical and the virtual, thus developing processes that respond to an ever-changing and unpredictable landscape. The intention is to shift away from colonial ideologies of land ownership, and associated systems of measure that prioritise parcelisation. The shift is to differentiate types of spatial knowledge and site formations. Knowledge systems are forms of classification developed to become more attuned to biological, chemical and ecological processes. The demarcation of land becomes a much more fluid and dynamic condition that prescribes a multiplicity of cohabitation in which the observer and observed intermingle.

Model for Reconfigurations

This design studio model considers worlding techniques that bring together a multiplicity of heterogeneous agents and actors, human and non-human, seen and unseen constituents from the past, present and future. A suite of entangled relationships is formed and continually refashioned in relation to what the landscape architecture discipline has traditionally termed "site". The act of worlding is itself a method of generating knowledge, a process "of knowing, being and telling", a co-becoming which is part of us (Bawaka et al., 2016), that occurs through an interwoven and continually shifting hierarchy between the observer, the observed, and the tools and techniques of observation. This kind of relational study can evidence multiple shifting positions and hierarchies of being and behaviour. This kind of relational study frames the climate crisis as a condition that describes a cycle of events and impacts we have collectively triggered, and that is also a future demanding action that accounts for all living systems and their interrelationships. This approach is a grounded one that does not frame the climate crisis as an abstract "over there" phenomenon.

A design approach of co-becoming as the production of space, place and morphology shifts knowledge systems away from an understanding of a single canon or representational convention. This shift demands academia's courage, particularly in the landscape architecture discipline, to take a radical turn that acknowledges a multiplicity of threaded knowledge lineages which can form multiple storylines, origins and interrelations. New techniques and approaches to communicate and respond through design are necessary to meet the challenges ahead. It has been proven historically that singular and definitive approaches only amplify inequities and vulnerabilities in the landscape (Demos, 2017). They are not sufficiently substantial to address the complexity of the climate crisis.

In this approach, indexical techniques are used as a subjective device of the landscape that define specific meaning within a context that furnishes the ability to embrace scientific, political and material information systems. These techniques inform multiple states of making in the landscape architecture design project.

This teaching and learning ambition positions the student, the academic and the academic institution as agents in the learning and world-making process. This approach offers students an explicit framework for developing skills that reveal and question explicit biases in their developed design project.

References

Barad, K. (2007). *Meeting the universe halfway: Quantum physics and the entanglement of matter and meaning*. Duke University Press.

Corner, J. (1997). Ecology and landscape as agents of creativity. In Corner J., & Hirsch, A. B. (Eds.), *The landscape imagination: Collected essays of James Corner, 1990–2010* (pp. 241–256). Princeton Architectural Press.

Country, B., Wright, S., Suchet-Pearson, S., Lloyd, K., Burarrwanga, L., Ganambarr, R., Ganambarr-Stubbs, M., Ganambarr, B., Maymuru, D., & Sweeney, J. (2016). Co-becoming Bawaka. *Progress in Human Geography*, *40*(4), 455–475.

Deming, M. E., & Swaffield, S. R. (2011). *Landscape architecture research: Inquiry, strategy, design*. Wiley.

Demos, T. J. (2017). *Against the Anthropocene: Visual culture and environment today*. Sternberg Press.

Gardner, H. E. (2006). *Multiple intelligences: New horizons* (Completely rev. and updated.). Basic Books.

Guattri, F. (2014). *The three ecologies*. Bloomsbury Academic.

Haraway, D. (2016). *Staying with the trouble*. Duke University Press.

Houghton, J. T., Jenkins, G. J., Ephraums, J. J., & Intergovernmental Panel on Climate Change. Working Group I. (1990). *Climate change: The IPCC scientific assessment*. Cambridge University Press.

Kahn, A., & Burns, C. (2021). *Site matters: Strategies for uncertainty through planning and design*. Routledge.

Reed, C., & Lister, N.-M. E. (2014). *Projective ecologies*. Harvard University Graduate School of Design: Actar Publishers.

Rowe, P. (1987). *Design thinking*. MIT Press.

Sitskoorn, M. (2010). Shape your brain, shape up. In H. Aardse & A. van Baalen (Eds.), *Findings on elasticity* (pp. 19–21). Lars Muller.

Tsing, A. L. (2012). On nonscalability. *Common Knowledge*, *18*(3), 505–524.

Whyte, K. (2020). Against crisis epistemology. In B. Hokowhitu, S. Larkin, C. Andersen, L. Tuhiwai-Smith, & A. Moreton-Robinson (Eds.), *Routledge handbook of critical indigenous studies* (pp. 52–62). Routledge.

Bibliography

Bateson, G. (1972). *Steps to an ecology of mind*. Ballantine Books.

Bohm, D. (1981). *Wholeness and the implicate order*. Routledge & Kegan Paul.

Maturana, H., & Varela, F. (1980). *Autopoiesis and cognition: The realization of the living. Boston studies in the philosophy of science*, vol. 42. Dordrecht, Holland, and Boston: D. Reidel Publishing.

Stengers, I. (2005). Introductory notes on an ecology of practice. *Cultural Studies Review*, *11*(1), 183–196.

PART 1
Material Ecologies

Narratives

The soil is red, dense and claylike. Over time, it slowly makes its way downhill, collecting and thickening at the edges of fallen logs and rocks. The soil is always accumulating, one way or another. Fresh gravel laid on the road above carries new seeds, and they join with those already embedded within the soil. *Dichondra repens* emerges everywhere, creeping outwards. Tree ferns and mountain ash (*Eucalyptus regnans*) tower overhead. Limbs and trunks regularly fall. The eucalypts are only just anchored in the soil by a taproot. So much of the tree exists above the soil, so little below. Upended trees expose the soil again, freeing it to move once again with its migratory seedbank. A non-linear accumulation and dispersal.

The soil here gathered from a lava flow 300 million years ago. First and enduringly inhabited by the Woiwurung and Boonwurrung people, settlers then occupied this fertile land to grow tulips, potatoes and deciduous trees. They converted wet sclerophyll forest into farmland reminiscent of their homelands. Claiming this wholesale reconstruction as an act of heroism, a taming of the "primeval forest" (Gembrook Nurseries, 1918) into ordered nurseries and orchards. The settlement formed around the Nobelius tree nursery. Trees sustained by the rich nutrient soil were grown here and circulated around the world (Afoot Round Mount Dandenong, 1908) on a phenomenal scale. At one time, 2.5 million trees were for sale, a proliferation of capital accumulated from the nutrients drawn from the soil. When the founder, Carl Nobelius, died, he passed on a fortune of £38,000 (Personal, 1922) to his children, yet another form of gathering from this soil.

More recently, the nature of this accumulating environment is seen through the prism of risk. The landscape is defined by competing overlays at various

DOI: 10.4324/9781003145905-2

levels of governance. The Victoria state government's "green wedge" policy aims to protect trees and indigenous vegetation by limiting removal. But this is in conflict with fire management and related home insurance policies. Retaining trees is discouraged – preference is bestowed on fastening the soil in place, keeping clearly defined property boundaries and avoiding limbs and trees falling on property. The view of the landscape as inherently dangerous echoes the fear of the untamed forest. Fear is increasing as various levels of government attempt to enforce individual responsibility for the performance, moving again to fashion this landscape with ideals of European security. Climatic perturbations further embed the view of landscapes of risk, as changes affect the lifespan and extent of this forest.

Definitions

In landscape architecture, material ecologies produced by designed and existing landscapes are often described in three ways: through their independent bodies of knowledge (geology, hydrology, biology); or by economic, environmental, political and social categories; or the built form of architecture. Definitions *of* materials abound. Take the example of soil:

> Soil is not just a base on which to build, but an essential creative tool for landscape architects. It creates form and functions as a link between the surface and the area below that can appear as a patch of vegetation, a pathway area or a building.
>
> *(Zimmermann, 2015)*

Definitions, by their nature, favour a focus on an understanding of materials as components. In contrast, site analysis and design processes in the field tend to understand material in a multivalent manner. Margolis and Robinson (2007) frame material as a system and technology that "defines material in terms of capabilities". This conception emphasises site logics, as defined by material performances, interactions and change over time. A more expansive positioning of "material" that questions what is inside this "system" and, in so doing, situates material performance within historical, economic and social contexts is also underway, exemplified by the work of Jane Hutton (2019) who offers an alternate reading of Central Park through the circulation of materials and their associated labour practices. This increased attention responds to contemporary discussions of what Rosi Bradiotti terms "neo-materialism". A process that "emerges as a method, a conceptual frame and a political stand, which refuses the linguistic paradigm, stressing instead the concrete yet complex materiality of bodies immersed in social relations of power". The conceptualisation of matter in relation to social, historical, economic and other forces is also outlined by Barad, De Landa, Latour and others. However, Todd (2016), as just one example, has noted these "new ontologies" closely represent indigenous perspectives without explicitly referencing them. The relationship between, and manner of referencing, these various ontologies require careful consideration and illustrates the responsibility to recognise indigenous sovereignty and knowledge as enmeshed within these relationships of power. A responsibility also to reflect also on the perspective of the soil itself, to ask what would it mean to "recognize an elemental, material agency distributed across bodies, human and non-human? Who or what would count as a 'stakeholder'? How would a 'public' be constituted?" (Khan, 2012).

We circle back to soil, now defined as a collective condition. A collection of minerals, leaf matter, microbes, fungi. Slowly solidified and continuously transmitting. It is not merely generative in a physical dimension alone. This is material as matter, as historical trace, as global circulation, as expressive, as transformable. This continual redefinition of the nature of "materialities" has in parallel influenced the understanding of the agency of the designer. The notion of agency has shifted from what McHarg (1995) described as "man having learned of the operation of the forest, or some part of it, is enabled to intervene in such a way as to increase its thermodynamic creativity". Agency has moved closer to Bohm's (2002) position, which argued for an understanding of the human perspective as situated *within* a material frame where "the body enfolds not only the mind, but also in some sense the entire material universe". Encapsulating an emphasis on extended design responsibilities to establish relationships with care.

To be part of, and to affect, material ecologies asks for approaches that allow positioning of oneself as a material agent. Seeking modes of action that consider connections, actors and agency, and how they are inextricably embedded in reciprocal, material, collective relations. To be in service to the material itself. These relations also imply an alternate formation of the studio environment, to allow for openness, interaction and transformation as a set of processes rather than hierarchical organisations.

Pedagogies

The work in this part explores landscape conditions through the lens of material ecologies. Climate crisis has brought into focus the connected, reciprocal and accumulative nature of these material interactions. Material processes continue to be amplified in unpredictable ways due to the multiple effects of intersecting issues. It has also revealed a legacy in the discipline of viewing material conditions through an extractive lens, then a management frame and finally as a risk landscape. Demonstrating pedagogies that reconsider, and decouple from, these related paradigms and extend beyond the studio into professional ethics and positions in relation to risk. Pedagogical approaches in this section trace a range of approaches from developing practices of co-formation through the indirect or direct alteration of material flows, situating material transformations and flows within a historical and sociological context and reorienting landscape stewardship from extractive and management frames of reference. Together, they present a series of ways of working through and in relation to the shifting material realities of the planet.

References

Afoot Round Mount Dandenong. (1908, February 15). *The Mildura cultivator (Vic: 1888–1920)*, p. 8. Retrieved March 24, 2021, from http://nla.gov.au/nla.news-article74851875.

Bohm, D. (2002). *Wholeness and the implicate order* (1st ed.). Routledge.

Gembrook Nurseries. (1918, March 28). *Dandenong advertiser and Cranbourne, Berwick and Oakleigh advocate (Vic.: 1914–1918)*, p. 3. Retrieved March 24, 2021, from http://nla.gov.au/nla.news-article88814647.

Hutton, J. (2019). *Reciprocal landscapes: Stories of material movements*. Taylor & Francis Group.

Khan, G. A. (2012). Vital materiality and non-human agency: An interview with Jane Bennett. In G. Browning, R. Prokhovnik, & M. Dimova-Cookson (Eds.), *Dialogues with contemporary political theorists (international political theory)* (1st ed., pp. 42–57). Palgrave Macmillan.

Margolis, L., & Robinson, A. (2007). *Living systems* (1st ed.). Birkhäuser Architecture.

McHarg, I. L. (1995). *Design with nature* (25th ed.). Wiley.

PERSONAL. (1922, April 28). *The Argus (Melbourne, Vic.: 1848–1957)*, p. 6. Retrieved March 24, 2021, from http://nla.gov.au/nla.news-article4657910.

Todd, Z. (2016). An Indigenous feminist's take on the ontological turn: 'Ontology' is just another word for colonialism. *Journal of Historical Sociology, 29*(1), 4–22. https://doi.org/10.1111/johs.12124.

Zimmermann, A. (2015). *Constructing landscape: Materials, techniques, structural components* (3rd ed.). Birkhäuser.

Bibliography

Reed, C., & Lister, N. M. E. (2014). *Projective ecologies*. New York, NY: Harvard University Graduate School of Design.

1

THE ANTHROPOCENE CHAMBER

A Pedagogic Experiment in Climate Change Communication

Rania Ghosn

Change, Climate, Representation

"Climate", writes the geographer Mike Hulme, author of *Why We Disagree about Climate Change*, "is weather which has been cultured, interpreted and acted on by the imagination, through storytelling and material technologies".[1] The argument put forward in this chapter is that climate – as it is imagined and acted upon – needs to be understood, first and foremost, aesthetically and culturally. Such a cultural understanding of climate reframes how the idea of climate change should be conceived: it is not only a calamity of the physical process – which must be stopped, fixed or solved – but also a predicament of the cultural environment. In this renewed climate imagination, many different forms of media – including visual art, installation, myth, performance and fiction – are needed to bring the matter to public concern, inviting us, designers and educators, to revisit the exigencies of climate change in the architectural curriculum.

For this experiment in pedagogy, the task began at the site of one of the established institutions in the mediation of 'nature and the environment'. In the spring of 2017, the MIT Department of Architecture issued an open call for Experiments in Pedagogy. The faculty was invited to take on topics and modes of inquiry that did not fit into the current curriculum, and to investigate new models, formats and topics of learning, design and research. The Earth on Display workshop proposed that the specific media of designers – drawings, models, material constructs – could make climate change (so often represented as in the unimaginable abstraction of graphs and charts) feel instead visceral, intimate and present. In particular, and at the site of a natural history museum, the workshop leveraged architectural media – historic narratives, material specimens and large diorama drawings – to foster an ability to relate, react and sense

DOI: 10.4324/9781003145905-3

contemporary ecological violence, and to cultivate an emotional vocabulary for articulating pain and urgency that is otherwise left unexpressed.

With support of the Harvard Museum of Natural History (HMNH), Earth on Display designed The Anthropocene Chamber and installed this temporary exhibit in the museum's Climate Change Gallery. In response to the prevalent language of climate change exhibits, replete with digital screens, climate models and expert video testimonials, the workshop privileged a strategy of material evidence and visual artefacts that gave form to what might otherwise be perceived as abstract climate matters. The diorama exhibition consisted of a series of eight large-sized drawings. Each drawing incorporated a museological specimen from the HMNH collection which was situated in a site drawing, making visible the specific geography and matter of concern.

The workshop proposed a fourfold methodology, which I outline in this chapter. The first step was to historically situate the natural history museum; its history, typology and its recent role in climate change communication. The second was to construct a theoretical and operative framework on the role of aesthetic practices in making public environmental systems and planetary scales of change. The third was to identify a series of specimens from the museum's collection and stage them within dioramas that visualise the landscape of such impacted geographies. The fourth and last phase, that of the public assembly, saw the dioramas brought together into the exhibition and the book.

Situate: The Museum of Natural History

> Objects, cabinets, remains: here is an assembling of wonders from a damaged planet, brought together in order to cultivate the arts of remembering effectively, so as to care seriously, to care for, to care with. Each essay is a provocation to curiosity in the sense of incitement to feel, know, care and respond.
>
> – Donna Haraway

Histories of museums and, more specifically, of museums of science and nature, generally trace their origins to the curiosity cabinets of Renaissance princes and scholars. Following the French Revolution, many previously private collections were claimed for the public, and new museums were established to give citizens ways of seeing the world. Considerable effort was directed towards making exhibitions educative for, and legible and entertaining to, the new mass public. In the introduction to his lectures, Charles Wilson Peale, a 19th-century artist, naturalist and founder of what became the Philadelphia Museum, explained why natural history held such crucial importance for Americans: they connect the vital educational function of a museum with the growth of the new republic and a democratic citizenry – the farmer, the merchant and the mechanic.[2]

The arrival of the Anthropocene has challenged the binary conception of Nature and Culture upon which the museum had rested, calling for the convergence of human, natural and geological histories.[3] It also means a renewed direction for the public educational mission of the institution. In the contemporary media landscape, the museum of natural history could build on the intrinsic interest in 'nature' among visitors, whether environmental activists or straight-out denialists, and channel that interest to communicate matters of climate change. According to Hulme, museums have an important role that requires institutions and artists to work with "the idea of climate change – the matrix of ideological functions, power relations, cultural discourses and material flows that climate change reveals as both a magnifying glass and as a mirror".[4] Through the looking glass, the institution reckons with its complex legacies, and the relations that empire, industry and laboratory have instrumentalised for various ends, not least those of the "white and male supremacist monopoly capitalism, fondly named Teddy Bear Patriarchy".[5]

Such deep entanglement of natural history with the extraction of knowledge and matter anchors the violence of the Anthropocene – often presented as exceptional and emergent – within the slow and *longue durée* violence of environmental history. The Dinosaur Wing at the American Museum of Natural History (AMNH), for example, continues to bear the name of David H Koch, co-owner of Koch Industries, among the leading polluters in the US and a major funder of climate science disinformation. Amid criticism from climate scientists, Koch stepped down from the AMNH Board of Trustees in 2016. Such ongoing corporate funding has launched a new activism against fossil fuel interests in the museum. The diorama "Will the Story of the 6th Mass Extinction Ever Include the Role of its Sponsors?" by Natural History Museum – a project of Not An Alternative, a non-profit collective of artists and activists – depicts a David H Koch Dinosaur Wing several hundred years into a dystopian future.[6] Drawing together climate's intricate relations of causes and effects, this cabinet urges AMNH directors to avoid giving cultural legitimacy to climate crisis perpetuation.

How, if possible, can we reclaim the museum as a terrain where a "common sense" of climate change is built and its social imaginary constructed? Rather than abandoning or destroying museums as a move towards liberation, the political theorist Chantal Mouffe proposed to engage the institution "with the aim of fostering dissent and creating a multiplicity of agonistic spaces where the dominant consensus is challenged and where new modes of identification are made available".[7] A more enabling politics, Mouffe argues, would transform the museum into agonistic spaces that foster new forms of public pedagogy on climate change. (Figure 1.1)

Construct: Making Climate Public

The quest for a new worldview requires an experimental method for conceiving and responding to the problem of climate change in the form of political arts. The Earth on Display workshop explored the gap between the importance of the politics of representation in ecology, and the narrow repertoire of emotions and sensations with which we approach these issues. The workshop proceeded in two conceptual steps. First, it drew on the work of Bruno Latour to articulate a framework that to shift worldviews on climate change from "matters to fact" to

FIGURE 1.1 Earth on Display, exhibition overview. Rania Ghosn.

"matters of concern", then, in a second step, and learning from Donna Haraway and Maria Puig de la Bellacasa, it explored how to shift those same "matters of concern" to "matters of care".

Climate change, Bruno Latour proposes, calls for a new worldview that counters the objectification of the Earth and the accompanying focus on inanimate, disembodied, undisputed reason. The new worldview centres on a political project of representation, of what world one wants to assemble, and with what entities she wants to live.[8] In times of turmoil, Latour's exhibitions and books, *Making Things Public* (2005), *Reset Modernity!* (2016) and *Critical Zones* (2020), are critical reminders of the agency of representation in bringing disputed things to public attention.[9] The framework of *Making (Climate) Things Public* allowed us to shift the debate away from matters of fact – that is, positivist solutions with its associated techno-fixes, such as green, sustainable, LEED certified – to matters of concern that position environmental matters at the core of natural histories and futures. "A matter of concern", Latour writes, "is what happens to a matter of fact when you add to it its whole scenography, much like you would do by shifting your attention from the stage to the whole machinery of a theatre".[10] In such an expanded worldview, media techniques foster the abilities to relate, react to and make sense of climate change, and to assemble a public around such matters of concern.

What forms should such matters of concern take? What might they sound or feel like? To address that, the workshop hosted filmmaker Fabrizio Terranova for a screening and discussion of his film *Donna Haraway: Story Telling for Earthly Survival* (2016).[11] A portrait of Haraway, the film portrait gives form to her theoretical and methodological SF: string figures, speculative feminism, speculative fabulation and the building of bridges between science and fiction. The film invites us to imagine new possibilities, of a world which goes beyond the critique of existing structures (such as those of capitalist witchcraft) to engage possibilities of life in the shadow of environmental ruins. *Story Telling for Earthly Survival* reminds us that we are clusters of stories: the way we move, the way we behave, is linked to the kinds of stories we have been told about the world, about our relations to each other. "We need other kinds of stories", Haraway implores as she faces the camera. Storying, otherwise, in Haraway's expression, experiments with different kinds of storytelling, bending the documentary genre by fusing the intimate everyday with the playfully surreal to produce new structures of telling and making the world.

Drawing on Haraway's work, Maria Puig de la Bellacasa deploys a feminist ethics framework to rework Latour's concept of matters of concern into that of matters of care.[12] The situated scholar can no longer rely on a disinterested gaze, asserts Puig de la Bellacasa. Troubling the notion of critical distance from scholarship, she details three dimensions of care: affect/affection, labour/work and ethics/politics. Each brings attachment, obligation and commitment into scholarly work. Matters of care, she argues, require that we engage with their "world-making effects", including the devalued and the unloved, ultimately bringing into the picture the "matter" of oppression and domination. To care for another, to care for a possible world, is to become affectively and ethically entangled, and consequently to get politically involved in whatever ways we can.

Represent: Political Arts of Climate

Science centres and nature museums have begun to produce displays on climate change and its impacts on the planet. The Carnegie Museum of Natural History dedicated its first major in-house exhibition in four decades to the topic of the Anthropocene and welcomed the world's first Curator of the Anthropocene. The Smithsonian National Museum of Natural History also opened its gallery on the Anthropocene in 2019 as part of a larger fossil/deep time hall. The media strategy of such exhibits often employs scientists to produce exhibition narratives and has mostly channelled the language of matters of fact. This has less hold on visitors' imaginations than displays in other wings of the museum, such as the 19th-century HMNH Ware Collection of Blaschka Glass Models of Plants.[13]

The implicit risk of climate change communication is decoupling knowledge from meaning and care. Digital screens play documentary or infographic videos, climate models and simulations and forecasting scenarios, to communicate fundamental shifts in the state and functioning of the Earth System. The Great Acceleration of the Anthropocene is cast in exponential diagrams of carbon dioxide emissions and species extinction. Disaster images abound – fires, spills, hurricanes. Throughout, human-induced climate change is perceived as a series of isolated natural disasters outside history and geography. It is a narrative of ecological violence, without history, which continues to abstract the effective causes and agents, implying a teleological trajectory to which all humans are equally culpable. They do not account for the anonymous and attritional events that are indifferent to the sensation-drive technologies of the image-world. Throughout, slow violence fails to be noticed. In his acclaimed *Slow Violence and the Environmentalism of the Poor*, Rob Nixon observes that the violence wrought by climate change takes place gradually and often invisibly, requiring different sorts of media to balance it. He adds, "How can we then convert into image and narrative the disasters that are slow moving and long in the making?"[14] How can the dirt and atrocities of the archive be recast to stage the many Anthropocenes? (Figure 1.2)

Earth on Display explored aesthetic strategies that could shift climate away from the epistemic authority of statistical data, and the populism of disaster images, and gear it towards forms of representation that couple knowledge with meaning, data with objects. A rapidly growing body of "climate aesthetics" explores this interplay between climatic knowledge and aesthetic experience. Such practices deploy a range of aesthetic formats to explore how climate evidence is made convincing in the eyes of the witnesses, often in the form of

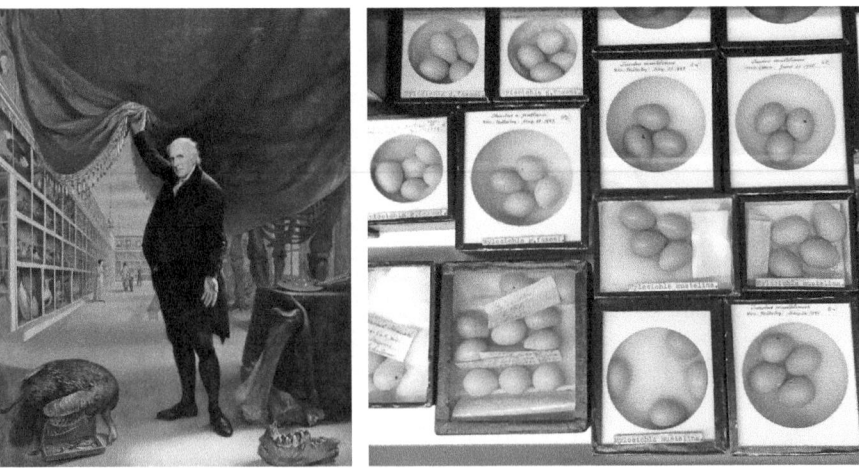

FIGURE 1.2 Left: The Artist in His Museum (1822). Charles Willson Peale. Pennsylvania Academy of the Fine Arts. Right: Bird egg specimen at Harvard Museum of Natural History. Rania Ghosn.

material collections such as Amy Balkin's *A People's Archive of Sinking and Melting*, which is a collection of materials contributed by people living in places that may disappear because of the combined physical, political and economic impacts of climate change, primarily sea level rise, erosion, desertification and glacial melting. Balkin's work makes explicit how the museum communicates beyond cognitive knowledge: it is a collection of thing-experiences, of unique and vivid objects that grab the attention and arouse emotions. Curatorial practices have increasingly been willing to explore the affective attributes of climate through objects. In Munich, the Deutsches Museum's Welcome to the Anthropocene, for example, engaged people in thinking about the Earth's past, present and future through objects, like a crocheted coral reef.[15] The conference and book *Future Remains: A Cabinet of Curiosities for the Anthropocene* gathered 15 objects – a pesticide pump, a jar full of sand, an old piece of calico – to offer clues to intertwined human and natural histories that shape our planetary futures.[16]

Materially situated and historically entangled, the specimen tells histories that make a nature's contested narratives. In the 1970s, researchers struggled to understand the rapid endangerment of the American bald eagle population. A once stable population of around 800 breeding pairs had fallen to a mere 60 pairs in less than 30 years.[17] Investigating this troubling phenomenon, the researchers examined collections like those held at the HMNH, surveying bald eagle egg specimens dating back decades. They observed that eggs from the 1940s onwards exhibited shells so thin they cracked under pressure of the incubating female. By 1970, female bald eagles seemed incapable of producing viable eggs and the shrinking number of viable hatchlings meant the American national bird faced almost certain extinction. What was the cause? A contaminant and endocrine disruptor, commonly known as DDT, threatened almost all birds of prey that came into contact with it by consuming infected fish that absorbed it from contaminated groundwater. The rising popularity of DDT as a pesticide between the 1940s and 1960s directly correlated with the thinning of bald eagle eggs, abnormal thyroid function, decreased fertility, a decrease in hatching success and alteration in immune functions. With evidence of this correlation from the natural history samples, the US banned most applications of DDT by 1972, and the population managed to recover, reaching pre-DDT levels by 2001.

There are thousands of stories like this one, told via collections held in natural history museums. The stories they tell are of the Anthropocene and the ecosystems humans have shaped, treasured and exploited as they have reshaped the Earth. In other words, the specimen creates material forms of object-based storytelling. The physical archive has become at once a testament to the richness of the planet and a roster of shrinking ecosystems, mass extinctions and climatic changes brought about by rapid and ill-considered expansion of human settlements and industries. The complexity of challenges facing humankind today can be understood with the help of such archives. In the design workshop, each student group chose a specimen from the HMNH collection – Museum of Comparative Zoology, Mineralogical & Geological Museum and Harvard University Herbaria. Derived from the Latin meaning "to look", the specimen is an illustration of something, from which the character of the whole may be inferred. Such a definition implies acts of discovering, observing and deducing, collecting and classifying, as well as drawing attention to the metonymic and synecdochic potential of the object itself.[18] Caring deeply, as Haraway suggests, "means becoming subject to the unsettling obligation of curiosity, which requires knowing more at the end of the day than at the beginning".[19] Concern and curiosity become the political register of this pedagogic experiment, which promoted an ethics of care towards previously neglected things.

Assemble: The Anthropocene Chamber

For the HMNH installation, students placed the Anthropocene specimen in a diorama drawing that made visible the ecological concern of the specific geography of concern, such as sand extraction in Indonesia, deforestation in Malaysia, the Fukushima nuclear accidents or the shrinking Aral Sea. The term diorama – from Greek *dia* (through) and *horama* (to see), and patented by Daguerre in 1822 – first referred to a theatrical form of visual art. It was later applied to natural history habitat displays. In "Earth on Display", dioramas plot windows into the Anthropocene, one site at a time; these views are populated with life forms and earth events that together narrate the processes that tie geography and history across scales. (Figure 1.3)

Beyond an emphasis on isolated specimens or species, the dioramas present a more complex account of spatial relations, always situated in a specific geography. In the egregious example of the disappearance of two dozen small islands in Indonesia due to black market sand mining – the second most consumed resource on the planet after water – the diorama Sand Extraction rendered visible the displacement of sand to infrastructural projects of bridges, roads and new land. To render visible the displacement of sand, an accompanying hourglass measured the speed at which sand was mined. The project "Mosquito Co-Evolution" narrated the story of the lethal transfer of infectious disease from animals to humans. Beginning with the transatlantic slave trade in the 17th century, when *Aedes aegypti*, today one of the most adaptable and widespread mosquito species, hitched a ride across the Atlantic, becoming a vector in the spread of yellow fever. A third project, "Deforestation Is Violence against Indigeneity", made visible the impact

FIGURE 1.3   Sand Extraction. Darle Shinsato.

of multiple extractive practices – coal mining, timber logging and palm oil plantation – on the incessant deforestation in the island of Borneo, home to thousands of indigenous people, some of whom are recruited as manual, indentured labourers in the mines. The project's installation included a 3D-printed model of a contemporary indigenous hat, itself a replica of a hat held by the Harvard Peabody Museum, to critique how Western institutional collections render dead living heritage in museum storage. A last example, "Fukushima 2100", took visitors on a tour of the Fukushima Prefecture through a series of instruments, and specimens from the radioactive fallout into air, soil and ocean. A Safecast Geiger counter, seaweed, anchovies and emergency iodine pills charted the technopolitical entanglements for life in the ruins of nuclear disasters. (Figure 1.4)

The Anthropocene Chamber assembled the series of diorama fragments into an "interior" experience of the Earth, as opposed to the stand-alone globe, which is permanently on display on the other side of the room. Contrary to the globe, the dioramas did not provide a totalising planetary overview of all things Anthropocene. Rather, the drawings emphasised relational approaches between fragments, in which a specific geography is entangled with the specimens on display. It was an assemblage that brought partial intimacy to claims of urgent universality. "Apprehending what is significant", notes political philosopher Jodi Dean, "may require adopting another perspective – a partial or partisan perspective, the perspective of a part … [from which] the whole will not appear as a whole".[20] The partisan perspective of The Anthropocene Chamber is an aperture onto agonistic public spaces transforming an abstract planetary model, difficult for all but the most learned to access, into a material and situated geographic matter that spoke to a multiplicity of different and differing audiences. (Figure 1.5)

FIGURE 1.4   Fukushima 2100. Jaya Eyzaguirre, Sebastian Kamau and Taesop Shin.

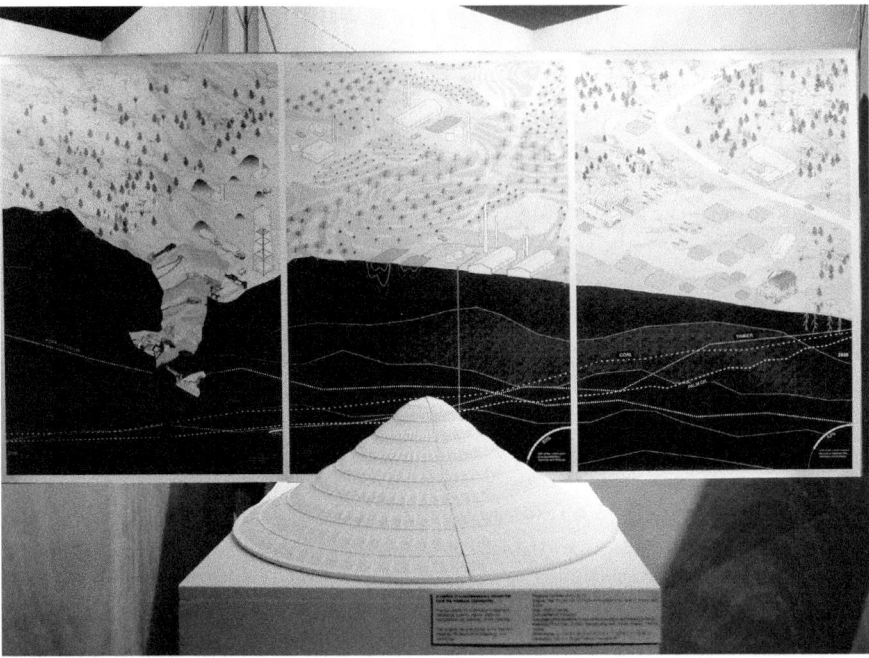

FIGURE 1.5 Top: Deforestation Is Violence against Indigeneity. Meng Fu Kuo, Semine Long-Callesen, Joude El-Mabsout and Mengqi He. Bottom: Woven hat, rim with pink plastic line, finial of thread and textile (1966 or earlier). Peabody Museum of Archaeology and Ethnology.

The workshop publication concluded with a manifesto that served as précis of the workshop's learning objectives, and as a potential roadmap for future similar experiments. The declaration included the following statements:

- Critique the museum with its own institutional tools of display;
- Acknowledge the museum's ties with a history of exploitation and extractive sponsorships;
- Counter abstract scale with material specimens;
- Chart specific geographies rather than a panorama of the globe;
- Expose long-term and slow violence;
- Decolonize figurative representations of the subaltern;
- Curate interpreted information to a non-expert audience;
- Care. (Figure 1.6)

The task of imaging and imagining the Anthropocene calls on a wide range of pedagogic experiments "to arouse a slightly different awareness of the problems and situations mobilizing us", which reach beyond the barbaric scenarios produced at the intersection between unfolding climate change and the capitalist attempt to turn extreme weather events into opportunities for profit.[21] For design pedagogy, the need to render climate change legible engages the discipline in its multiple attributes and capabilities; it is at once analytical, representational and material, as it is critical, aesthetic and experiential, and throughout affective and public. The workshop was one such design experiment to make climate change public by connecting political ecology with aesthetic experience in the form of object-orientated democracies. As such, it extended an invitation to rethink our intellectual conceptions and hierarchies; as Donna Haraway puts it, we are urged towards a new "response-ability" – that is, to acquire an emotional apparatus capable of sensing and experiencing the Earth and the crisis caused by extractivism and industrialisation. Ultimately, the climate change exhibit solicited affective strategies that called on the capacity of both students and visitors to care and respond: a political commitment that affects the way we produce knowledge and action in the world.

FIGURE 1.6   Mosquito Co-evolution. Angeline C Jacques.

Notes

1 Mike Hulme, *Weathered: Cultures of Climate* (London, UK: SAGE Publications, 2017).
2 Steven Conn, *Museums and American Intellectual Life, 1876–1926* (Chicago, IL: University of Chicago Press, 1998), 35.
3 Dipesh Chakrabarty, "The Climate of History: Four Theses", *Critical Inquiry* 35, no. 2 (2009): 197–222.
4 Mike Hulme, *Why We Disagree about Climate Change: Understanding Controversy, Inaction and Opportunity* (Cambridge, UK: Cambridge University Press, 2009), 362–3.
5 Donna Haraway, "Teddy Bear Patriarchy: Taxidermy in the Garden of Eden, New York City, 1908–1936", *Social Text*, no. 11 (Winter 1984–1985): 20–64. See also Rainer Fuchs, ed., *Natural Histories: Traces of the Political* (Köln, Germany: Verlag der Buchhandlung Walther König, 2017). The exhibition, with works by Joseph Beuys, Marcel Broodthaers, Candida Höfer and Mark Dion, provides an example of nature as a politically and historically constructed concept.
6 Natural History Museum, http://thenaturalhistorymuseum.org
7 Chantal Mouffe, "Artistic Strategies in Politics and Political Strategies in Art", *Dissidence* 10, no. 2, The Hemispheric Institute (2013), http://hemisphericinstitute. org/hemi/en/e-misferica-102/mouffe.
8 Bruno Latour, "Waiting for Gaia. Composing the Common World through Arts and Politics", November 2011, www.bruno-latour.fr/sites/default/files/124-GAIA -LONDON-SPEAP_0.pdf See also Rania Ghosn and El Hadi Jazairy, "Gaïa Global Circus: A Climate Tragicomedy", in *Climates: Architecture and the Planetary Imaginary*, eds, J. Graham et al. (New York, NY and Zürich, Switzerland: Columbia Books on Architecture and the City and Lars Müller Publishers, 2016): 52–60.
9 Bruno Latour and Peter Weibel, eds, *Making Things Public: Atmospheres of Democracy* (Cambridge, MA: MIT Press, 2005).
10 Bruno Latour, "What Is the Style of Matters of Concern?" (Spinoza lectures, April– May 2005, Department of Philosophy, University of Amsterdam), 39, http://bruno -latour.fr/sites/default/files/97-SPINOZA-GB.pdf.
11 https://earthlysurvival.org/
12 Maria Puig de la Bellacasa, *Matters of Care: Speculative Ethics in More than Human Worlds* (Minneapolis, MN: University of Minnesota Press, 2017).
13 In such denialist times, however, it might be important to reassert that to question the prevalence of the scientific epistemic object and the affective nature of climate data does not imply an anti-science or an anti-expertise agenda.
14 Rob Nixon, *Slow Violence: The Environmentalism of the Poor* (Cambridge, MA: Harvard University Press, 2013), 3.
15 Jennifer Newell, Libby Robin and Kristen Wehner, eds, *Curating the Future: Museums, Communities, and Climate Change* (New York, NY: Routledge, 2017).
16 Gregg Mitman et al., *Future Remains: A Cabinet of Curiosities for the Anthropocene* (Chicago, IL: University of Chicago Press, 2018).
17 Fraser, JD, SK Chandler, DA Buehler and JKD Seegar, "The Decline, Recovery and Future of the Bald Eagle Population of the Chesapeake Bay, USA", in *Eagle Studies*, 181–7, eds. BU Moyberg and RD Chancellor (Berlin, Germany, London, UK and Paris, France: World Working Group of Birds of Prey, 1996).
18 Helen Gregory, *Un-Natural Histories: The Specimen as Site of Knowledge Production in Contemporary Art* (PhD diss., University of Western Ontario, 2016).
19 Donna Haraway, *When Species Meet* (Minneapolis, MN: University of Minnesota Press, 2007), 36.
20 Jodi Dean, "The Anamorphic Politics of Climate Change", *e-flux Journal,* no. 69 (January 2019), https://www.e-flux.com/journal/69/60586/ the-anamorphic-polit ics-of-climate-change/.
21 Isabelle Stengers, "The Cosmopolitical Proposal", in *Making Things Public: Atmospheres of Democracy*, eds, Bruno Latour and Peter Weibel (Cambridge, MA: MIT Press, 2005), 994.

2

THINK LIKE A RIVER

Designing from the Riparian Zone

Jane Mah Hutton[1]

Rivers flood. Slopes erode. Materials decay. To design and build is always to negotiate with powerful forces, difficult site conditions and factors beyond one's control. One can hope that the storms don't come, that the waters don't rise, and that it all works out as planned. But the storms always did come, and now getting acquainted with change is undeniably a necessary life skill. But the paradigms through which the construction industry operates tend to lean on a falsely stable state. North American settler culture is founded on binary thinking, seeing the more-than-human world as something to be controlled, exploited and defended. These binaries permeate the language, policies and theories that surround the culture of building, and they manifest in the physical constructions that architects and landscape architects design and the drawings and models they make. At the same time, awareness of a changing climate has made *change* a central concept in design education and theory: we know things are complex, and the problems massive. When change is catastrophic and the stakes existential, how is it most useful to practise seeing change as a designer? This short chapter describes a two-week modelling design exercise in an undergraduate architecture studio at the University of Waterloo School of Architecture. The studio was called *Think Like a River: Designing from the Riparian Zone* and grappled with those questions.[2] In this exercise, students studied fluvial processes and experimented with 1:1 modelling techniques that bring together field observations, demonstrations of change and design inquiry.

Spending time in rivers and riparian zones is not a bad way to ponder change. Rivers are forces manifest; they shape terrestrial environments, moving material across the continents, integrating channel, overland and groundwater drainage. And at the same time, people shape rivers: to extract energy, control floods, drain

DOI: 10.4324/9781003145905-4

lands, mine resources and urbanise adjacent parcels. To observe water dynamics, as Alpa Nawre demonstrates through a study of vernacular water systems in India, is also to access an understanding of complex sociocultural relationships (Nawre, 2018). It is easy to think of rivers as and tangible and fixed blue lines on a map, yet as Anu Mathur and Dilip Da Cunha remind, there is no such thing:

> Why is it that despite waters *everywhere* precipitating, seeping, soaking air, soil and vegetation, collecting in interstices, pores, terraces, cisterns, and aquifers, evaporating, transpiring, and sublimating, we see water *somewhere*, confined within or behind lines and generally colored blue in maps?
>
> *(Mathur and Da Cunha, 2014)*

How lines are drawn, and which words are spoken influence how we understand; Jane Wolff's *Bay Lexicon*, a dictionary of San Francisco's water–land boundary, demonstrates the intermingling of language and perception when it comes to complex water-and-landscapes. And when modelling rivers, how are we to explore and demonstrate forces rather than depict them in a static form? For inspiring models, we can look at Bradley Cantrell's research on environmental simulation and sensing (TED Archive, 2017, 03:15–05:21), Cantrell and Justine Holzman's co-edited book, *Responsive Landscapes*, and studios led by Christophe Girot and Gramazio Kohler Research at the ETH (Girot, 2021) But what can be done with just a few days and minimal materials?

Our studio windows at Waterloo Architecture overlook the Grand River, sometimes calm, sometimes a raucous ice-jam, whose flow once powered mills nearby. Draining 6,800 square kilometres of southern Ontario into Lake Erie, the Grand River is also a political entity. It is centreline of the Haldimand Tract, a 950,000-acre parcel that in 1784 was promised forever to the Six Nations of the Grand River; within 40 years the Crown had sold off the majority of the parcel, and this land remains actively contested today (Six Nations Lands and Resources Department, 2015). The studio's focal sites – a small municipal dam, a former aggregate pit planned for subdivision development, and a riverside park run by Six Nations Tourism (Six Nations Tourism, 2021) – reflect the complexities of the Grand River as colonial landmark, industrial power and spine of urbanisation.

Peering, dipping toes and sometimes wading into the cold May river, students observed one of these processes to start this two-week exercise:

1. Flow velocity, or how water runs
2. Turbulence, or how water swirls
3. Channel erosion, or how water carves
4. Sediment transport, or how water carries sediment downstream
5. Disruption, or how water and sediment behave around human constructed elements

6. Floodplain deposition, or how water deposits sediment during floods
7. Bank erosion, or how overland flow carries debris into the river
8. Riparian vegetation distribution, or how plants behave in relation to changing water levels
9. Inhabitation, or how different species move through and occupy the riparian zone

Along the river they looked for and documented evidence of these particular processes, observing particles of different sizes being organised in the river flow, the non-linear work of erosion along a gullied bank, and how riparian plants held on to riverbanks or gave way. They cross-referenced their observations with river science literature, looking for relationships between what they saw and what was written.

From these observations and further research, students explored the process further by designing and building a dynamic model, given a few suggestions: to privilege *demonstrating* the process rather than *representing* it (so to show erosion by eroding something rather than by creating an image of erosion); to consider a model made *in situ* – in the river itself or engaging with existing forces (like an introduced or existing water source); and to privilege using simple materials – rocks, sand, mud – that bring their own physical dynamics and won't end up in a landfill. The week was part science fair, part land art: rocks, sand, mud and some carefully gathered plant samples filled the studio. Students traipsed back and forth to the river with nets, underwater cameras and rubber boots, and a few themes emerged.

Students tried different approaches. Some made the river the model itself, dropping lightweight materials onto the surface and documenting their behaviour as they flowed in particular patterns. Others used underwater cameras to record otherwise invisible movements of sediment or plants. Others "performed" forces by experimenting with making models by the physical tendencies of the materials themselves. With channel erosion research in mind, Caroline Brodeur and Philip Carr-Harris propelled plaster into a sand bed, observing the forms that came from the gravity, viscosity and momentum of the material (Figures 2.1 and 2.2). Alifa Frebian, Ericson Ho and Jieyu Wang, interested in the invisible annual deposition of sediment along the floodplain that comes with annual flood cycles, constructed a device to record the movement of particles of water. In a box with a clear plexiglass bottom, they swirled water in different ways with additions of soil, sand and gravel, and recorded their movement and settling patterns; they then documented these patterns with cyanoprints, compositing trials of different methods of disturbing the water (Figure 2.3).

Observing how water moves over different rock configurations, producing hydraulic jumps, eddies and riffles, Vincent Chuang, Lily Tran and Yi Ming wanted to simulate some of this movement and designed a "kit" composed of

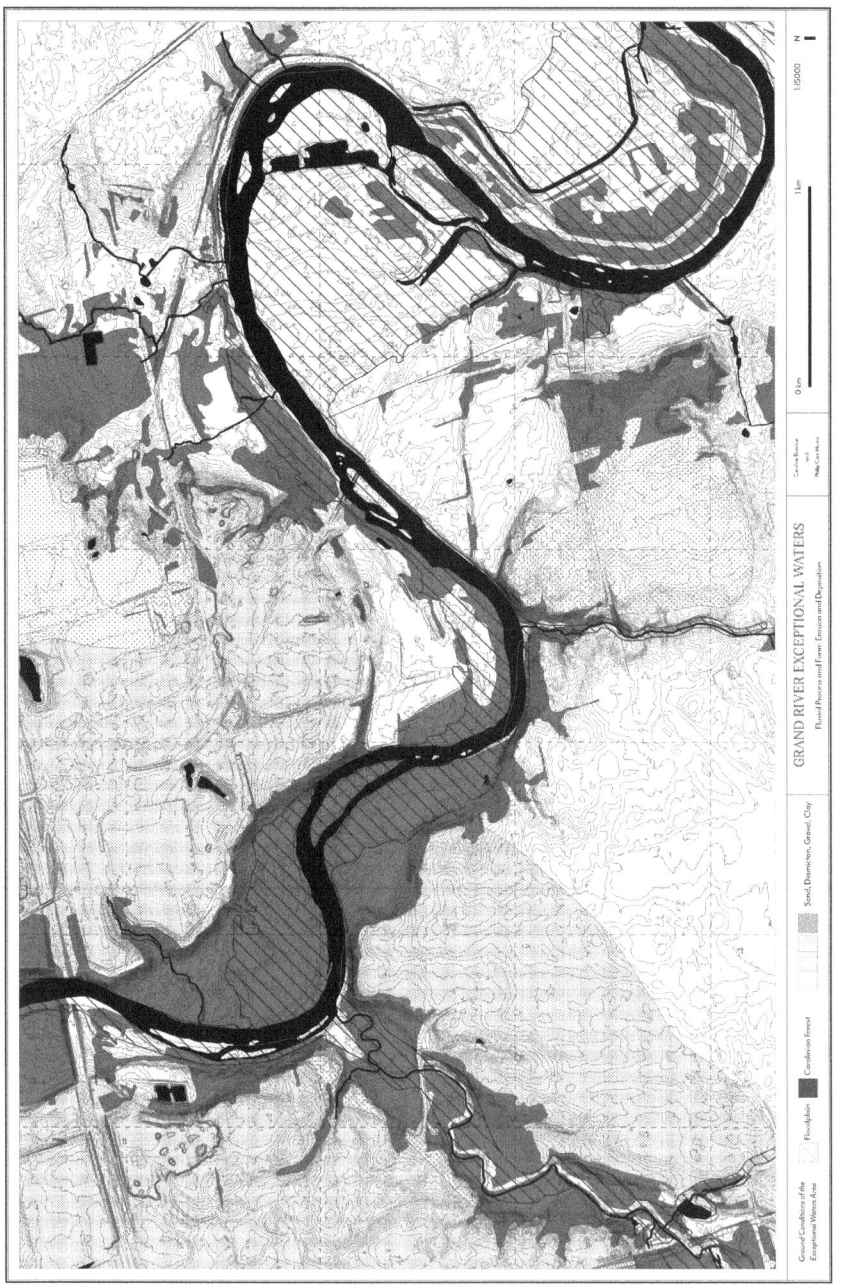

FIGURE 2.1 Grand River Exceptional Waters Map (2018). Caroline Brodeur, Philip Carr-Harris.

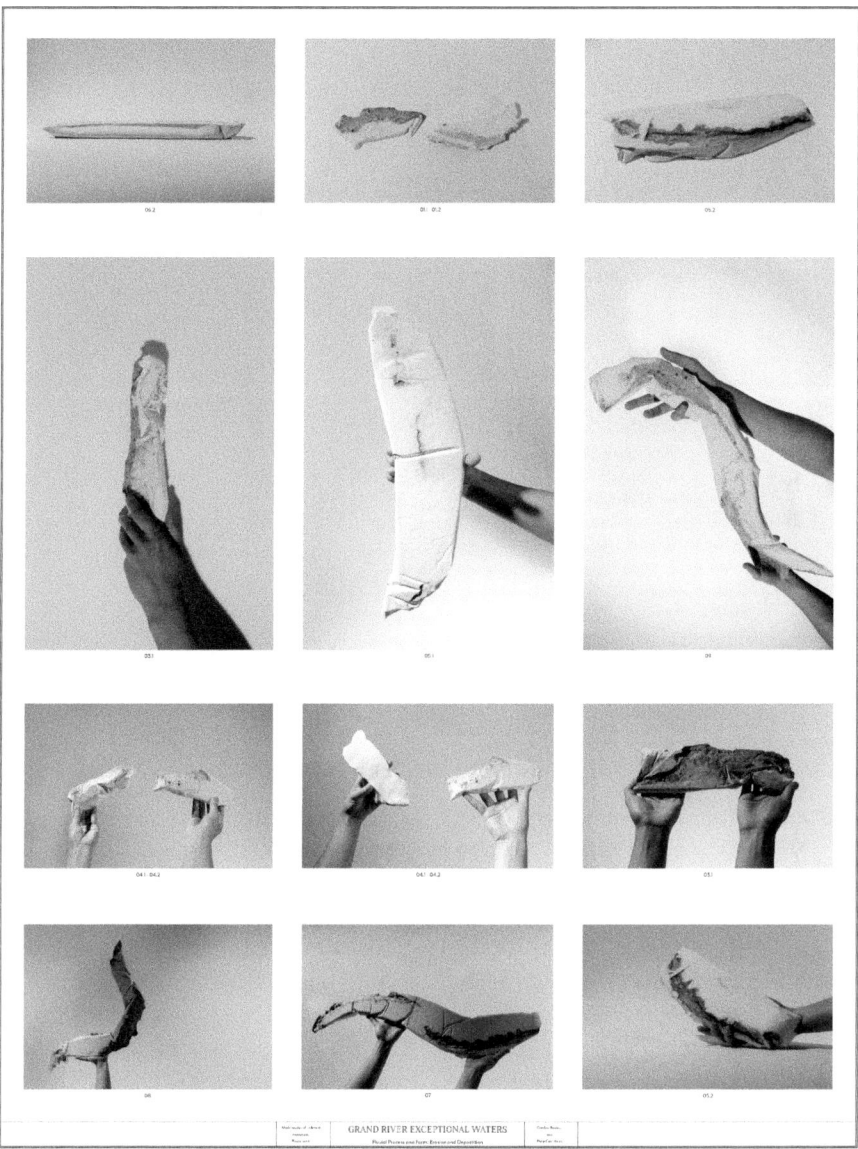

FIGURE 2.2   River erosion and deposition models: sand, plaster (2018). Caroline Brodeur and Philip Carr-Harris.

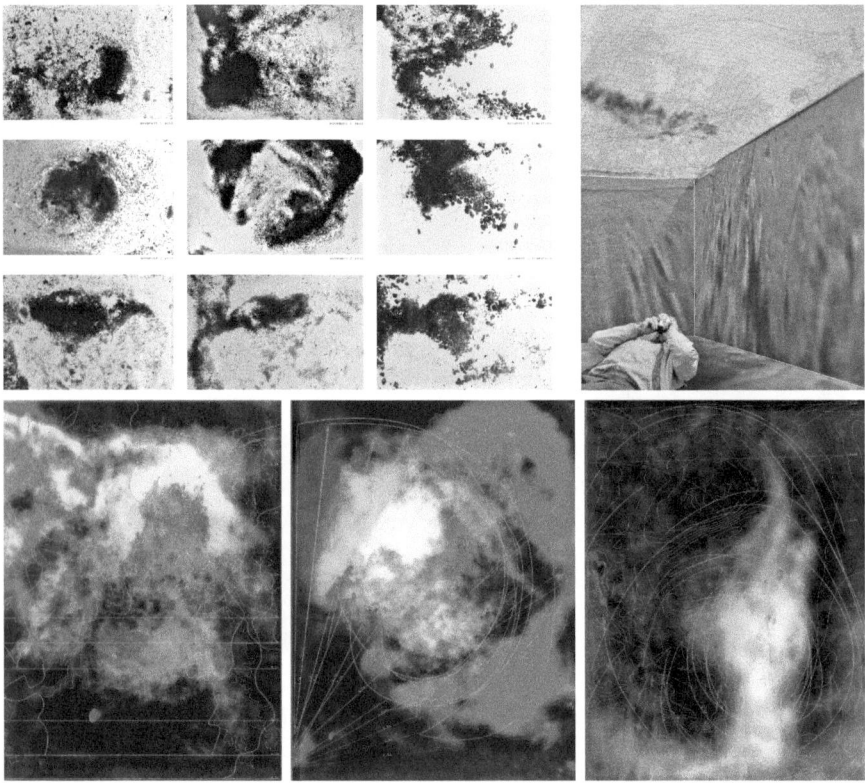

FIGURE 2.3 Top: Deposition studies: soil, sand, gravel, water, plexi (2018). Bottom: Deposition studies composite; cyanoprint (2018). Alifa Frebrian, Ericson Ho and Jieyu Wang.

a sloped channel (enclosed by two plexiglass walls) that had interchangeable shaped pieces referencing rocky river bottoms. By mixing and matching these pieces and running water down the slope, they were able to watch and record the dynamics of flowing water. (Figure 2.4). Another group, Leela Keshav, Michelle Li and Rushali Patel, focused on soil formation through cycles of flood deposition and organic material decay. They explored this convergence of plant material and sedimentation by casting gathered plants and site sediments in plaster, and then recast in glycerin tiles; these tiles were suspended and arranged by site elevation (Figure 2.5).

Like the river itself, most experiments were both clear in some ways, murky in others. The models helped us to be suspicious of single-state landscape drawings and models. They prompted discussions about how we make models, how we understand complicated principles, and about the relationship between models,

FIGURE 2.4 Top: Hydraulic jump, eddy, riffle model; foam, plexi, paint, water (2018). Bottom: Hydraulic jump, eddy, riffle model in action (2018). Vincent Chuang, Lily Tran and Yi Ming Wu.

simulations and "truth". There was joy in working with simple, easily acquired materials that could be scattered around after the work was over, as ephemeral as the processes under study. Many experiments failed gloriously and productively – both physically, as rocks and sand and wax spilled onto the studio floor – and also conceptually as we recognised how different the forces of the river are from ones modelled at a smaller scale and with different materials. The models involved a lot of conjecture; it was a challenge to both seek an understanding of the scientific principles which have precise definitions, while also using the models as idea generators, open for imagination. We caught ourselves making questionable or causal scientific interpretations ("this happened in my model and so this would happen in the river"), but these led to interesting discussions about what could be learned with minimal means, and what can or should not be extrapolated. A quick follow-up exercise asked students to alter their models to imagine designing with the force they were studying – to make a space where the force might be felt by a visitor. The final two projects involved research on the social forces in the river, and a site-focused design proposal working with the confluence of river and social forces. In some cases, these preliminary models became useful working tools during the semester, taken apart and remade to test out new ideas. The models challenged the notion of a designer's total control and created an opening for humility in grappling with something as powerful and complex as a river.

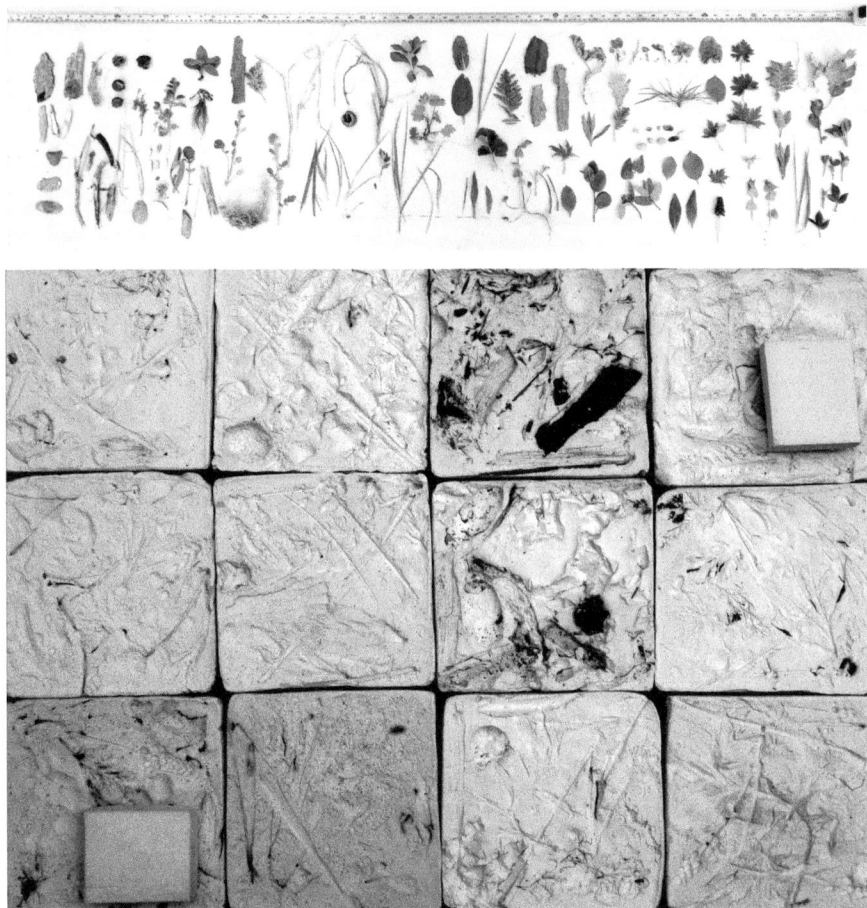

FIGURE 2.5 Top: Riparian vegetation study (2018). Bottom: Riparian vegetation and sediment castings; plaster, glycerin (2018). Leela Keshav, Michelle Li and Rushali Patel.

Notes

1 Based on a studio run in 2018 and 2019 with co-instructors Rick Andrighetti, John McMinn, Christie Pearson, Lola Sheppard and Scott Sørli, and teaching assistants Maighdlyn Hadley, Dani Kastelein, Jason McMillan and Morgan Wright.

2 Thanks to all the students who participated in the 2018 and 2019 studios, those whose work is shown here, guest reviewers. Thanks also to Justine Holzman for sharing materially exuberant models from her teaching at the University of Toronto Department of Landscape Architecture; they constituted an inspiring model for this exercise. For more on the Grand River from Waterloo Architecture, see Carly Kandrack, Guiding the Grand, M.Arch thesis.

References

Cantrell, B. E., & Holzman, J. (2015). *Responsive landscapes: Strategies for responsive technologies in landscape architecture* (1st ed.). Routledge.

Girot, C. (2021). *Robotic landscapes*. ETH Zurich Architecture Department. https://girot.arch.ethz.ch/tag/robotic-landscapes, accessed February 4, 2021.

Mathur, A., & Cunha, D. D. (2014). *Design in the terrain of water*. Applied Research & Design.

Nawre, A (2018). Between community and culture: The criticality of landscape infrastructure reuse in India. *Journal of Landscape Architecture*, *13*(3), 54–63. https://doi.org/10.1080/18626033.2018.1589141.

Six Nations Lands and Resources Department. (2015, February). *Land rights: A global solution for the six nations of the Grand River*. http://www.sixnations.ca/SNGlobalSolutionsBookletFinal.pdf.

Six Nations Tourism. (2021). *Six nations tourism*. https://www.sixnationstourism.ca/, accessed February 4, 2021.

TED Archive. (2017, July 19). *Creating living infrastructures | Bradley Cantrell* [Video]. YouTube. https://www.youtube.com/watch?v=-vQZPllqw-Y, accessed February 4, 2021.

Wolff, J. (2021). *Bay Lexicon*. McGill-Queen's University Press.

3

EDIBLE ECOLOGIES

Zaneta Hong

As landscape architects, urban planners and designers, we are becoming increasingly mindful of the material practices that are embedded in our disciplines and to an awareness of how manufacturing and construction are interrelated at multiple scales, influenced by multiple species and conducted across multiple territories. Our knowledge and skillsets provide us with agency to systemise people, communities and environments, to alter atmospheres, hydrospheres and lithospheres, and ultimately, to shape and alter living matter into corporeal realities and speculative futures. This multifaceted, multidimensional enterprise acknowledges the interconnectivity of systems and ecologies that exist in the design and planning of built landscapes. And for both practice and pedagogy, a substantial emphasis is placed on translating the discipline's production of designed spaces and artefacts with material affordance and cultural substance. Landscape architect and theorist James Corner states that

> The processes of which ecology and creativity speak are fundamental to the work of landscape architecture. Whether biological or imaginative, evolutionary or metaphorical, such processes are active, dynamic and complex, each tending toward the increased differentiation, freedom and richness of a diversely interacting whole.

In other words, landscape architecture is the discipline that is uniquely positioned to synthesise the immaterial with the material, to catalyse a design from individual components and assemblies into well-informed spaces and shared experiences.

However, those experiences of the everyday and the very elements that surround us are continually being threatened, endangered, and in many cases, are on the brink of extinction. The effects of climate change in our environments have been rhizomatic, cascading through networks of materials and energies,

DOI: 10.4324/9781003145905-5

biologies and chemistries, economies and cultural exchanges. Consequently, this has affected every minutia of our world in diverse and unpredictable ways, and most notably through the quality and composition of our everyday foods. As changes transform the future of food and the farming industry, the current and future state of all biotic matter becomes imminent and comprehensive. The United Nations Intergovernmental Panel on Climate Change (IPCC) released a 2019 report identifying the climate crisis as one linking food consumption and population growth; and it has been widely forecasted that by the year 2050, the world's population will surpass nine billion people. This has become a call to farmers, nutritionists and restaurateurs, as well as landscape architects and urban planners, to tackle climate change and global warming with not only the support of local communities and federal government investments, but with greater influence towards rethinking our collective practices and daily habits.

A future of overpopulation concomitant with mass urbanisation, forced migration and increased energy usage, each of these environment stressors will test humanity on land use planning and food security issues. In order to support a global food supply, productive agricultural lands need to expand at the very expense of rainforests, wetlands and other delicate ecosystems. Currently, 40 per cent of the Earth's land mass has already been cleared for agriculture use alone. Seventy per cent of the world's fresh water has been redirected for farming purposes. The extensive use of fertilisers has doubled the amount of nitrogen and phosphorus in the environment, causing widespread water pollution and significant habitat and biodiversity loss. Thirty per cent of total greenhouse gas emissions comes from agriculture, its greatest single source. An evolution in farming practices and technologies including advanced mechanisation, irrigation and cleaner transportation is a start for better food futures; however, the sustainability of food systems is incredibly complex. A systems-wide transformation of our current food network and supply chain – from food production to food distribution – will require more than better water and soil management, crop diversification and regenerative agriculture.

In the last quarter-century, 571 of the world's seed-bearing plant species have become extinct; and a main concern escalating within food security matters lies in the fact that only 150 food crops are cultivated for human consumption, out of a known 300,000 edible plant species. Among the most important food crops are three plants in particular – rice, maize, wheat. These "Big Three Crops" contribute to more than half of all caloric consumption for humans worldwide. Since the 1900s, approximately 75 per cent of plant genetic diversity and 90 per cent of crop varieties have disappeared, mostly in part by the replacement of genetically uniform, high-yielding varieties. Although testing and research has been widely published on genetically engineered variations of commodity crops and hybrids (e.g., heat-resistant, heat-tolerant, drought-resistant, disease-resistant, etc.), these agricultural innovations alone will not carry the weight of the world's growing population, nor the indeterminacies brought on by natural disturbances.

A vital aspect to preserving agrobiodiversity and other cultivated ecologies are strategies that contribute to resilience in the face of climate change. With a rise in extreme weather conditions including hurricanes, typhoons, tornadoes, floods, droughts, wildfires, fungus, pests and blights, etc. each of these have devastating and long-term impacts on food production, including the contamination of water bodies, loss of harvest and livestock, increased susceptibility to diseases and the destruction of agricultural machinery and infrastructure. However, yet another important piece to understanding how land degradation along with plant extinction is occurring at a rate 500 times faster than normal is understanding the effects of human activity and influence over this past century (Figure 3.1).

Edible Artefacts

A landscape affords us with a unique memory and identity to place, where the human body and its corporeal sensibilities can distinguish the material qualities of one environment to another. Considerations of affect, comfort and enjoyment, each provide an opportunity and a commonplace medium for landscape architects to employ through design. Also, within this framework, landscape architects can simulate a range of sensorial experiences and responses, well beyond servicing the visual and tactile domains alone.

What does a site taste like?

What does a season or year taste like?

One could say that every minute changes a plant's life. Through the basic interactions of weather–climate, soil–subsoil and terrain, a flower, grapevine or barley seed can forever be altered (Figure 3.2).

When one takes a sip of coffee, tea, beer or wine, one is consuming something that is grounded in a particular geography, a product of a specific time and landscape. The natural ingredients, which are highly sensitive to their environmental conditions, commingle to craft edible pleasures and create a complex and delicate balance of chemicals that produce distinct aromas, flavours and textures. Whether considering vegetables, meats and grains or alcohol, coffee and confection, food is one of the most quintessential of necessities to our health and wellbeing. Food also plays a significant role in our bodily nourishment and our daily routines – from cultivating, gathering and purchasing to preparing, consuming and discarding. Whether considering food as a domestic ritual, commercial endeavour or object of desire, it reveals our cultural evolution of place and memory, where kitchens and dining rooms, fields and gardens have become the landscapes that represent our identities, traditions and preferences to different cuisines and geographies. Food, in essence, defines who we are as individuals, as communities and as citizens; and its cultural and ecological value can undeniably be seen as complex and manifold (Figure 3.3).

The concept of terroir resonates authenticity and integrity for edible ecologies; an idea that values a traceable lineage and a timeliness of ingredients. Subtle varieties of flavour can differ from region to region, but also from field to field, from harvest to harvest and from crop to crop. A flavour profile can not only transform one's own understandings of landscapes, from one place to another, but also the time in which ingredients coalesced and distilled into one finished

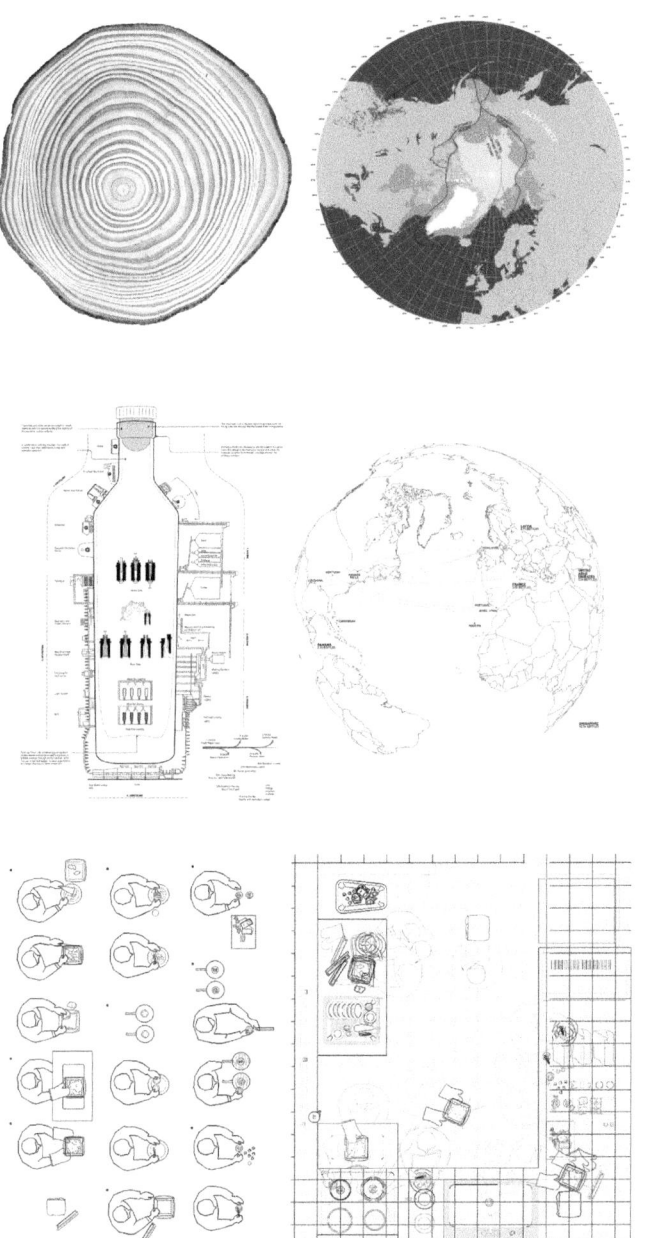

FIGURE 3.1   Top: Douglas fir, *Pseudotsuga menziesii*, material traceability. Taryn Wiens. Middle: Glass, manufacturing and distribution. Aleksander De Mott, Xitong He, Jingwei Jiang and Alex Kiehl. Bottom: Operative landscapes. Hangxing Liu.

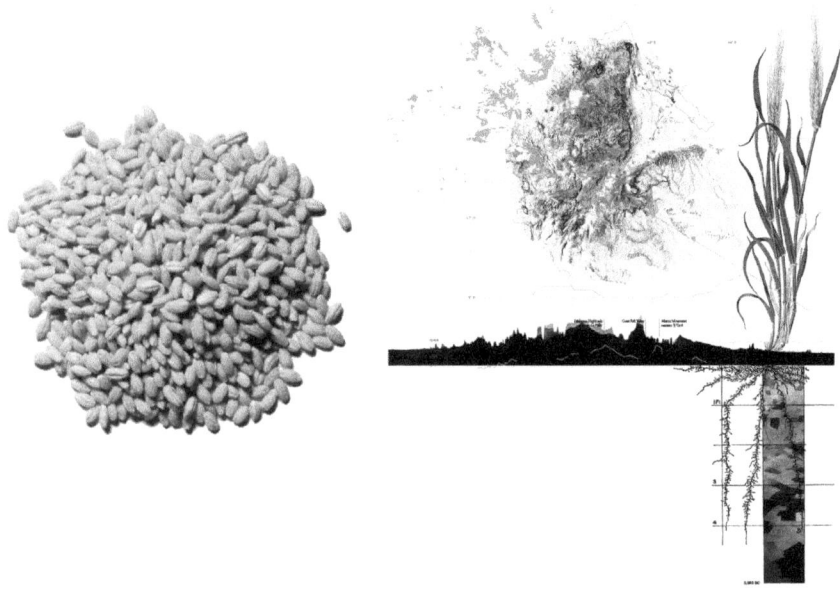

FIGURE 3.2 Barley (domestic), *Hordeum vulgare*. Barley is a member of the grass family and is a major cereal grain grown in temperate climates globally. It was one of the first cultivated grains, particularly in Eurasia as early as 10,000 years ago. Barley has been used as animal fodder, a source of fermentable material for beer and distilled beverages, and an ingredient of various health foods. In 2016, barley was ranked fourth among grains in quantity produced (141 million tonnes) behind maize, rice and wheat. Kazufumi Shimomura.

product. The differences within this one finished product become the foundation to what defines a food's unique set of properties or identity, a credit to the rudiments of its agricultural heritage including specialised methods of cultivation, harvesting and processing. In turn, these traditions and customs become a distinct recipe – an edible artefact – that defines an industry, a brand and a vintage.

Yet, with the onset of rising temperatures, intermittent droughts and erratic weather patterns, these climatic events have become everyday challenges to productive landscapes around the world. Even with advanced weather forecasting, extensive planning and preparation, the food industry is vulnerable to ecological disturbances. One notable event in recent years is the wildfires that have distressed the California wine industry. Although the region has become accustomed to stretches of wildfires over the last decade, most recently, 67,484 acres within Napa and Sonoma counties were decimated by the 2020 Glass Fire, what has been considered one of the most destructive wildfires in the state's history. Along with the resulting devastation of hundreds of buildings destroyed and damaged, most of the counties' autumn wine harvest was irrecoverable from

FIGURE 3.3 Cocoa (Cacao), *Theobrna cacao*. The seeds of the cocoa (cacao beans) are used to make chocolate liquor, cocoa solids, cocoa butter and chocolate. Historically, chocolate makers have recognised three main cultivar groups of seeds used to make cocoa and chocolate: Forastero, Criollo and Trinitario. Eighty per cent of chocolate is made using beans of the Forastero group, the most ubiquitous variety. Forastero trees are significantly hardier and more disease-resistant than Criollo trees, resulting in cheaper cacao beans. The most rare, expensive and prized is the Criollo group, the cocoa bean used by the Mayans. Hangxing Liu.

flames and smoke. Blackened grapevines, charred earth and damaged wineries have devastated this US$40 billion industry and have threatened not only its current state of production but its future viability (Figure 3.4).

At its core is the most prized artefact, Cabernet Sauvignon, a crop that reached US$1 billion in gross value in 2019. Two notable grape varieties have prevailed over the centuries; one variety that flourishes along the California valley floor and another in the hillsides. The flavour profile from the valley floor variety is lush and refined, finely crafted with hints of blueberries, ripe plums, black cherry, liquorice, mocha and violet. A dusty, bold and mineral-driven flavour from the hillside variety is accentuated by hints of blackcurrant, black cherry, wild berry, spice box, anise, espresso, cedar and sage. The environmental complications stemming from the Glass Fire, including remnants of contaminated soils and smoke taint (wildfire smoke that lingers in the air for an extensive period of time, causing chemical compounds to blemish the skins of wine grapes) have transformed the artefact's terroir into an ill-fated smoky and smouldering aftertaste. Although some wildfires are the products of natural phenomena,

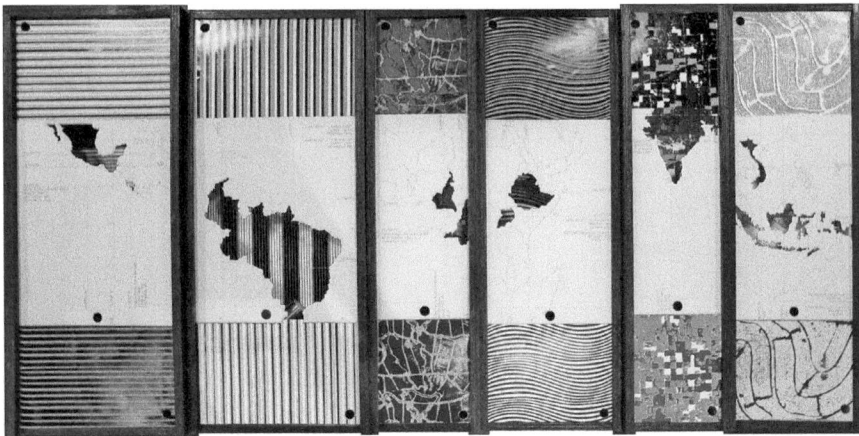

FIGURE 3.4 Coffee, *Coffea arabica* and *Coffea canephora*. The Coffea plant ranks as one of the world's most valuable and widely traded commodity crops and is an important export for several countries including those in Central and South America, the Caribbean and Africa. The seeds (coffee beans) of some Coffea species are used to flavour various beverages and products; and the fruits, like the seeds, contain a large amount of caffeine and have a distinctly sweet taste. The two most popular species are *Coffea arabica*, which accounts for 60–80 per cent of the world's coffee production, and *Coffea canephora* (also known as Robusta), which accounts for about 20–40 per cent. Palak Shah.

the likes of strong winds and lightning storms, their ignition for the most part have been recently attributed to human influence, e.g., power transmission lines, pyrotechnic devices, campfires, cigarettes, etc.

Edible Traces

When one purchases and consumes a food product, one is naturally not investing or directly ingesting the landscape or climate of a particular region; terroir cannot be seen as a monetised commodity, although it is frequently marketed as such. Rather, in the act of consumption, one is tracing in as much as sustaining a network of edible ecologies – ecologies that are entangled with landscapes, people and matter. In a recent research and design studio, students were asked to evaluate the gravities of today's material consumption, or rather overconsumption, of natural resources. The studio offered a framework to study material systems that exposed the interconnectivities that lie between design, construction and edible ecologies, providing a lens through which to understand and interpret lifecycle assessments and design decisions including material selection, specification and manufacturing. Research findings became recipes for generating

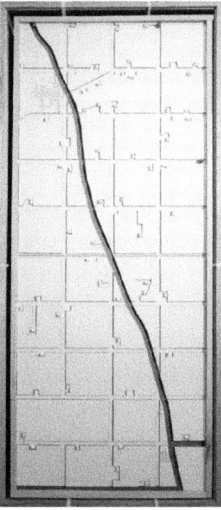
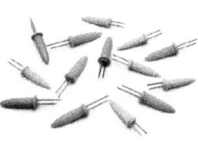

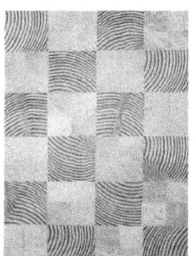

FIGURE 3.5   Corn (maize), *Zea mays convar.* saccharata var. rugosa. Corn is a cereal grain first domesticated by indigenous peoples in southern Mexico about 10,000 years ago. Corn has become a staple food in many parts of the world, with the total production of maize surpassing that of wheat or rice. However, little of the maize is consumed directly by humans; most of its use is for corn ethanol, animal feed and other maize products such as corn starch and corn syrup. Sugar-rich varieties called sweetcorn are usually grown for human consumption as kernels, while field corn varieties are used for animal feed. Other corn-based food uses include grinding into cornmeal or masa, pressing into corn oil, and fermentation and distillation into alcoholic beverages like bourbon whisky. Alex Kiehl.

designed artefacts; and the *cooking* of these artefacts equated to edible embodiments of cultural, socioeconomic and environmental values assigned to materials and food (Figure 3.5).

A selection of common materials (e.g., copper, cork, glass, oak, etc.) influenced by both the construction and food industries, coupled with a series of mainstream crops (popular and widespread plants, often commercially produced and consumed by a large majority of the global population and mainly influenced by global markets and economic trends) instigated discussions on the consumption of resources and the interconnectivity of material ecologies. Extracted from the United Nations Food and Agricultural Organization (FAO) list of the most heavily produced individual crops in the world, 12 mainstream crops were highlighted for further investigation: barley, cocoa, coconut, coffee, maize, rice, rye, sorghum, soybean, tea, wheat and wine grape. Each mainstream crop provided an opportunity to delve deeply into an edible ecology, and an opportunity to trace the material cycles and flows that have come to shape a food industry, a productive landscape, as well as our physical bodies. Information was measured, layered and digested in ways that communicated site-specificity, as well as time-specificity. Grounded in scientific research and design methods, information on each edible ecology was blended, strained and stirred through mappings, ontographies, models and prototypes. Each of these forms of representation offered novel systems-based relationships and (in)edible experiences that were consumed throughout the semester (Figure 3.6).

Investigating alliances between two distinct ecologies – one material, the other food – steered students from the misconceptions of a designed artefact or designed landscape as an autonomous object or environment. This holistic perspective forefronted the fact that living organisms do not operate in discrete or detached boundaries of space and time. Knowing that material ecologies are interconnected, and that they can be extraordinarily impacted by the myriad indeterminacies and inequities of environmental and climatic fluxes, as well as socioeconomic factors operating at the local to global scales, accepts how we

FIGURE 3.6 Tea. *Camellia sinensis* var. sinensis and var. assamica. Camellia sinensis is an evergreen plant that grows mainly in tropical and subtropical climates. Tea plants are propagated from seeds and cuttings; about 4–12 years are necessary for a plant to bear seed and about three years before a plant is ready for harvesting. Tea is generally divided into six different categories based on processing including white, yellow, green, oolong, black and post-fermented (dark). Aleksander De Mott.

as humans, and the matter that surrounds us, are but of a collective ecology in constant limbo.

A central tenet for the studio was embracing design objectives stewed in eco-literacy. We considered how nature sustains our lives, how we live with the uncertainty of living processes, and how unpredictable external factors including disturbances and variable weather patterns can alter supply and demand, incite economic volatility and wrench detrimental impacts on material ecologies enmeshed with constructed landscapes. Shepherding an ecological approach to design prepared students to address complex sites and turbulent issues; and the concepts manifested through the making of artefacts and landscapes acknowledged that design rarely ever serves a singular purpose, rather it addresses the many programmatic requirements and environmental agendas in which we are continually challenged with today. To endure the ways of being and becoming requires us, as landscape architects, to value sustainability and ecological responsibility with an empathetic and mindful practice.

Bibliography

Bennett, J. (2010). *Vibrant matter: A political ecology of things.* Duke University Press Books.

Cohen, J. J. (2014). *Prismatic ecology: Ecotheory beyond green.* University of Minnesota Press.

Corner, J. (1997). Ecology and landscape as agents of creativity. In G. Thompson & F. John Steiner (Eds.), *Ecological design and planning* (pp. 80–108). Wiley & Sons.

Corner, J. (2006). Terra Fluxus. In Charles Waldheim (Ed.), *The landscape urbanism reader.* Princeton Architectural Press.

Food & Agriculture Organization of the United Nations. http://www.fao.org/faostat/en/#data/QC.

Hill, K. (2004). Shifting sites. In Carol Burns & Andrea Kahn (Eds.), *Site matters* (pp. 131–155). Routledge.

Ingold, T. (2011). Landscape or the weather-world?. In T. Ingold (Ed.), *Being alive: Essays on movement, knowledge and description.* Routledge.

McGrath, B. (2018). Intersecting disciplinary frameworks: The architecture and ecology of the city. *Ecosystem Health and Sustainability, 4*(6), 148–159.

Mobley, E. (2020). Wildfires have ravaged Napa Valley. Will California's wine industry survive? *National Geographic Magazine,* 9 October.

Morton, T. (2018). *Being ecological.* Pelican, Penguin Books.

Tsing, A. L. (2017). *The mushroom at the end of the world: On the possibility of life in capitalist ruins.* Princeton University Press.

Wallace-Wells, D. (2019). *The uninhabitable earth: Life after warming.* Tim Duggan Books.

4

CONVERSATION WITH FORMAFANTASMA (SIMONE FARRESIN)

FIGURE 4.1 Cambio Formafantasma: Andrea Trimarchi and Simone Farresin.

Could you talk through the Cambio exhibition? How does it relate to your thinking about material ecology, and how you might define ecology?

I will start with Cambio and how the project started and then how it references back to our work and some of the questions we have in the studio (Figure 4.1).

DOI: 10.4324/9781003145905-6

Hans Ulrich Obrist asked us to do an exhibition at the Serpentine. But his request was different from many other museums and galleries because they weren't necessarily interested in presenting products or outcomes. They wanted to see the exhibition as a starting point: not as the final presentation but as a beginning. And they were interested in the multidisciplinarity aspect of our work.

We started thinking about the location of Serpentine in Hyde Park, London – that it was in a park and surrounded by trees, but also that close by was the location where the Crystal Palace was built for the first international exhibition. For us, these combinations of elements guided the Cambio exhibition. We thought that the Great Exhibition was an important event because it really shapes the understanding of design as a discipline. It did so in the sense that for the first time, the achievement of industrial production was presented, together with what was probably not yet named design, but what was considered to be design. And also, all the links of how contemporary production is shaped were already there. For instance, exploitation by European countries, moving of resources and the presentation of these resources together with industrial means.

Through selling tickets to the Crystal Palace, the Victoria and Albert Museum was founded, and Henry Cole was asked to rethink and redesign the art schools of the kingdom to link them to industrial production. You see here a very clear agenda: there was the understanding that industrial means could be incredibly relevant for development. That was firmly linked to both environmental exploitation and human exploitations, alongside building of institutions now exploring design through a study collection. That is, students could be developing their taste and their abilities based on a collection but, the *implications* of those designs were not explored.

The realisation of the implicit role design has in production, and the use and applications of materials, is an area design should explore more. This is important both in terms of understanding the implication of design, and also in developing a more holistic perspective. So that we don't consider just the needs of the user but also the infrastructure upon which the design relies. What I mean to say, if I need to deliver a service or a product, inevitably I'm going to support a system of extraction, distribution and refinement of resources. With Cambio, we're trying to say – if we are aware of the system, our choices, and maybe even our questions to our possible commercial partners, might be different.

We are also questioning this idea of the user. Who really is a user? Is it the person who's going to use our services or products or architectures, or is it also the people who will build those architectures, the companies who will extract those materials needed to build those architectures? Or maybe user is not the right word anymore. Important actors in this discourse are also all those humans that inevitably we're going to affect with our own practice. Cambio raises a lot of questions in this respect, taking the governance of the timber industry as a subject. But it also tries to give, if not direct answers, then I would say implicit design briefs, and I would say methodological ways of approaching these objects.

Looking at both Cambio and Ore Streams and to really talk about these sets of relationships and their entanglement. What is it about the current situation, the climate crisis, that you think these kinds of methodologies can assist with when looking at a broader set of problems?

> There are several responses in both projects about the way we approach these complex questions, or even hyperobjects, both within our work and within education. First of all, design here is applied to understand entanglements. In the case of Ore Streams, for instance, between all those different figures, they're shaping industry but rarely actually entering contact. Here is the first problem design can address: the problem is how the idea of efficiency, in terms of production and economic growth, compartmentalises and separates knowledge. This creates fragmentation, which results in you not feeling responsible as an individual or even as a corporation, as a company, as a design studio and so on. It also fragments your understanding of the bigger picture.

In the case of Ore Streams, I would say we took a very traditional research method through interviews and reading published papers. What we tried to do was map out, for instance, the complexity of electronic waste recycling, looking at legislation, economic research, NGOs working on a global level, and recyclers and producers. Our goal was to understand individual knowledge and position, and see what, on a very pragmatic level, can be sorted out now. How can we improve the design of electronic products so that it can be more reusable and recyclable? I would say this responds to the basic needs of today. It's the urgency of today. But on the other side, one important thing is to think about solutions on multiple timescales.

One difficulty we see in education is the tendency to either work on a metaphysical level, so that you work on the deep roots or the system, and thinking on those problems in a visionary way, or alternatively to try to solve problems immediately. But resolving those problems often ends up with very short-term solutions, or not even solutions but what look like solutions. As designers, it is important to tackle this complexity with multiple scales of intervention; to think about what is needed now, what is needed tomorrow and what is needed maybe in 50 or 100 years.

Time is an element that needs to be considered. The other element is scale. Both within our own practice and when we teach or within the department that we just started, we try to look at a given, let's say, problem or subject, as a body, and design an intervention as acupuncture. It is not about one solution, but about a multiplicity of triggering points that helps the definition, the outlines and the scales of the body. It is never about the big solution or the big intervention. It is always about respect and trust and recognising that a multiplicity of small triggering points would help cure the problem.

In terms of scale, is it also thinking that some of the acupuncture might be policy?

Absolutely. Those multiplicities or levels would work both in Cambio and Ore Streams, but one level is on the level of the governance. The question here is, why should the architects and designers be interested in policymaking? And are they the right people to be involved in that? Yes and no. What I mean to say with no is that, obviously, we are not here to say designers should improvise and become policymakers. Nevertheless, especially when we talk about certain specific subjects it is obvious why designers should be involved in policymaking, in the case of electronic waste for instance. It's because we are aware of the structure of the production, development and engineering of objects. And if, for instance, we look at European policymaking, while there are several very interesting actions at the moment, whenever you look at policymaking at the level of design, the outlines are rather limited.

If a designer is not necessarily involved in policymaking, at least the designer should be aware of what is happening in policymaking. Because it makes you much more aware about the industry you're working with, and it might also inform your decision-making. I'm thinking, for instance, about the case of Cambio and the timber industry. At this moment we are trying to collaborate with a company producing furniture in wood, and basically, we want to look at their production in relation to forestry, to the forest. To return this relationship which has been taken apart in recent years. Working on a policy level, or being aware of policymaking, will have a direct impact on the way even the business model of that company might change in the future.

What you're describing are material ecologies but also the new potentials of how we practise in relationship to the climate crisis, is one that we understand in relationship to its political and economic imperatives, and thinking about that, not only from the local, or acupuncture point, but also in relation to planetary impacts.

> Absolutely. The attempt with Ore Streams, and in part with Cambio to make a very practical example: if we design things here and now, well, this here and now will not be the reality of that product. That product will spread branches and expand all over the planet, and the awareness of the different contexts where the product will go might affect our own way of thinking about it. For instance, recycling, there is the reality of recycling in Western countries, and there is the reality of recycling in other than Western countries, that needs to be taken into account.

When we started looking at strategies of making the product more repairable or recyclable in Western countries, and when we looked into what we call developing countries, well, you realise the necessities are very similar. It's looking on a more planetary scale, then the problem might help you understand almost archetypical problems.

Could you talk about setting up the Geo-Design programme, your ambitions for that, and the structure of it?

> One reason I am interested in education is because we are addressing some issues with our more research-based projects that are, for us as designers, extremely relevant. But we cannot necessarily address issues in our daily, more commercial practice. Education responds to this, recognising that we can address these subjects with our research-based practice, but not necessarily with our working-world practice.

We started thinking that another generation will be more effective with the things we care about, and education can respond to that need. At the beginning we try to put into practice immediately, our own interests in the research projects. That's often not possible. And we understand, again, it is a matter of scale and time. If our time is limited because of our lifetimes, well then there are other people who can continue to address certain issues. Thinking about education, it seems our observations are actually about letting others continue, letting others contribute to things. From this perspective, the Geo-Design Department wants to consider design as an act that is the outcome of an authorial perspective, an individual perspective, but it is also a collective way of thinking about subjects.

The Geo-Design programme was introduced in 2018 by Joseph Grima, the head of Design Academy in Eindhoven. And there was a Geo-Design exhibition platform, and there is also a Geo-Design Department, and Joseph asked us to lead the department. He left us free, to a certain degree, to interpret the term Geo-Design. What we like about it is that clearly the term "geo" inevitably references the planetary scale of design. It makes a clear declaration that design cannot live without this consideration.

First of all, the Geo-Design Department believes the department itself should develop as a studio, where knowledge is shared. Individuality is respected but knowledge is shared, and it is built upon. And it is, exploring some of the issues we are exploring within our studio. It is thinking about the planetary scale of design intervention and the consequences of design, and the infrastructure upon which design performs.

The department is trying to address complex issues with a multidisciplinary approach. What we're trying to do is to have students from multiple disciplines join the course, and also to have mentors who are not necessarily design mentors. For instance, in the third trimester we will have a scientist, a biologist and a conservationist join the course. The mentoring team will be an architect and a conservationist. The students will look into working on a specific location in the Netherlands on a project to reintroduce wilderness in the country. The question will be how design can intervene in such a place. The challenge will be understanding if design can contribute by developing design ideas for other than human species, and whether it is even possible to do so.

It seems you're proposing that the individual is important in terms of their rela-
tionality to the planetary or … It's suddenly shifted the hierarchical possibilities.

> It's a very pragmatic decision. How many design students are develop-
> ing their industry projects and the research is kept for themselves? I don't
> think we can necessarily apply a scientific method, or model, to the design
> discipline. But we can at least refer to it in terms of sharing a certain
> body of knowledge that can be appropriated by others. I think that's a very
> reasonable way of working. Then the individual outcomes are extremely
> important because design cannot apply the scientific method fully because
> it is a humanistic science. The idea of sharing research is also an extremely
> valuable way of working. How to do that? That's a different question. We
> are working on that!

The educational environment that you're suggesting – it's a laboratory type of envi-
ronment where it's peer learning, collaborative learning. Would you say the pedagogi-
cal structure to be very similar in that it allows for continuous knowledge transfer of
knowledge year to year or semester to semester?

> Everything is in a state of becoming as we try out things. The first
> important thing is to clarify the objective with the students. We must
> have a pedagogical structure which is transparent, and with transpar-
> ent aims. It seems obvious, but often students understand what they do
> differently if they are aware of the reasoning behind why you're doing
> a certain thing.

We start with very basic tools. For instance, students are developing an online
platform where content is shared. Opening a conversation between students and
mentors about how that content should be shared. Raising questions – Should it
be edited or not? Should we set up a format? Then in another trimester, we will
work on taking the outcomes of this into a publication that will be for, not only
the department but also for the outside world. There are multiple levels of editing
of content. If this works, and probably one assignment for next year, will be to
work upon the research of the previous year.

The students in the current studio are working on the Cambio archive. We
use Cambio as an exhibition but also an educational programme, inviting mentors to look into our archive and develop assignments. Then the students work
on those assignments, and the touring exhibition will implement some of the
works of the students. We're trying to make sure there is a constant feeding of
one element to the other, and at the end of this year, or maybe at the end of two
years with the first graduates, there will be a moment of evaluation and of reassessment of what we're doing and if it's working.

The work implies there's a constant shift in subjectivity, both in your research and
also within the teaching pedagogy, could you expand on that?

A lot of our work is what it is because of reflections on the responsibility of being a designer. Some questions emerge simply by working. When we were working on Cambio, the exhibition starts with a very up-close observation of materials, with any information within it, and then the more the exhibition goes ahead, a shift in perspective and even timescales occur. We felt it was important to have at the end of the exhibition this work that we developed with Emanuele Coccia, the philosopher, because it is a reversed perspective. It reversed what was done before, which is extremely anthropocentric in perspective. That exercise is an extremely important one: to try to develop a more empathic understanding of the needs of other than human creatures. In our work there is often this shift in perspective, and often there is not a resolution of this multiplicity of perspectives.

For example, students are working on some of the issues in Cambio and thinking about the forest as an environment that includes productions but also other services that trees and forests provide to other creatures as well as humans in terms of CO_2 intake and so on. Students are thinking of everything that can be done on, I would say, a production level, an extraction level, without the extraction of timber. Which other things can we do with things that trees provide? That is a very practical way of interpreting the necessities of trees, for instance, of trees that do not necessarily need to be exploited as a commodity.

There's a generosity in the multiple perspectives. If you're using it as the source for the students, it allows them to also enter into multiple places or ways of thinking. Is this a mode of open source?

Open source is implicit in the research model of sharing research. The idea of open source often implies scalability, and applications in different environments. That is extremely interesting, but the idea of having ideas and research open source is much more interesting, because it allows different people, different communities, to take knowledge and bend it towards what is interesting for them in that specific place instead of thinking of open source as a system that is created in a place and is then distributed globally.

I like that you use the word generosity because, indeed, what is also interesting is that possibilities open up when you try to be generous and to embrace the needs of multiple users, or multiple communities. Instead of becoming a frustrating process of further and further limitations, if you're a good designer, or a good architect, or a good thinker, you see all possibilities open up instead of constant limitations growing around you.

Your work challenges the idea of advocacy, that we can advocate for the many.

It's for the many, but also for what is important to advocate for right now, and what is important to advocate for tomorrow. We're constantly obsessed with this idea of scale, but also time. The idea of scale has been explored a lot, but the idea of what design can do over time is less explored, because it is all about the human scale and what we can do now. No matter what our practice, it's always in the service of economic expansion, and economic expansion means actions now. But if we do things with the idea of being generous in building a more sustainable, ecological world and way of living on the planet, then inevitably the timescale becomes much wider. It's not about what you can achieve today.

There's a sense of empathy in the work. And that consequently constructs differ-ent value systems that are being brought to the foreground. When we think about empathy and value systems, there's a novelty in what artefacts are generated from. The relationship actually shifts and novelty becomes, in the way that we under-stand the artefact or the spatial condition, actually becomes transformed. It allows for new possibilities or new assemblages to emerge. Would you agree?

The question on the materialisation of ideas, it's a very big question that should always be addressed on a contextual level. Again, it is important to also read our practice for what it is, and so, for instance, the outcomes of Ore Streams: they are what they are. Also, because we were working with a museum that had a design collection based on furniture.

It is important to recognise that because then you read the work for what it is. This collection together with our interest, for instance, made the artefacts, or the visualisations what they are. In the case of Ore Streams, we knew we could have engaged in whatever projects we wanted, but at the end we had to design some furniture pieces. With Ore Streams the production of those objects became the excuse of the Trojan horse: it allowed us to engage with those industries like recyclers, people were sourcing secret boards to extract. A lot of people interpret those objects, for instance, and I want to be specific with Ore Streams because I think it is a very good example of a visualisation of the outcome, of the results.

The work is what facilitates the research. The research is the outcome, and the object is the process to get there. The objects and the artefacts explore ideas on a visual level. It is about what they do physically – using recyclable materials and so on – but it is also how they look and the ideas that they explore on a visual level. And this is extremely difficult to discuss even in the context of a museum because the relationship of how objects are presented and how design research is explored, is often very binary. Because of how design has been exhibited, por-trayed and used, people enter with very clear ideas of these relationships.

And this is why I was mentioning this very clear relationship created on many different occasions, but specifically with the Great Exhibition and the found-ing of the Victoria and Albert Museum. We use that as an example of that very clear relationship between all these elements that makes design the discipline it is. Often we do not have the occasion to discuss this relationship between the artefacts and our research process in depth, which is a pity. For instance, Cambio has that outcome because it is an exhibition, and the spatial condition of the Serpentine dictated what the exhibition was. The two rooms at the centre, and the sort of corridor structure that goes around it, already created a curatorial structure.

There are always limitations and possibilities in the media and in the environments we work with, too often those ideas are not really taken into consideration. When you read an exhibition, you should also really read the material in the space, the conditions of the space.

5

SHIFTING GROUNDS

Curating Creative Instabilities in Design Studio Pedagogy

Chris Reed

Change in the environment is inevitable.

Complex adaptive systems ecology posits that the environment is always in a state of change – and, in fact, it is organisms' and ecosystems' ability to change and adapt that both characterises the norm and defines ecosystem health. As early as 1971, ecologists CS Holling and MA Goldberg noted that conceptions of dynamic ecosystems ecology should trigger new thinking in landscape planning and management to better accommodate this understanding (Figure 5.1).[1]

Climate dynamics can be seen as an extension of complex adaptative systems ecology, but in ways that are increasingly complex and include humans and our constructions (physical, regulatory and sociopolitical). In fact, today's climate crisis is in part a result of a failure to act in concert with the Earth's rhythms and its biological and atmospheric functions. Our understanding of the environment now rightly includes human-constructed and human-influenced systems and places, but they are having devastating effects on the efficacy of ecosystems worldwide. Typical responses – which include armouring, protecting, defending buildings and lived fabrics – just exacerbate the problems, and often result in inequitable responses.

> *What I think of as history on this land – the events that occurred and the narratives told of them – can never be complete or single-voiced. Each of us participates in it. We contribute to it as players, as witnesses, as narrators, as producers and consumers, in an ongoing past-to-present.*
>
> Lauret Savoy, Trace: Memory, History, Race and the American Landscape

The climate crisis is also heightening an awareness of the intertwined relationship between people and the environment – and exacerbating the disproportionate

DOI: 10.4324/9781003145905-7

FIGURE 5.1   Shifting sand and water currents, Nantucket Island. Google (n.d.). [Shifting sand and water currents, Nantucket Island]. Retrieved 23 June 2020, from URL: https://www.google.com/maps/place/Nantucket,+MA/

effects of climate impacts felt by low-income communities of colour, which are already suffering from disinvestment, environmental and social injustices, racism, etc. (in part a result of racist policies and planning regulations of the past). In fact, self-described "earth historian" Lauret Savoy argues that understandings of landscape and environment (and by extension climate) can never be dissociated from people who inhabited and explored these lands and must be expanded to include muted or deliberately buried histories and people.

All of this suggests a need to step back from the immediacy of solving problems to explore how to first recognise and then work with the dynamics that are in play – environmental, social, cultural and economic, political, regulatory. Here, the design studio can serve as laboratory for design research, design investigation, design exploration – for tapping into investigations that explore *in*stabilities rather than static conditions; that recognise and embed multiple and diverse social and cultural histories, stories and needs; and that inform richer, more emergent possibilities that embrace the fluid environments in which our work participates.

Climate crisis requires new ways to think and work, across disciplines, across jurisdictions, across boundaries and property lines, and that span from design to policy to advocacy and education. This means not assuming fixed points or fixed goals – rather, understanding that environmental dynamics might inaugurate a lighter and more temporary or at least iterative approach to designing land, with a sense of how landscapes might unfold in time. It also means that the morphologies of landscape and landform, for instance, are never quite complete, always thought of in a state of transition. The broader goal is to create new adaptive futures through hybridisation of agendas, performance-informed interventions, and to embrace the idea of open-ended futures – that these can be directed but not controlled, deliberately left often to engage conditions and circumstances to come.

> *How can public and civic architecture go beyond engaging its users with amenities and instead fundamentally realign and reorient the devices of power toward users' aspirations, toward creating community, framing the conditions of shared citizenship, of res publica – the commonwealth of diverse social and cultural experiences?*
>
> Okwui Enwezor, "Gestures of Affiliation"

This is also an opportunity for creative design exploration, exploration that integrates the various scientific and social intelligences of the process, peoples and places. Here we might consider a dual focus on socio-ecological informants to the climate crisis, and the material flows/physical conditions that can be manipulated to respond to it: sediment, construction materials, construction and demolition processes, ecological palettes and processes, water, heat, shade, etc. (Land tenure is one starting point, for instance; invoking indigenous understandings, whereby land is thought of as a common resource, where an individual only takes what she or he needs, and that occupation of that land is considered temporally and lightly – even instigating seasonal migrations across land in response to weather, shelter, food resources, etc.[2]) But we must also continue to consider the projective nature of what we do, the cultural nature of our individual and collective design project. As designers, we will never be the best scientists or policymakers, the best environmental advocates or sociologists. But we do have the ability to integrate these disciplinary imperatives, and to project and imagine and inspire new worlds. Here I turn back to the writings of Graham Swift, whose novels are both poetically descriptive of the in-between, land and waterscapes of the Fens in eastern England and also aspire to an art form that both takes us away from everyday toils and inspires imaginings of new futures yet to come.

So how to approach teaching and studio pedagogy to mine these fertile grounds? For me, the process is centred around deep and creative research paired with iterative design methodologies that begin with the conditions and dynamics in play, and only eventually make their way to the physical and conceptual ground of the studio site. And while these questions sit alongside some of the work we may do through design research in practice, the studio environment allows us to pull back even further, to be more reflective, less constrained by a particular set of agendas or briefs or scope limitations.

Over a decade of studio teaching at Harvard's Graduate School of Design has touched on this thinking in very different ways and has evolved to take on a rich myriad of issues and challenges. Early core studios introduced a landscape-based approach to urbanism and integrated early fluid simulation modelling for urban river and stormwater flows as one component of dynamic landscape systems that could frame and form urban districts, largely in temperate cities like Boston and New York. Later option studios advanced both interests in dynamic modelling and simulation on the one hand, and urban infrastructure on the other, with a succession of studies at California's Owens Lake (denuded and changing ecologies, related to the metropolitan infrastructure of Los Angeles, through "projective processing" techniques); at Houston's eastern Buffalo Bayou (flooding and climate change within the context of disinvested black and brown communities); and involving the conversion of Santa Monica's municipal airport to an arid-adapted climate infrastructure (incremental park, food production and biodiversity catalysts). More recent investigations include multidisciplinary work in a historically black community in Miami, disproportionately affected by disinvestment, racism and the ongoing impacts of climate change; and dynamic coastal environments of the east coast, and the ways in which resetting timeframes and timescales might help us better recognise the waves of environmental succession and human migrations (and peoples) that have inhabited such a place.

Typically, initial investigations in these studios ask broadly: what are the forces in play, at multiple scales of space and time? Typically, this work is diverse, deep and intersectional, and examines issues from policy to physical conditions and dynamics to social environmental issues and processes. Yet this is not mute analysis; student teams are asked to uncover complexities and missing links, contradictions or gaps, that could lay foundations for further investigation. Issues of landscape, climate, social and racial equity, identity and urbanism are all in play, and the work is both quantitative (including geographical, measurable) and qualitative/cultural in nature. Students are asked to develop a combination of deliverables (maps, booklets, animations) that tell multidimensional stories about the place, organised around individual topics, that can eventually be shared across the studio cohort. Here multidimensional refers to spatial and scalar issues, but most importantly implicates and integrates the temporal, working across multiple time scales from the geological, to the ecological to the momentary (Figure 5.2).

Subsequent exploratory design investigations stem from the very specific conditions and forces uncovered through the initial research. They step back from the particularities of an individual site to address broader conditions (physical and otherwise) that may be encountered across the territory of a studio site. This could be anything from how underwater scaffolds interact with water flows, offering ripe habitat for shellfish growth or sedimentation, for instance, to how gentle manipulation of earthen surfaces in arid climates might instigate a subtle but important change to hydrologies – and, by extension, to plant growth and succession. Here design is invoked as a creative, projective and interactive act – investigating landscape and urban processes and protocols (how do these systems work, and by rules do they behave?) and programmes as they relate to physical form. The intention is to explore design form almost too early in the process, almost not wholly informed – but as a deliberate inquiry in its own right, and then to test through wide-ranging iteration. This often happens through one of two lenses:

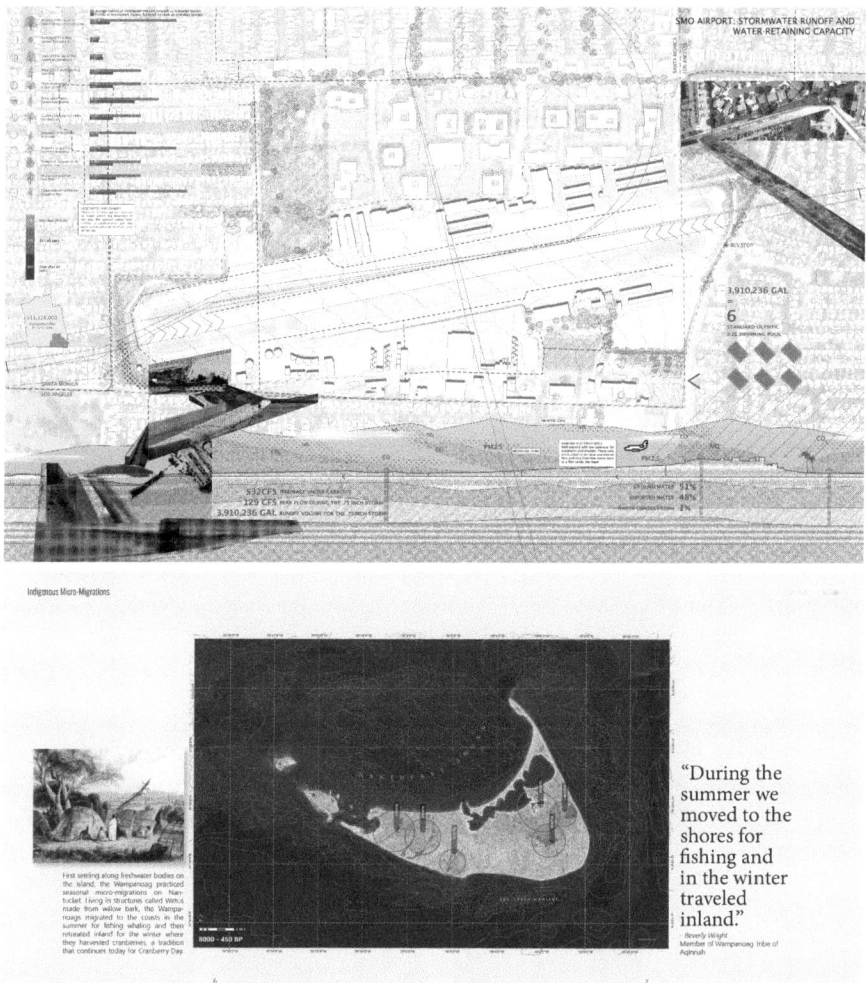

FIGURE 5.2 Top: Flood and Wildfire/Disastrous Dualities, analytical research drawing, SMO PARK+ studio, Harvard Graduate School of Design. Haoyu Zhao and Xiaoji Zhou. Bottom: Indigenous occupations, Nantucket Island, Offshore … Adrift … studio, Harvard Graduate School of Design. Yvonne Fang and Kymberly Ware.

- Seriality, or investigations of processes and patterns, their intersections and opportunities for creating new worlds (multiple and alternate morphologies in relationship to formative and/or evolutionary processes and dynamics, often using fluid simulation software) (Figures 5.3 and 5.4)
- The mash-up (Sean Canty's term), or investigations of multiple agendas, multiple informants, and their hybridised opportunities (in most cases the

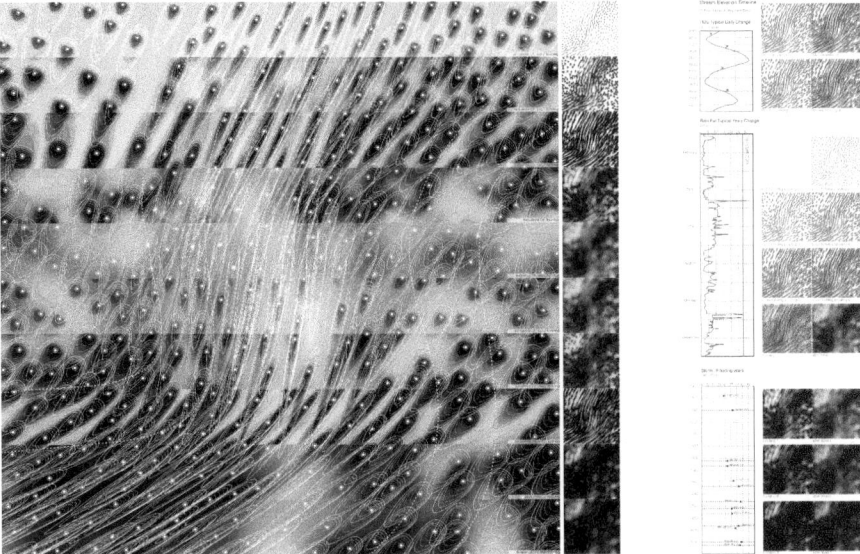

FIGURE 5.3 Fluid projections/flow simulation studies, Re-Tooling Metropolis studio, Harvard Graduate School of Design. Xun Liu and Ziwei Zhang.

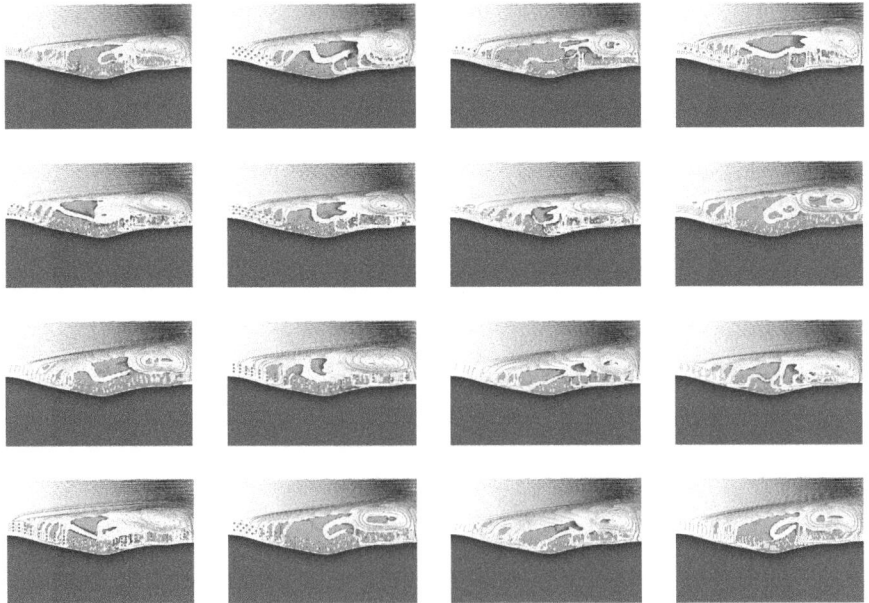

FIGURE 5.4 Water flow and sedimentation simulation studies, Away … Offshore … Adrift … studio, Harvard Graduate School of Design. Caleb Nash and Gena Morgis.

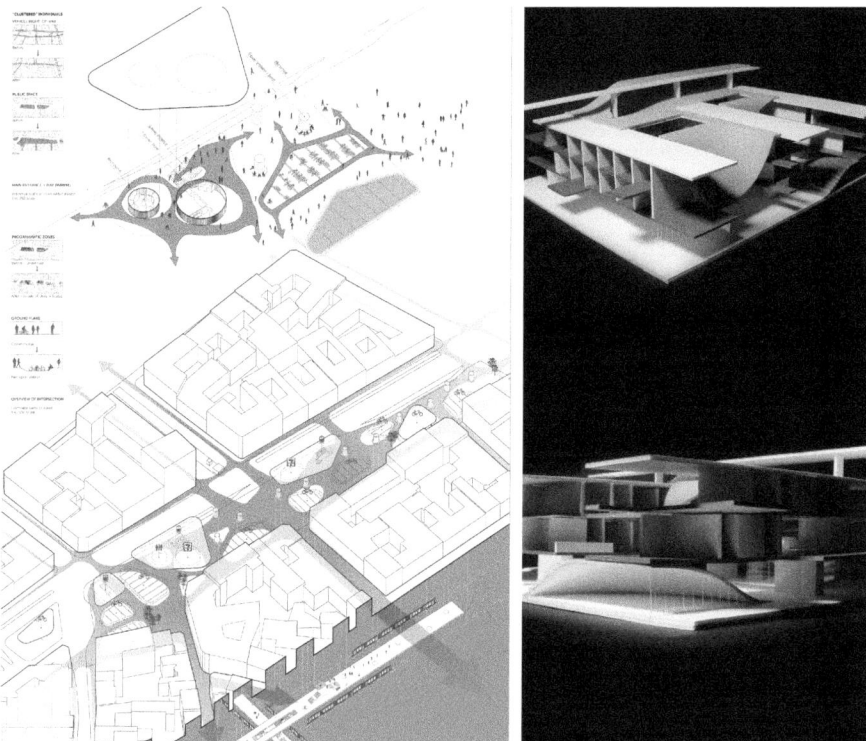

FIGURE 5.5   Mash-up of Gifu Kitagata (Gifu, Japan; SANAA) and Noørreport Station (Copenhagen, Denmark; Gottlied Paludan Architects + Cobe) Multiple Miamis studio, co-taught with Sean Canty, Harvard Graduate School of Design. Sam Atkinson and Hiroki Kawashima, Ting Liang and Zishen Wen.

hybridisation of two designed entities, a building or designed infrastructure with a designed landscape) and new hybrid forms and programmes that could emerge (Figure 5.5)

The idea is to allow for wide-ranging design exploration, situated in the milieu of the studio site but unencumbered by the specificities of a particular site's complexities. In doing so, students are asked to develop operational diagrams that describe the interrelationships and dynamics in play, and the various design responses and catalogues of parts that might be strategically rearranged or tactically deployed in response to any number of situations or conditions.

Finally, work turns towards the development of site-based strategies that engage a multitude of factors but are scenario-based, built upon sets of assumptions that could change and do change over time. Here students are asked to apply their design investigations to designated sites or sites of their own choosing, to scale and re-scale, to organise and connect, to ground their explorations in the contexts and dynamics of a particular place. In doing so, students are asked to consider a number of factors:

- What are the rules of engagement, how do things work and operate, how could they play out in time?
- How to consider questions of initiation, where and how to start?
- How to implicate time, flexibility and adaptability – in part through scenario-making, a nod back to complex adaptive systems theory that allows for multiple productive futures

Proposals mix the physical and temporal, as well as the enabling infrastructures, policies and regulations that could allow them to emerge over time. Students describe multiple potential futures through scenario-making dependent on a range of conditions, if–then sketches that can respond and adapt to changing circumstances, and that leave parts open to future curation

Profiles and prompts of three recent studios, taught at Harvard's Graduate School of Design and open to landscape architects, architects and urban designers, offer three points of comparison within three very different social and environmental settings. *SMO PARK+* is situated in the arid, metropolitan landscape of Santa Monica and Los Angeles, on an airport site about to be retired, and focuses on plant ecologies, and demolition and construction materials and methodologies. *Multiple Miamis* focuses on the inner city, African-American and Afro-Caribbean neighbourhood of Overtown in subtropical Miami, and focuses on water, heat and shade and basic necessities like quality housing and open space. *Away … Adrift … Offshore …* is situated on the island destination of Nantucket, offshore from temperate New England, and examines potential climate futures of coastal waters, sediments and habitats. All three mine the muddy waters of climate change, urbanism and questions of what it means to project new fluid and equitable futures for these places and the people that inhabit them.

SMO PARK+ Climate Infrastructure, Arid Landscape, Spring 2020

Santa Monica Municipal Airport (FAA code SMO) is a 227-acre general aviation airport about 6 miles from the Pacific Ocean that largely handles private and corporate flights. After many years of battles between the FAA and the city, the federal government agreed to close the airport in 2028 and the city will convert it to parkland. Despite Santa Monica's relative wealth and prosperity, though, the city is struggling to imagine how it will afford to build and operate a new, large public park. More generally, Metro Los Angeles continues to burn and flood at an increasingly alarming rate due to the rapidly accelerating effects of climate change, homelessness is at a critical level and the whole state of California continues to put pressure on food, water and energy supplies in ways that exacerbate the state's climate- and environment-induced challenges.

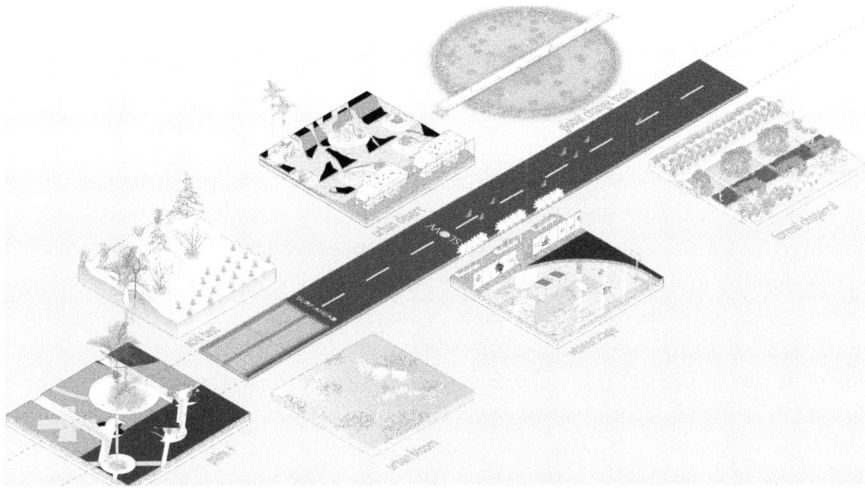

FIGURE 5.6 Freakish Grounds, A Botanical Collection for a Hotter LA (open space re-use framework for Santa Monica Municipal Airport), SMO PARK+ studio, Harvard Graduate School of Design. Karissa Campos, Hannah Chako and Zoë Holland.

This studio is set within this complex environmental, social, economic, regulatory and jurisdictional context, and will explicitly explore the nature of what a contemporary public park can and should be. The studio will take on questions of the role of the public realm within an arid environment, and in an era of climate change. Students will be asked to develop multifunctional park proposals for this large-scale urban site that integrate climate mitigation and adaptation strategies associated with critical food and water supply, excessive heat, storm and drought/flood and wildfire cycles, biodiversity loss and clean energy production and distribution – with potential secondary agendas for social housing and revenue generation (Figure 5.6).

- Part 1 Climate Impacts/Infrastructural Misalignments
- Part 2 Seriality: Process and Pattern
- Part 3 Adaptivity: Climate Park

Multiple Miamis, Fall 2018 (co-taught and co-conceived with Sean Canty)

Multiple Miamis looks to imagine new and diverse urban strategies that recognise a multiplicity of environmental, social, cultural and economic starting points, and that redefine what an inclusive and civically minded urbanism could be. The work will focus on an area of Miami that was once productive agricultural fields and later a vibrant black community, that was torn apart by racist zoning policies and targeted interstate highway construction, has recently witnessed high poverty rates, vacancy and land speculation; and is subject to multiple effects of climate change. Design and strategy will emanate from the mash-up, in which diverse cultural and disciplinary starting points will be fused in open-ended, formal and strategic explorations. They will build on initial investigations of hydrologic, demographic, infrastructural, typological, economic and productive foundations, past and present. And they will inform multidimensional speculations that may focus on affordability, housing and public space; on multifunctional infrastructures and productive landscapes; and on access and mobility-related development initiatives – all of which deliberately take on questions of what is a just and equitable urbanism in 21st-century America. Multiple Miamis will also cultivate conversations on the nature of contemporary urbanism, on race and identity in America, on the possibilities for transit, affordable housing and landscape infrastructure strategies that could transform the city, and on the opportunities and challenges faced by the 21st-century American city. The studio will embrace the idea that projects can be rooted in multiple starting points, in multiple cultural and environmental traditions; and that they can emerge over time as a result of catalytic and transformational processes that continue to work at multiple scales of time, space and policy.

- Part 1 **Multiplicity**: a large and diverse variety, complex in nature or effects
- Part 2 **Mash-up**: a mixture or fusion of disparate elements
- Part 3 **Mutations**: a change in form or nature

Away … Offshore … Adrift … Shifting, Spring 2021

Nantucket, meaning "faraway land or island" or "sandy, sterile soil tempting no one" in Algonquin, is an island 30 miles off the south-eastern coast of Massachusetts, formed through glacial processes and ice melt, and continuously re-shaped by strong ocean currents, winds, storms and human constructions and impacts. It served as both home and seasonal farming and fishing grounds for the Wôpanâak tribe (meaning "People of the First Light"), and it came to be a haven for an extensive black community, whose members could find stable work around the wharves, far from mainland racist attitudes and laws. It has since become a tourist destination and summer playground for the elite – but this simplistic characterisation denies the substantial year-round and seasonal workforces, racially and ethnically diverse, who power the robust service-sector economy.

Climate change is already bringing rising seas, regular flooding, and coastal erosion to many parts of the island, and threatening areas in and around the main harbour town that are low-lying, close to eroding bluffs and to shifting sands. In this context, many are asking questions about how to counter these effects, by armouring, raising and protecting the various cultural, infrastructural, ecological and personal/private resources on the island. While we recognise the importance of these efforts, this studio will step back and reassess our collective starting points. In doing so, we will ask different questions – questions of how to first recognise and then work with the dynamics that are in play – environmental, social, cultural and economic – as starting points for exploring alternative futures

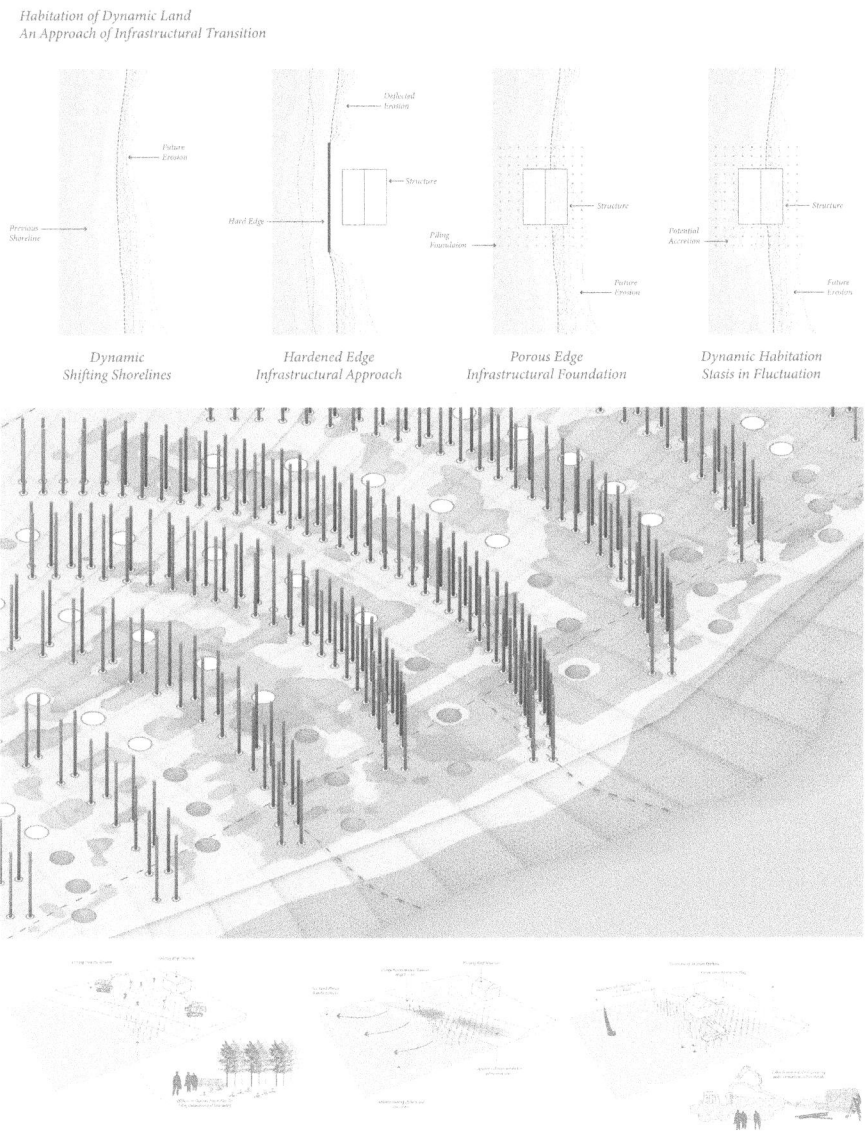

FIGURE 5.7   Inhabiting Instabilities concept strategy diagram, deployment study and project conceptualisation boards, Away … Offshore … Adrift … studio, Harvard Graduate School of Design. Caleb Nash and Gena Morgis.

for the island. How can a rethinking of the relationships between stability and instability lay new fluid grounds for more adaptive solutions? How can we render invisible processes and people visible and central to the conversations about land occupation, landscape and cultivation dynamics and sustainable work practices? How might we think about new forms of occupation and regulation/law relative to land, land use and our place and time in and on it?

- Part 1 **Migrations**: movement or periodical passage from one place to another
- Part 2 **Shifting Grounds**: changing the form and nature of surfaces, soils and land and water areas – and of the fundamental conditions and beliefs that underlie them
- Part 3 **Provisional (Re-)Occupations**: the interim (re)uses or (re)settlements of land, and the temporary or conditional activities in which one engages to realise evolving futures

Coda

In short, none of these studios take direct aim at only solving the problems at hand – they are deliberately speculative, acknowledging the cultural ambitions that are (or should be) at the heart of all we do as designers. While they certainly look to have students explore the very broad and very real range of issues and dynamics that may be in play on very complex sites, they also shift the foundations of the work from the simple frame of problem–solution to something that is more hopeful, more optimistic, more opportunistic, perhaps. They look to instil a different mindset and agility in the students and in the work, that embraces fluidity, lightness of touch, complexity, shifting grounds and social and cultural multiplicities. This approach also fosters a porosity of ideas and informants to the work, opening up our focused disciplinary explorations to extra-disciplinary concerns and forces. In short, this approach productively destabilises our sense of certainty, and of fixed solutions and boundaries, in favour of explorations that embrace open-endedness, the interrelationships between and among things, and the wider landscapes of the imagination (Figure 5.7).

Notes

1 CS Holling and MA Goldberg, "Ecology and Planning", 1971, republished in Chris Reed and Nina-Marie Lister, *Projective Ecologies,* 2014.
2 These observations are taken from class-based discussion with Beverly Wright, former Chair of the Wampanoag Tribe of Gay Head, as well as collective student reach in the spring 2021 studio "Away … Offshore … Adrift …" at the Harvard Graduate School of Design.

6

CLIMATE CORE

A Roadmap for Climate Education in the Built Environment

Jesse M Keenan

There is evidence that climate change has been on the shortlist for the collapse of a great number of civilisations (Keenan, 2021). But the world faces a slew of challenges, none of which are perceptively proximate to the immediate demands of professional education in the built environment. This chapter offers a perspective on why climate change is critical to the future of professional education in the built environment, and how curricula might be designed to address this grand challenge. As such, this chapter outlines the curriculum of a core course in climate change that is accessible to a wide range of degree programmes in the built environment (hereinafter, "Climate Core"). Educators are tasked with identifying both relevant subject matter and those skills that will define future demands for professional literacy and competency.

As the impacts of climate continue to manifest in greater frequency and intensity, there is a corresponding political and economic demand from all sectors to advance both climate mitigation and adaptation. Many public and private stakeholders recognise that the physical and transition risks from climate change impact both the effectiveness and sustainability of their underlying missions (Martinez–Diaz and Keenan, 2020). In this regard, the production and consumption of the built environment is both a source of risk and opportunity in mitigating and adapting to climate change (Keenan, 2014, 2015). Designers, planners and managers of the built environment bear a unique burden in playing catch-up to climate change, in part because of the immediacy of the physical risks associated with ongoing impacts. The long useful life associated with material interventions in the built environment also impose an immediate challenge to the path dependencies associated with buildings, infrastructure and land use that will shape the options future generations have to mitigate and adapt (Keenan, 2020).

DOI: 10.4324/9781003145905-8

Herein rest the ethical and moral dimensions of intergenerational equity that have mobilised students and faculty of the built environment to engage with climate. In terms of prevailing ideologies in today's design schools, there are two contrasting viewpoints that will shape the trajectory of climate education. One frame suggests a challenging retrospective evaluation of historical people and designs in the context of today's social norms. This would suggest future generations would likely – by extension – find today's consumption and design of the built environment to be entirely without merit, in terms of both symbolic and material performance (Cohen, 2020; Anguelovski et al., 2020). Here designers are motivated by the fictional judgement and shame of future generations that define much of the contemporary methods of progressive social norm setting (de la Fuente, 2020). Another viewpoint grounded in the liberal division of art and science seeks a greater connection with empirical science, acknowledging the underlying limitations of social behaviour that cloud scientific consensus, and corresponding design decisions and management actions that are unnecessarily positivist, deterministic and insular (Cole, 2019; Boarin et al., 2020).

This chapter attempts to mediate these two ideological frames, recognising there are strengths and weakness to each, and also advocates the underlying proposition that the strongest elements of each frame can reinforce a curriculum and corresponding practices that engage a truly diverse range of considerations in advancing social and environmental welfare. Therefore, the first step in processing a climate curriculum is to recognise the political and ideological dimensions that shape an ever-shifting baseline of social and environmental values. By extension, future landscape architects will need to see the world through the lens of sustainable real estate, and future developers will need to understand the role of biomass accumulation in the thermodynamic performance of energy-efficient buildings and communities. In this sense, the various professions of the built environment need translators just as much as they need specialists (Fiedler et al., 2021). This is a translation of not only science, but also the associated values that shape production of the built environment.

Global Change, Not Just Climate Change

In the search to address climate change in higher education, faculty are quickly overwhelmed by the vast challenges defining global change. By viewing climate change through a narrow prism of science and applied science, one loses the opportunity to draw productive connections and relationships that bear on practice. One cannot fully engage the study of a low-to-no-carbon energy transition without understanding the implications for social inequality (Nielsen and Farrelly, 2019). Likewise, the economic and environmental trade-offs of adaptation will not make much sense in many urban areas without understanding the environmental justice legacies of historically marginalised communities. Social vulnerability in that context of climate change is incomplete without understanding the finite limitations of an ageing society (Haq, 2017). By the same token, the externalities of biodiversity loss and unsustainable consumption have long been removed from the built environment, but climate policies such as carbon taxes and supply chain certification are quickly changing that.

Finally, the very nature of labour and work in the built environment is changing radically thanks to technology and artificial intelligence (AI) (Pan and Zhang, 2021). Technologies can bring greater analytical prowess to species selection and building material optimisation, but generative forms of AI threaten to undermine the work of junior ranks of designers who train and mature through the production of detail and specification (Sönmez, 2018). The convergence of new forms of labour with new economic values of a post-climate built environment will be extremely destabilising. But those competent enough to know the alignment of these new post-climate values with the value of their own work will have a competitive advantage.

Labour Force and Professional Practice

When it comes to climate change and global change, the labour force needs two types of professionals – specialists and translators. Specialisation within the built environment is comparatively robust. Energy modellers and disaster risk planners have plenty to stay on top of. But housing architects and public landscape architects are increasingly calibrating their practices to accommodate greater oversight and accountability in the environmental performance of their designs. Whether this is through building codes, engineering standards or client preferences, the standards of practice are changing. With greater technical specificity, particularly in sustainability, comes a broader array of specialists who compose increasingly complex design and engineering teams (Shafique and Mollaoglu, 2020). For instance, physical risk exposure associated with precipitation, humidity and heat now requires external expertise that can negotiate the assumptions in codes and standards (for example, ASHRAE 90.1) with relevant ranges of data from various types of global change models and products (Lian et al., 2020).

Here the division between specialist and translator is blurred. But all relevant parties must have the literacy to understand the confidence and value of these streams of information as they inform their designs, material specifications and operations and maintenance. This is the real value of being a translator. So, while there are new practices that arise as specialists, much of the challenge is to mainstream climate as a means to calibrate existing forms of practice. All these new modes of practice and analysis force a new calibration of project delivery models and risk-sharing among stakeholders (for example, model underestimation and mistranslation). If anything, it amplifies the value of integrated project delivery (Larsson and Larsson, 2020).

While some of the motivation may be expanding new lines of service that offer value-added contribution, there are also push factors associated with changing standards and duties of care that will require degrees of compliance to maintain insurable practices (Mihaly et al., 2018). This is already a hot topic for architecture firms working in highly litigious urban markets in Florida, where local customs and performance are misaligned in increasingly prohibitive ways for many practices. If climate competencies are not developed in emergent professional classes, then the business-as-usual alternative is a quickly emerging a class of uninsurable practices.

Climate Core

Developing a common core curriculum across the various degree programmes of the built environment is both desirable and practical for a few reasons. First, it economises the limited time and space accredited degree programmes have in their existing curriculum. One must pack a lot into a relatively narrow timeframe. The good news is that accreditation bodies are increasingly steering towards required climate-related curricula, even if licensure bodies are slow to respond (Cole, 2020). There is also an economisation of faculty time, not just student time, as very few faculty have the expertise to teach a Climate Core. This means schools must invest in interdisciplinary built environment faculty with demonstrated expertise in climate change, not just a passing interest.

So, what makes up a Climate Core curriculum? The best place to start is with a basic education in the philosophy of science. It is important to highlight how ideas as to what constitutes science and the scientific method have evolved over time, as well as the extent to which current consensus-based paradigms are premised on a slew of important assumptions and experiences. From post-positivism to falsification, students must be able to articulate evidentiary standards, and position such empiricism within relevant scientific methods for their fields of study. This is not a means to substitute empirical science for design research as a method – rather, it is a mere complement (Luck, 2019; Aburamadan and Trillo, 2020).

With this foundation, students can contextualise observational science associated with the basic physics and chemistry driving our contemporary understanding of climate change and its impacts. There are many resources that define the basic elements of climate literacy (NOAA, 2020). You start with radiant forcing, W/m^2 and the planetary energy budget, and go from there. Understanding how and where greenhouse gases proliferate, and the spatially distributed nature of heat within atmospheric, terrestrial and oceanic systems, are key. This should open up a broader discussion on decarbonising the built environment and the world's economy, and who bears the burdens of doing so.

Along the way, it is important to highlight the mechanics of climate modelling and how climate relates to weather. Seeing climate through the lens of weather is always helpful for students in understanding associated scales of time and space. This also presents the opportunity to introduce students to attribution science, which is important for properly contextualising natural variability and climate change within any given extreme event.

From here, students can launch into observational science associated with various climate change impacts, from ocean dynamics to ecological range shifting (Fletcher, 2020). This is a foreground for further studies in the science and social science of adaptation science. When teaching adaptation science, it is always helpful to start with a basic understanding of biological adaptation, including phenotypic and genotypic adaptation. This can be further positioned through studies of complex adaptive systems. Here, the opportunity is not only to highlight emergent behaviours, but also to understand the empirical and philosophical limits of reductionist science.

Moving from biological to social behaviour, studying the social science of adaptation provides students with an opportunity to understand the psychological, sociological and anthropological mechanisms that shape how people understand and respond to environmental change. How people prepare for and respond to climate impacts is heavily shaped by belief systems, resource allocation and a broader notation of the adaptive capacity of individuals, groups, organisations and institutions. Understanding how people and civilisations have adapted and maladapted to prior climate change is critical for understanding our current predicament.

From here, the applied science of adaptation can be evaluated more acutely in terms of decision science, risk management and economics. Disaster risk reduction is often the first step in seeing how people relate to risk and uncertainty, particularly in the context of material design. At this juncture, climate change has to be positioned for its capacity for both shocks and stresses. Very often, through a lens of disaster risk reduction, students tend to conceptualise climate as a series of extreme events without understanding the "stress" that incremental change has on humans and ecological systems.

It is also important at this early stage to introduce the evolution of the basic concepts of adaptation and resilience. This allows students to see that adaptation is not simply about managing risk. Rather, they can see adaptation is also, by definition, focused on the opportunities that may arise, incidental to processes of transformation. The applied science of adaptation varies a good bit by discipline and sector. Here you centre on the body of knowledge memorialised in the IPCC and move to apply that knowledge to cover various infrastructure sectors. This is a great opportunity to introduce the various categorical variants of resilience (Meerow et al., 2016; Davidson et al., 2016). There is no single form or definition of resilience. Rather, there are many categorical variants of resilience, including engineering resilience (most common in the US), disaster resilience, ecological resilience and community resilience (Keenan, 2018). Each variant has its own body of theoretical, analytical and empirical knowledge. Engineering resilience is largely descriptive and is an accessible entry point to understand its relevance in energy, transportation, water and other infrastructure sectors.

Conversely, community resilience is largely normative. It can be best understood when studying the complexities of social and organisational learning as a means to build capacity for recovery and stabilisation. When studying community resilience, it is often useful to introduce the idea of development indicators for measuring the qualitative and quantitative processes of both resilience and adaptation. In the context of community resilience, this is a great opportunity to introduce the broad field of study and practices associated with human and public health and climate change. In particular, public health indicators are some of the most empirically robust ways to understand and measure community resilience.

Studying adaptation and resilience is theoretically daunting for students given the tremendous diversity of scholarship and practice associated with these concepts. The most important thing to get across is that any given intervention advanced in the name of resilience and adaptation has to be evaluated in terms of stakeholder orientation, time horizon and the extent to which there are concurrent adaptive and maladaptive possible outcomes. Each outcome may lead to both synergies and conflicts between different types of categorical resilience. For instance, building a concrete flood wall might advance the engineering resilience of homeowners, but it likely undermines the ecological resilience of altered coastal ecologies. Likewise, if such localised ecological impacts also impact fisheries, it might also undermine community resilience for a local fishing community. In the aggregate, this might be an example of a maladaptive engineering resilience intervention. The bottom line is that resilience is not an absolute good. Here, students begin to set up analytical trade-offs, including opportunity costs and the distributional implications of costs and benefits. They also begin to understand the path dependency of action and inaction. If you are going to teach adaptation, you are also going to have to teach maladaptation.

For students to engage with all this complexity, they are going to need some background in environmental science and policy. It is always helpful to start with ecology. Urban ecology, in particular, provides a useful set of analytical and representational tools for understanding the complexity of human interaction with ecological processes. Thereafter, more formal study of ecosystem services and ecosystem accounting is useful for quantifying – albeit incompletely – these processes. This provides a tangible way to understand how ecosystems are part of the design and management process of the built environment. The best place to start is with land use. Students need a sense of the scale of human management of land, and the role such landscapes play in broader orientations to urbanisation. Land use also provides fundamentally important modes of analysis relating to the first order impact and sustainability of any given design intervention.

From land use, the next scale of analysis relates to the flow of energy and ecological systems. Students should be familiarised with the science and applied science of thermodynamics and energy. This can be extended very broadly, for example, from building and landscape performance to material selection. It is at this stage that students can be introduced to lifecycle analysis and environmental lifecycle analysis, including carbon accounting. This is a good opportunity to introduce the broader field of sustainability science, with a critical eye for the anthropocentric limitations inherent in much of this theory and practice.

This opens up a much broader module within Climate Core about environmental design and technology. It varies a good bit depending on one's specialised degree programme. However, what unites various degree programmes is a basic understanding of the processes and strategies associated with sustainable urbanism – from affordability and accessibility to albedo of landscapes and diffusion of natural light (Loh et al., 2020). What is important is that students learn the various measures of environmental impact and how to benchmark and optimise accordingly. This comes with the recognition that optimisation is rarely possible, and that deeper questions concerning consumption, population growth, models of economic growth and planetary urbanisation often evade critical evaluation (Pearce and Ahn, 2017).

Students have to be able to understand the political and economic forces that shape investments and disinvestments into climate-sensitive buildings and landscapes. This requires a basic understanding of real estate, infrastructure and public finance. It is also incumbent to teach the basic operations of disaster capitalism and carbon economies. When it comes to finance, the basics really are not all that complex. It is about the time-value of money, the flow of cash, and a few key metrics for understanding economic discounting and investment return analysis, including investments made in the name of environmental stewardship. Students of the built environment rarely understand who pays for (and who benefits from) the built environment. Because adaptation and mitigation are not free, it is incumbent that students have sufficient financial literacy to understand, among other things, where their designs and plans can add value.

Value-added innovation requires an understanding of the institutions that incentivise good and bad behaviour. This requires basic knowledge of domestic law and policy about air, water and toxins regulations. This knowledge should be contextualised through the various models of climate governance, from the federal government right down to planning and public works departments in local governments. Here, students can begin to study adaptation planning, which is a specialised form of environmental planning that includes a variety of considerations from flood plain management and hazard mitigation, to managed retreat and climate vulnerability assessments and inventory management. Engaging issues of planning also allows for a broader internalisation of the interaction of climate with housing, labour and mobility, and the allied challenges of racial, social and economic inequality.

A big part of the practice of adaptation planning is the process of engaging the public. Students should be familiarised with basic models of participatory planning, and existing practices utilising participatory planning for everything from vulnerability assessments to project mobilisation. Learning how to develop consensus, and to what ends, is critical for being an effective agent of the built environment. With these tasks and practices comes a certain responsibility to understand the ethical and moral dimensions of climate change.

In this regard, it is incumbent on students to know the historical and empirical basis for environmental justice and climate justice, as both a field of empirical knowledge and a body of political practices. The challenge is for students to move from rhetorical use of equity to more finite analytical understanding of distributional equity and procedural justice. Who pays and benefits are central questions in both climate mitigation and adaptation. Likewise, who decides who pays and who benefits is central to recognition theories of climate justice, as well as basic tenets of procedural justice. With a sensitivity to these dimensions of social behaviour, students are more adept at understanding the impacts of their designs and management interventions in the built environment.

Skills Development

To engage and apply this broad-spectrum knowledge of knowledge in a Climate Core, students need stress-tested analytical ethics that allow for a measure of reflexive analysis on the nature and impact of their own professional practices. By extension, this requires a capacity both to evaluate multi-scalar relationships, and to identify trade-offs and conflicts across scales. All these decisions must be further contextualised to identify path dependencies. The challenge for many designers is to design the adaptive capacity of buildings and landscapes to be able to adapt to future environmental conditions and programmatic uses (Keenan, 2014, 2020). Students must be able to balance immediate lifecycle analysis associated with energy, carbon and environmental impacts, with redundancies of a designed adaptive capacity responsive to a range of potential unknown futures. At the end of the day, managing all this complexity requires communication skills and creativity – something that often gets lost along the way. But as design and management teams become more complex, these skills will be important for specialists and for translators.

It is also worth reflecting on what analytical and decision-making skills students need so they can apply this substantive knowledge. First, students need a basic understanding of statistical analysis and inference. They must be able to analyse and manage risk, and also uncertainty and deep uncertainty. To do this, they must be able to source, interpret and properly apply an exponentially expanding universe of environmental data. They must also be able to position qualitative and quantitative data within the appropriate decision support models. In doing so, they must be sensitive to cognitive biases, and be able to identify deterministic logic and normative rationale. Ultimately, as climate data proliferates, they must be able to recognise stationary forms of data in data, institutions and their own designs.

Sequencing and Course Development

Given the exhaustive nature of the survey, Climate Core could be positioned within a two-part, two-semester course. As previously referenced, programmes rarely have time and space to fit in much coursework. Therefore, the challenge is to use Climate Core as a means to accomplish other non-climatic required tasks associated with professional practice, ethics and environmental requirements. In ideal terms, Climate Core could be sequenced after core introductory classes, and before elective sequences begin to open up. Climate Core would require some basic disciplinary knowledge, but the goal is to open students up to a wide range of conceptual and analytical knowledge so they can apply this new knowledge within their elective studios and paper-writing electives.

Conclusions

Studying climate change is both deeply depressing and supremely liberating. Within the built environment, there is a healthy tension between the preservationist tendencies of resilience, and the transformations necessary to manifest the goals associated with adaptation and mitigation. As society seeks to facilitate deeper transformations that rewrite societal relationships in everything from environmental sustainability to economic growth, there is an infinite array of risks, uncertainties and opportunities. Students of the built environmental intuitively sense this change but lack the knowledge and the skills to engage with these processes of global change.

Developing a core curriculum in climate change is necessary both to capture this change in social norms, and to develop minimal levels of professional literacy and competency. As designers and managers of the built environment, students have the opportunity to be specialists and translators of new forms of knowledge already driving new practices and service lines. Indeed, students with an education in the knowledge and skills outlined in this chapter will be desirable in a variety of sectors and industries that need a new workforce of climate translators.

Beyond the self-serving labour dimensions of Climate Core, there is also a more fundamental proposition at play. Society relies entirely on future generations to design, develop and manage the built environment in a way that mitigates climate change and takes advantage of the many opportunities to advance a more sustainable and just way of life. Through developing a common core in climate change, schools of the built environment have an opportunity to positively contribute to society and the environment in ways that will be a true model of university education in the years to come (Monroe et al., 2019).

References

Aburamadan, R., & Trillo, C. (2020). Applying design science approach to architectural design development. *Frontiers of Architectural Research*, *9*(1), 216–235.

Anguelovski, I., Brand, A. L., Connolly, J. J., Corbera, E., Kotsila, P., Steil, J., & Argüelles Ramos, L. (2020). Expanding the boundaries of justice in urban greening scholarship: Toward an emancipatory, antisubordination, intersectional, and relational approach. *Annals of the American Association of Geographers*, *110*(6), 1743–1769.

Boarin, P., Martinez-Molina, A., & Juan-Ferruses, I. (2020). Understanding students' perception of sustainability in architecture education: A comparison among universities in three different continents. *Journal of Cleaner Production, 248*, 119237. doi: 10.1016/j.jclepro.2019.119237.

Cohen, D. A. (2020). Confronting the urban climate emergency: Critical urban studies in the age of a green new deal. *City, 24*(1–2), 52–64.

Cole, L. B. (2019). Green building literacy: A framework for advancing green building education. *International Journal of STEM Education, 6*(1), 1–13. doi: 10.1186/s40594-019-0171-6.

Cole, R. J. (2020). Mainstreaming carbon zero in architectural education: Within a decade? *Buildings & Cities*. Retrieved from https://www.buildingsandcities.org/insights/commentaries/mainstreaming-architectural-education.html.

Davidson, J., Jacobson, C., Lyth, A., Dedekorkut-Howes, A., Baldwin, C., Ellison, J., Holbrook, N., Howes, M., Serrao-Neumann, S., Singh-Peterson, L., & Smith, T. (2016). Interrogating resilience: Toward a typology to improve its operationalization. *Ecology and Society, 21*(2), Art 27. Retrieved from https://www.jstor.org/stable/26270410?

de la Fuente, P. P. (2020). Guilt-tripping: On the relation between ethical decisions, climate change and the built environment. *Urban Planning, 5*(4), 193–203.

Fiedler, T., Pitman, A. J., Mackenzie, K., Wood, N., Jakob, C., & Perkins-Kirkpatrick, S. E. (2021). Business risk and the emergence of climate analytics. *Nature Climate Change, 11*(1), 87–94. doi: 10.1038/s41558-020-00984-6.

Fletcher, C. (2020). *Climate change: What the science tells us*. 2nd ed. John Wiley & Sons.

Haq, G. (2017). Growing old in a changing climate. *Public Policy & Aging Report, 27*(1), 8–12.

Keenan, J. M. (2014). Material and social construction: A framework for the adaptation of buildings. *Enquiry: Journal of Architectural Research, 11*(1), 18–32. doi: 10.17831/enq:arcc.v11i1.271.

Keenan, J. M. (2015). Sustainability to adaptation and back: A case study of Goldman Sach's corporate real estate strategy. *Building Research & Information, 43*(6), 407–422. doi: 10.1080/09613218.2016.1085260.

Keenan, J. M. (2018). Types and forms of resilience in local government planning: Who does what? *Environmental Science and Policy, 88*(1), 116–123. doi: 10.1016/j.envsci.2018.06.015.

Keenan, J. M. (2020). The positive, negative and neutral outcomes of designed adaptation in the built environment. *ZARCH: Journal of Interdisciplinary Studies in Architecture and Urbanism, 15*(1), 154–163. doi: 10.26754/ojs_zarch/zarch.2020154821.

Keenan, J. M. (2021, March 8th). What's the most climate-safe place in the world? *Gizmodo*. Retrieved from https://earther.gizmodo.com/whats-the-most-climate-safe-place-in-the-world-1846409071.

Larsson, J., & Larsson, L. (2020). Integration, application and importance of collaboration in sustainable project management. *Sustainability, 12*(2), 585. doi: 10.3390/su12020585.

Lian, Z., Liu, B., & Brown, R. D. (2020). Exploring the suitable assessment method and best performance of human energy budget models for outdoor thermal comfort in hot and humid climate area. *Sustainable Cities and Society, 63*, 102423.

Loh, S., Foth, M., Caldwell, G. A., Garcia-Hansen, V., & Thomson, M. (2020). A more-than-human perspective on understanding the performance of the built environment. *Architectural Science Review, 63*(3–4), 372–383.

Luck, R. (2019). Design research, architectural research, architectural design research: An argument on disciplinarity and identity. *Design Studies, 65*, 152–166.

Martinez-Diaz, L., & Keenan, J. M. (Eds.). (2020). *Managing climate risk in the U.S. financial system.*U.S. Commodity Futures Trading Commission.

Meerow, S., Newell, J. P., & Stults, M. (2016). Defining urban resilience: A review. *Landscape and Urban Planning, 147*, 38–49.

Mihaly, E., Franczek, W., & Selman, A. P. (2018). Legal liability of design professionals for failure to adapt to climate change. *Journal of the American College of Construction Law, 12*(2), 87–90.

Monroe, M. C., Plate, R. R., Oxarart, A., Bowers, A., & Chaves, W. A. (2019). Identifying effective climate change education strategies: A systematic review of the research. *Environmental Education Research, 25*(6), 791–812.

National Oceanographic and Atmospheric Administration (NOAA). (2020). *The essential principles of climate literacy.* Retrieved from https://www.climate.gov/teaching /essential-principles-climate-literacy/essential-principles-climate-literacy.

Nielsen, J., & Farrelly, M. A. (2019). Conceptualising the built environment to inform sustainable urban transitions. *Environmental Innovation and Societal Transitions, 33*(1), 231–248. doi: 10.1016/j.eist.2019.07.001

Pan, Y., & Zhang, L. (2021). Roles of artificial intelligence in construction engineering and management: A critical review and future trends. *Automation in Construction, 122,* 103517. doi: 10.1016/j.autcon.2020.103517

Pearce, A. R., & Ahn, Y. H. (2017). *Sustainable buildings and infrastructure: Paths to the future.* 2nd ed. Routledge.

Shafique, F., & Mollaoglu, S. (2020, November). Transformational leadership and team performance in green construction project teams: Development of study model and measurement tool. In David Grau, Pingbo Tang and Mounir El Asmar (Eds.), *Construction Research Congress 2020: Project Management and Controls, Materials, and Contracts* (pp. 115–124). American Society of Civil Engineers.

Sönmez, N. O. (2018). A review of the use of examples for automating architectural design tasks. *Computer-Aided Design, 96*, 13–30. doi: 10.1016/j.cad.2017.10.005.

PART 2

Generative Lineages

Narratives

An extended root reaches skywards, incrementally higher with each new frond. The tree fern slowly grows taller. Small tendrils at the base express its rhizomatic nature. Unfurling its fronds from the fibrous trunk, the tree fern stretches out. Tightly wound, the spiral begins to extend. Unfurling, in one direction and then crossways, then crossways again as, lastly, tiny leaves unroll. The fronds sway gently in the breeze, releasing tiny spores that float away. Epiphytes attach, rough ferns grow underneath, and insects feed. The 2020–21 La Niña brought so much rain, funnelled through the trunk-root. Two full sequences of frond growth lifted the lacy canopy higher than ever.

It has been given a name – *Dicksonia antarctica*. *Dicksonia* in honour of James Dickson, 1738–1822, a Scottish nurseryman; *antarctica* denoting "southern" or from the Antarctic regions.[1] Its antecedents are frozen in drawings and specimens collected through the operations of the British empire. An "exchange" plant of the Royal Society, the tree fern was widely circulated from Australia (particularly Tasmania) to England. Caught up in a "fern fever" that never ends, tree ferns are harvested by cutting them off at the trunk, wrapping, shipping and replanting the trunk. Many plants are still exported every year. There are reported instances of tree fern "laundering"[2] which refers to moving ferns from Tasmania to Victoria to avoid laws that prevent their harvest. Fern fever was then inverted and re-projected back onto the grounds of the tree fern through a tourism explosion in the Dandenong Ranges in the late 1800s and early 1900s. The fern's continuous extraction is an expression of the ongoing circulations set in motion by frameworks of knowledge that prioritise naming, categorising and coveting the fern.

DOI: 10.4324/9781003145905-9

However, these processes are not perceptible to, nor needed by the fern. It continues to extend, fronds wither and drop, new sequences of growth occur, epiphytes find home in its shade and the fern endures. The fern does not "know" to which class it belongs. It is separate from the types of knowledge that inform perspectives on it. Botany provides a guide for understanding its relationship with other plants, biology a guide for understanding its transformations and other "ologies" relate narratives about the transmission of food and energy. Multiple distinctions are held within each, and together they constitute an aggregated understanding of the tree fern.

Forms of knowledge production dislocated the fern from its context – the habitat and relations that allow for its flourishing. These forms make further physical dislocations reasonable. But even a single "species" is constituted of the many. Many organisms within and without constitute an evolving whole. Botanical knowledge allowed for a great understanding, comparison and organisation of plant species. But they also divided the plant from its habitat; biodiversity loss as a nested issue of the climate crisis. It is essential to re-encounter plants used in everyday practice as biodiverse in their own right as well as within a diverse habitat.

Definitions

Landscape architecture has the benefit of diverse disciplinary lineages (gardening, architecture, art and science) from which to project into multiple variations of practice. That landscape architecture is a combination of art and science is commonly found as a claim in design programme marketing material. These lineages originated and continue to operate within colonial, imperial and capitalist logics. In addition, the tendency of an educational institution is also to organise, classify and disseminate. We contend with a boggy ground of intertwined systems of knowledge, classification and expression.

These lineages carry forward. They imply ways of thinking, and tools and techniques for seeing and making. They imply steps, sequences and frames of and for knowledge. Extending into the present where the commodification of biology as a form of capital is relevant to the way in which landscape and landscape design are valued through an economic lens. Colonial classifications implicate the work of landscape architects as they have informed the way in which landscape architectural knowledge is framed within the institution.

An example of acknowledging multiple lineages was outlined by Lister and Reed who identified three ecological genealogies – natural sciences, humanities and design thinking and practices.[3] This acknowledgement opened a space in which lineage can be both articulated and animated. In this way a recognition of the conceptual frame through which knowledge is constructed allows for processes of alternate organisations. A framework for knowledge that is itself ecological similar to that developed by Warburg in the field of art history "a knowledge in extensions, in associative relationships, in ever renewed montages, and no longer knowledge in straight lines, in a confined corpus, in stabilized typologies".[4]

Considering foundational knowledge through a process of declassification opens them to transformations, reformation and repositioning in relation to how they interact with one another. It offers a generative potential of confluence, contradiction and cross-pollination. This is a continuous process of worldmaking, extending and flourishing. In this way, specialisations multiply the worlds, each limited, together generative.

The climate crisis is a disruptive agent to the discipline's groundings. It exposes the genus of practices that have been formative to the physical condition of the climate crisis, and now extend into the way the "crisis" tends to precipitate "solutions-based thinking" in design disciplines. The continuous amplification of climate effects further cracks open foundational issues, indicating a responsibility to acknowledge and critique the inherent logics of existing tools and techniques, emphasising the studio environment as the place of interrogation, exposure, integration and transformation, creating a potential for cross-pollinations of lineages that extends into pedagogical structures.

Pedagogies

"Generative Lineages" explores practices that engage in disruption, inflection, extension and questioning. The disciplinary grounds are not defined as static entities; rather, the chapters discuss processes whereby complexities can be acknowledged, and modes of teaching practice can be proposed that allow for inflection and generative engagement in multiple directions. This may be through amplifying the embedded multidisciplinarity of programmes.

This section identifies a range of specialisations in the discipline that introduce diverse lineages of knowledge as generative design research processes. These sample specialisations are demonstrated through the various contributors who identify different forms of practice and academic exploration. They use collaborative approaches and are informed by the infiltration of knowledge from other disciplines which expand the field. The scope of these practices and pedagogies produces novel outcomes and, consequently, new areas of contribution to the discipline. The responses to this provocation urge that it is the time to resist solutions-oriented work; rather, it is the time to examine perspectives and positions that extend ecological thinking into the institutions and modes of practice. The increasing size, scope and opportunity for landscape architectural work in the context of the climate crisis will require continuing articulation, inflection and development of strands of specialisation within the discipline. From a pedagogical perspective, this necessarily entails approaches that re-evaluate what substantiates core knowledge in the discipline, and that enable cutting across diverse forms of knowledge within the studio.

Notes

1 See https://www.anbg.gov.au/gnp/interns-2003/dicksonia-antarctica.html.
2 Tree fern "laundering" to the UK. See https://www.theguardian.com/society/2002/mar/06/highereducation.biologicalscience.
3 Reed, C & Lister, NME (2014). *Projective Ecologies*. New York, NY: Harvard University Graduate School of Design.
4 Didi-Huberman, D (2004). Foreword, in Philippe-Alain Michaud (Ed.), *Aby Warburg and the Image in Motion*. New York: Zone Books.

7

HOPE IN RESTLESS PEDAGOGY

Rosetta S Elkin

The cascade of planetary warnings and policy failures that contour our recent past are only useful insofar as they might inspire a more prepared and thus equitable future. On 11 October 2000, more than 300 million gallons of viscous coal burst from a slurry dam, destroying fresh river streams across Kentucky, killing millions of fish and altering drinking water resources for human communities. The spill was not an "act of god" as the mining company declared; rather it exposed the corruption of environmental policy.[1] By 2005, Hurricane Katrina followed suit, echoing the social ramifications of unequal policy, when it was framed as "an act of nature".[2] The following year, Al Gore's pursuit of global warming related the cascade of crises around the country to rising temperatures, warming seas and melting ice caps. Global warming revealed to the general public that the vulnerability of infrastructure was not godly and the velocity of a storm was not natural; rather, each disaster was the consequence of human carelessness. Fast forward to the next decade, as the worst fire seasons burned in Alaska, California and Texas, inspiring the term "gigafire", a word to describe the complex burning of millions of acres so intense it generates another climate. By 2012, the world also invented the first "superstorm" as Hurricane Sandy dimensioned the Atlantic coast from the Caribbean to Rhode Island. These are just a few examples of the lag of environmental regulation, the inability of policy to keep up with the times. Notwithstanding the lag, professions including landscape architecture do little to address the warnings and the misfortune, evidenced by the number of professionals eager to access funding packages, enter rebuild competitions and proffer design solutions that tend only to camouflage static engineering protocols. The problem is no longer that the climate is changing: it is that humanity is not.

This chapter is a plea to understand and accept the past, and leave it there in order to use the present to motivate the future. What this means is that we might adapt rather than "fight" climate change. Adaptation requires restless pedagogy.

DOI: 10.4324/9781003145905-10

At present, the professions are forced to follow the protocols of regulation, policy and governance. But the institution is not. Therefore, the decades of unrest experienced across the planet must be met by an equally restless pedagogy – a pedagogy trying, testing and aspiring to revise the profession. A pedagogy is required that is committed to a future inscribed by change, that shapes how we interpret and respond to, and show up for, the present.

Restlessness embraces change and might be embarrassed by stasis. To be restless is to display continual curiosity, adaptation and growth. Yet, restlessness begins to appear out of place when set against landscape architectural traditions that resist or even disregard change. I can accept the stasis in relation to 20th-century models of modernism, a relic of design in the Holocene. What I cannot accept are static planning procedures that lie prone to the cascade of planetary warnings and policy failures of the 21st century. Witness the inertia of our most predictable park systems, well-intentioned plazas and bespoke masterplan exercises. Witness the pressure of cultural preservation, as pumps and walls are grafted to the rubric of restoration. Witness the rise of sterile cultivars that replace species, so that urban street trees are prevented from seeding, fruiting or otherwise reproducing. Each detail is reliant on outdated standards that actually work against change. What can a raised seawall, a flood-proof square or a layer of dredge really do for a community except *conceal* change? To conceal change is to lie to the public and is a form of climate denial so prevalent it seeps into pedagogy. If professionals believe the landscape shakes and floods with more intensity, or that mudslides, permafrost thaws and fires rage with new frequency, then why not expect landscape architecture to change in turn?

I am committed to a restless pedagogy because it includes collaboration between humans across generations, it thrives as a form of co-creation, it acknowledges the activity of the earth, and it has little use for hierarchy. It encourages a reading of transformation that might not always have clear winners and losers. I am committed to restless pedagogy because I no longer trust received procedures and traditions that lock out equity, or import millions of plastic pots, rely on pacifying water systems, or mulch urban parks with dollar bills at the request of a risk-averse client.

Restless pedagogy is a direct means to advocate for more equity in our standards that range from planting specifications to environmental policy. It addresses the ways professional standards can be informed by research. But it takes effort, and it takes time. It also might not look the same; it might be messier and create a kind of vulnerability because climate impacts are so multiple. As educators, if the response to the crisis is to annex climate dynamics to your research, vie for new funding, insert a "climate" module to a standard-issue seminar or invite a guest to describe how climate affects the field, then you will be left in the neoliberal dust, so to speak. To capitalise on crisis is not only a denial of change itself, it is a blatant form of prejudice against the generation inheriting your mistakes. Perhaps climate change might play an active role in creating our institutions, just as prejudice has played an active role in the systems of education, health care, ownership, employment and virtually every other facet of life since European colonisation of the Americas. Perhaps the creativity of pedagogy might be allied to life as we *don't* know it.

Allow me to contextualise my position. I am a landscape architect with a deep respect and admiration for plant life. I am also a professor stirred by how increased climate risk is affecting the ways we nurture and share knowledge. As a consequence, one of the first questions I usually field is how I reconcile my interest in plant life with the injustices brought on by climate change. The framing suggests plant life is only a diversion to the weightier climate-based issues that *actually* affect humankind. I have also been accused of *overlooking* human life, implying that studying plants takes away from the significance of racial justice, community impoverishment or gender politics, for instance. In defending my research to fellow plant-dependent humans, I explain that studying plants requires a slowing down, a deceleration that helps me take stock of living on the planet. Rather than work *on* plants, I work *with* plant life in order to engage seriously with being human. Plants bridge the overly siloed tendency to separate design from theory, often framed academically in the detachment of a history seminar from studio culture, for instance. Such constricted ideologies not only divide our worlds and destroy hope, they isolate the past from the present. This is why I resist teaching plants as a technical course, a side-gig to the more intellectual content of the episteme. But I also impart plant knowledge because plants are the surest way to acknowledge the connection between the pedosphere and the atmosphere, the land and the climate.

Restless pedagogy expands by building powerful relationships through non-material agency, revealing that opportunities are concealed in the thick profile of the landscape. To demonstrate what this means, I offer the following three concepts. Consider the student work that is paired with each concept. The methods are grounded in landscape architectural tradition, yet they emerge as a means to track and follow change. Each is isolated, yet the reader will soon notice that they mingle, and cannot be secluded or extracted easily from one another. It is my hope that they will inspire others to think about how to transform landscape education.

1. Policy is under your feet

Environmental regulation and policy are unfortunately trapped in planning, a superficial projection across the thick changeability of the land. Restless pedagogy suggests policy is a sectional dynamic, a distinct subdivision that might be reoriented to reveal the aliveness of landscape structure. The aliveness of the land is currently lost on a public and a profession with their heads bent to the canopy. Standing in a forest, billions of microorganisms, dormant seeds and rootlets are invisible to the collective, despite the fact that their extents far surpass vulnerable crown spread, ownership lines or zoning guidelines. Following suit, models of conservation transform the landscape from a fertile, changing ground which nurtures the future, into a collection of items to be admired, seemingly crystallised in time, much like an artefact locked in a museum or an animal held in captivity. A well cited example of the friction inherent to plan-based environmental policy is

Pando, a giant *Populus tremuloides* organism valued for its spread and its age, despite the fact it is ageless because of its root-sprouting capacity, its life underground. The only way to conserve Pando is to conserve the movement of its roots, since individual ramets are short-lived. Yet, human appreciation still tends to augment with age metrics – champion trees and centennials – making youth less significant and therefore more disposable: the bole radius of a giant sequoia, the extraction of genetic heritage in monumental olive trees, and the physical fences that attempt to enclose millennial oaks. While some practices aim to move away from "saving" the products of nature to embrace the mobile processes of the environment, plant life remains in an impoverished middle ground.

What might emerge from a sectional consideration is the power of longevity, expressed as changeable lifecycles. Consider that longevity remains a mystery in plants, since some seeds only survive a matter of decades while others seem ageless. But to take care of the land under our feet is to advocate to conserve dormancy. For instance, a Judean date palm was germinated from a 2,000-year-old seed, making it the oldest known viable seed. But the tree is only a few years old. Seeds survive across the planet by establishing contingency plans for an unknown future, banking on the thickened section of the soil. Restless pedagogy accounts for sectional movement (Figure 7.1), conserves seed dispersal (Figure 7.2) and explicates landscape change (Figure 7.3). Hope emerges in acknowledging that transformation is not absent, but often radically concealed from notice.

2. Design relationships

With less and less at hand, or underfoot, I notice that students are further than ever from the land, from the plants upon which we depend and from one another. We are linked to one another through our shared experiences of dislocation and displacement. The reliance on scientific certainty and technological prediction increases because we are in an era of low social trust. This means it is easier to trust data than each other. Therefore, a challenge emerges in the gap between the natural sciences and the social sciences, between disciplines that reason and

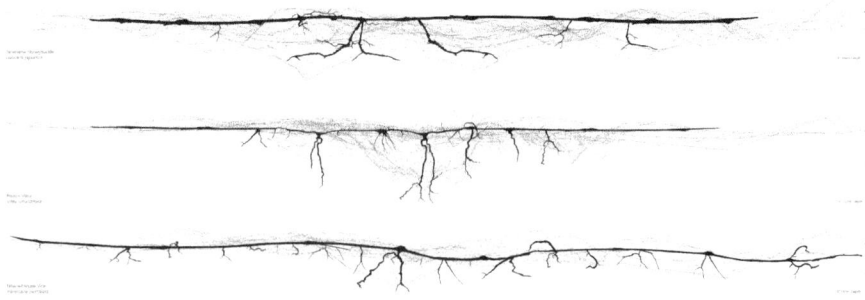

FIGURE 7.1   The aliveness of the soil rendered by movement in the rhizosphere (2014). Canepa, Cox and Li.

UXO
existing monitoring
well

FIGURE 7.2 Serial sections show the thickening of the seed bank, in relation to a landform model (2012). Deuschle, Rodriguez and Zhang.

theorise humankind, and those that study global, physical and geomorphic processes.[3] The assertions distance humanity from the land, abstracting our relations.

For instance, it takes 15–20 years to release a new crop to farmers, such that any smallholder farm necessarily invests in and supports scientific research, postponing experimental or indigenous practices due to insurance restrictions.[4] What smallholder farmers really own is crop insurance, not land or plants, because landowners no longer have any influence on species selection. Once scientised, plants are more easily commodified, traded, replaced and felled. Paying attention to how

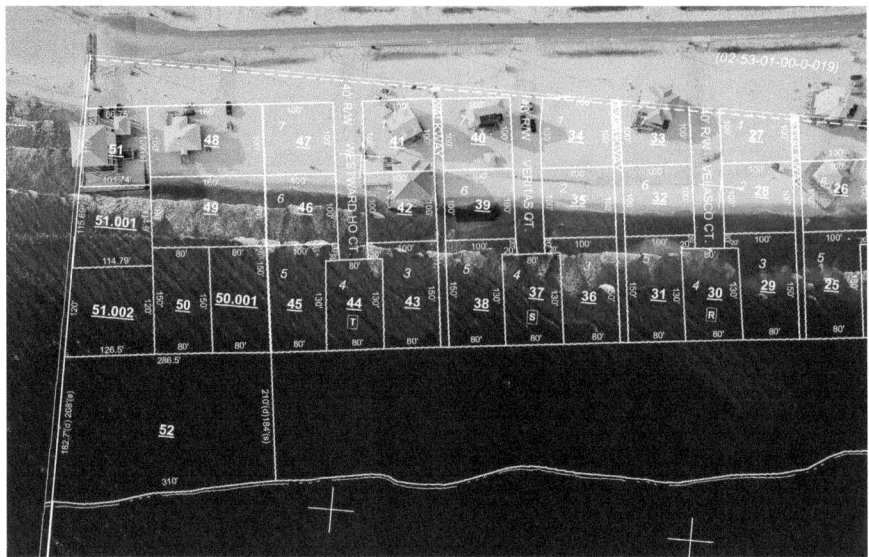

FIGURE 7.3 A fragmented understanding of change creates friction between ecology and policy (2019). Stein and Tsang.

relationships are fragmented reveals the possibility that relationships radicalise what the sciences normally manage to study, and certainly radicalises what is communicated or shared with the public. The elaborate exchange between cells, organisms, bacteria and society suggests science cannot be split from its social context. It also suggests that landscape – at once culture and nature – might be a powerful bridge.

Thus framed, restless pedagogy suggests landscape architects design durable relationships, imagining bonds between one another, between species and with the land, to engender a different kind of power. Designing relationships can be achieved without abstraction, in real time. Students learn to discover the landscape on their own terms, to engage with propagation, classification (Figure 7.4), and first-hand experience (Figure 7.5). The co-creation occurs between who is studied and who is being studied, blurring instruction and instigating a supraprersonal form of knowledge capture.

3. Non-materiality is real

Restless pedagogy critically engages the human: the viewer, the critic, the recipient of design intentions. Consider the role of models in the shift from studio to fieldwork. How can the actions of users be qualified and quantified if durable relationships are acknowledged as design? The actions embedded in a return to the landscape are created by engendering a familiarity with the thick, living soils, the aliveness of plants and the determination of moisture, for instance. As a result, landscapes create a critical exchange as space becomes a co-producer of interaction rather than a background of action. Curious minds open to learning from

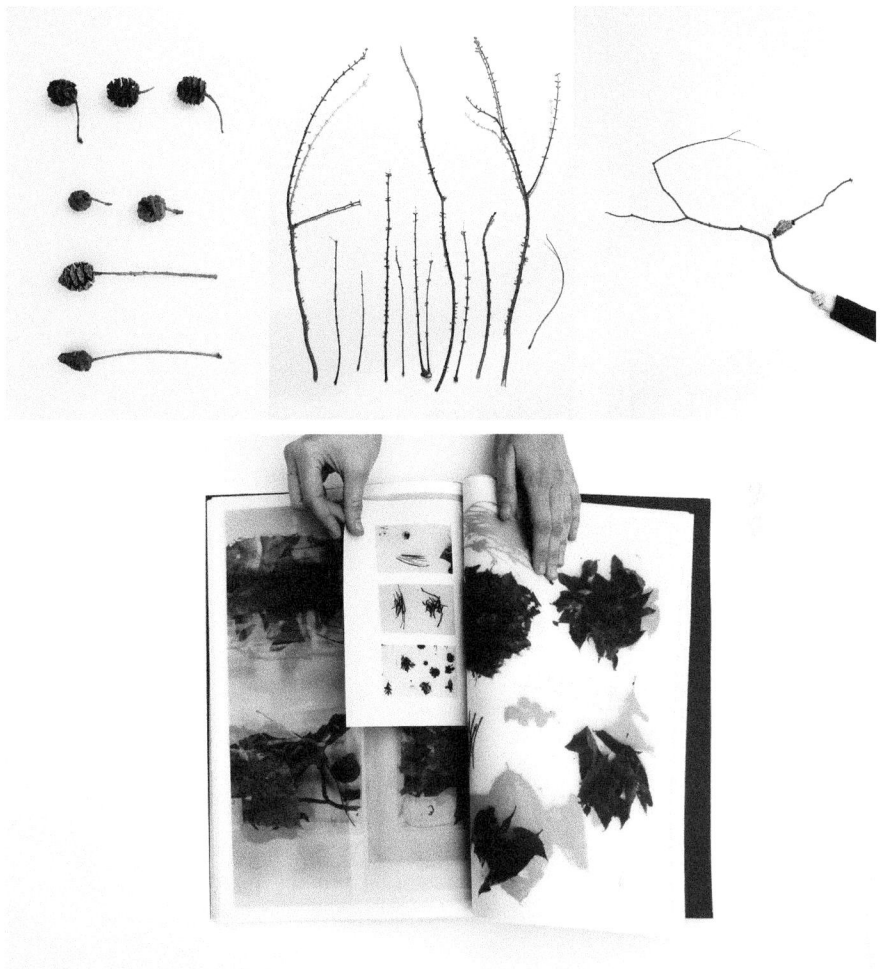

FIGURE 7.4 Top: Fieldwork practices move disciplinary theory to practical application through first-hand experience (2018). Armida and Aranza. Bottom: Curious minds open to learning from different formats and altered engagements (2018). Alcega, Benitez and Collard.

engagement simply because the ambition to study landscape embraces a world fundamentally indifferent to formal geometry. For instance, grasslands cannot be reduced to carbon sink statistics, million-tree projects cannot offset deforestation statistics and forests cannot be summarised for their timber resources alone. Rather, landscape holds non-material significance, which is no less "real" because it resists formal outlines. A more inclusive set of relationships bridges science and theory, humans and nature, culture and technology, because the thickness of the planet requires both thinking and doing. This inclusive ecology is political because it has no authoritative precedent, no imposition into social structures, and no a priori

FIGURE 7.5 One way to ensure science is not isolated is to expose students to creative studies of the living formation (2016). Rosetta S. Elkin.

scientist that has to remove herself from the experiment. I highlight that living stabilises political ecology, because "ecology" expands beyond nature-based issues, harmonising the insistence that human politics is the most prevalent transformative force. As a corrective, non-materiality substantiates the agency of ecology by articulating the concealed behaviours, slower advances and intelligence of living organisms that still produce tangible results.

For instance, the small flowering annual *Corydalis aurea* has expressed preference for dispersal by ants (myrmecochory) over humans. Experiments in deciduous forests show that black shiny seeds planted by ants produce 90 percent more seedlings than those planted by hand.[5] The dispersal advances through manifold investments that promote movement, across openings in the forest floor, through gravel and duff, with intentionality. Proliferation is hinged on connections that do not have material form but emerge as discernible outcomes when the golden yellow flowers materialise each spring. Design emerges as a common ground between abstracted, theoretical space and physical, material space. What emerges is consensus, a model of political practice. Consensus is not only unanimity, it is the effort involved in creating reliable multi-agent landscapes.

Manifold species are appreciated by restless pedagogy, whether they are ephemeral, aggressive, transitory or otherwise. In other words, the changing climate acts as a catalysing force to increase relationships, overcoming the deception of a "productive" species (an unfortunate label that might be retired for its exclusive claims of higher value to humankind alone). Non-materiality is established in recognition of cartographic informality (Figure 7.6, top) and the activity of the earth (Figure 7.6, bottom), and it challenges the distractions of data (Figure 7.7). Restless pedagogy teaches students to see beyond the most obvious encounters, grounding non-material value in the context of design.

If the structure of institutional education can overcome static pedagogy, then there is hope. The hope I am referring to encourages confidence in change, inspired by optimism for the future of the profession.

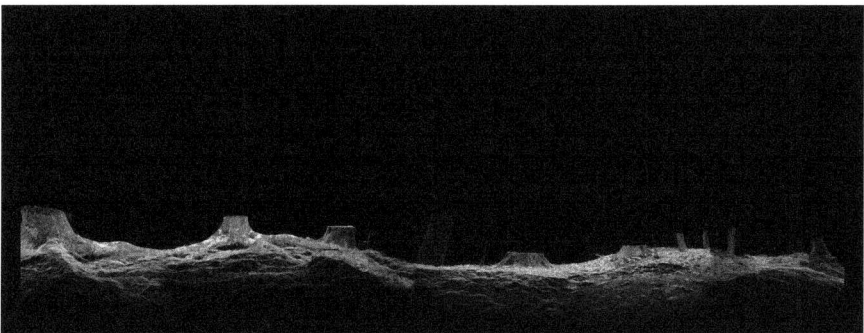

FIGURE 7.6 Top: Informal wayfinding across the US–Mexico border – design on Department of Interior correspondence (2015). Michelle Franco. Bottom: The temporal relationship between above and below ground (2017). Berdichevsky, Haines and Ling.

FIGURE 7.7 The critique of tangible evidence is explicated by an overwhelming amount of data, presented as a studio review (2018). Rosetta S. Elkin.

Opportunities enlarge in collaboration because students are no longer entering academia with a casual relationship to change. Rather, decades of crisis define their relationship to the failures of policy, the sluggishness of governance, or unrestrained and unequal urbanism. The lag of environmental regulation is no longer an excuse. Restless pedagogy is a counterpoint to the patchy acceptance of change and demonstrates to the next generation that we are not only "in" this together, but we inhabit the same earth. There is power in that acknowledgement, so restless pedagogy concedes that year after year the syllabus must change, the readings must change, the framing must change, the references must change, the methods must change, in order to begin healing our *dis*location – *dis*- is a Latin prefix meaning lack of, apart, away, in a different location. The hope found in restless pedagogy is that it embraces kindred locations, is close and near because the precarity is more personal than ever.

Notes

1 Martin County Sludge Spill occurred after midnight on 11 October 2000. See, for instance, Peter T Kilborn, "A Torrent of Sludge Muddies a Town's Future", *New York Times* (1923–Current File) (New York), 2000.
2 The consequences of the designation "natural disaster" were critiqued shortly after the storm by geographer Neil Smith: https://items.ssrc.org/understanding-katrina/ theres-no-such-thing-as-a-natural-disaster/

3 This disciplinary distance is articulated by Dipesh Chakrabarty in 2009, by mobilising the parallel discussions between globalisation and climate change, and the distinctions between natural history and human history: "The Climate of History: Four Theses" *Critical Inquiry*, 2009.
4 The United States has 311 million acres of agriculture protected by crop insurance. See for instance: https://cropinsuranceinamerica.org
5 Peter Thompson and Stephen Harris, *Seeds, Sex and Civilization: How the Hidden Life of Plants Has Shaped Our World* (New York, NY: Thames & Hudson, 2010).

8

A CONVERSATION ABOUT LANGUAGE

*Teresa Gali-Izard, Luke Harris, Cara Turett
and Bonnie Kate Walker*

Translation

Translation is an act of love, care and empathy. It requires a close reading and the capacity to move from one language to the other. It implies an understanding of the context, and it is never perfect. Translation is an impossible endeavour.

These four design studios in the past decade have explored and translated the hidden potentials of sites. In landscape architecture, the act of translation gives voice to invisible creatures, processes and relationships that occur around us. In our attempt to translate, we have learned how to read: the marks, the paths, the presence – sometimes imperceptible – of the inhabitants of places. A continuous, dynamic overlapping of milieus is deployed in the shared environment defined by climate and geology. In our design studios we seek to explore those milieus, their relationships and hidden potentials.

Translation enables a common conversation. This chapter is a conversation between a professor, Teresa Gali-Izard, a Professor of Landscape Architecture at ETH Zurich, and members of the Office of Living Things, a landscape design collective consisting of her former students from the University of Virginia School of Architecture (UVA) who now work with her at ETH Zurich. Since teaching is a call and response, together we reflect on this methodology as it has evolved.

Drawing

Critical in the process of translation is the act of drawing. Drawing is a common language that allows for the encounter of different cultures, backgrounds and actors. It is a medium through which every concept can be represented at different scales.

DOI: 10.4324/9781003145905-11

Through drawing we build bridges between landscape architecture and other fields, such as architecture or engineering, but also between different forms of knowledge, including traditional practices and the historical systems of ancient civilisations. New and unexpected relationships appear through the lines that translate different worlds because these lines talk to one another; they contain the information. Drawing a line is the first decision of the design process.

As students engaging with this approach for the first time, it was difficult to draw something that we did not fully understand. In this methodology, students are challenged to draw as a means of discovering something. This puts students in an uncomfortable position. In a studio project, the role of the initial drawings is to begin to systematically grasp the complexity of a site – to figure out the existing relationships, dependencies and enmeshed phenomena. To enter into these dynamics as a designer, we have to try to comprehend the information by drawing it – and then respond to the drawing. It is the conversation between site and studio, between specificity and abstraction. As a student, the first line is the hardest thing to draw. This step of translating what is happening on a site into drawings means you are designing from day one.

In each of the studios described below, we develop a common drawing language that is shared among all students. Each studio builds on the ones before it, and this language has evolved towards an abstract symbology that contains units of measure and an accurate dimension of key parameters.

The invention of a shared drawing language enables the studio to become a collective project of knowledge-building. The group of students work together to encompass a wider range of intelligences, with someone drawing climate, someone else drawing the topography, and so on. This becomes the common foundation for transformation. In the end, instead of competing interpretations of the site, the projects take part in a collective conversation.

Studios

These studios are an intellectual exercise; a limited and framed approach to a question, with very specific constraints, that the students explore in depth. We do not pretend to include all the possible actors and points of view, because we are very aware of the complexity around us, and because we seek to contribute with a very particular agenda to an open conversation. The outcome is one among others. Our thesis is that by making visible the invisible, and by giving the voice to forgotten processes and living creatures around us, we will discover new design logics, new architectural proposals and new forms to inhabit the planet.

We want to give space and time to find unexpected relationships that can only appear by translating logics other than our own into a shared language.

The "Unfinished Studio", taught in the autumn of 2014 for first-semester master's students at UVA with Leena Cho, was the first of a series of studios where we tested the methodology of translation. We challenged the contour line as language for the landform, instead exploring languages from historical drawings, the fields of geology and geomorphology, methods of survey

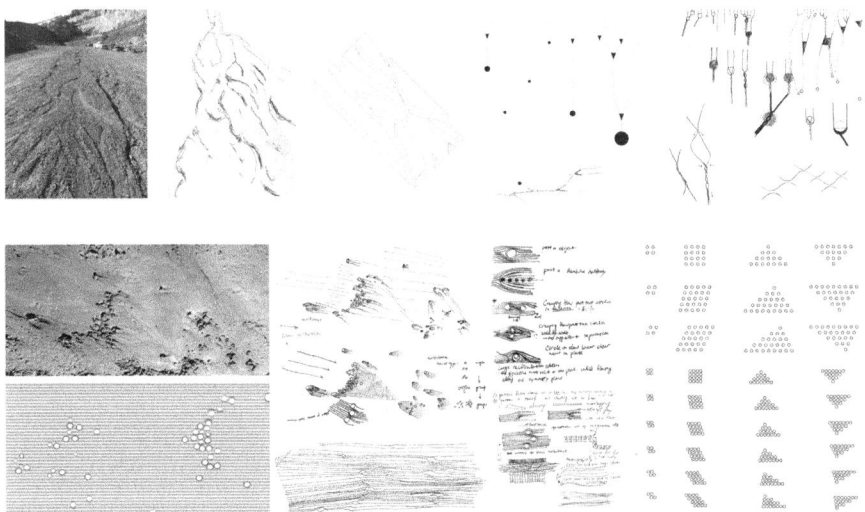

FIGURE 8.1   Documentation from quarry visit (2014). Top: Claire Casstevens. Bottom: Xingyu Yang.

and digital programmes that produced three-dimensional surfaces as meshes or continuous surfaces. The understanding of the rules behind the landforms brought us to the dynamics of climate and their capacity to shape land based on its geological conditions. Focusing on water as an agent of shaping landscapes, the students were asked to make an intervention in an existing quarry (Figure 8.1).

Students had to discover water as an agent through which to imagine the potential future of the site. Surprisingly, when visiting the site, the students reacted by discovering only the close context around them. Their reaction mimicked the behaviour of a child in their early years of discovery. Children discover their environment while growing and developing their psychomotric apparatus. This organic approach to understanding the environment, which starts with our immediate surroundings and expands over time while growing up, seemed embedded in their bodies, and we decided to let them discover in their own way. Many times, practising landscape architecture resets our most basic relationship with our environment. Our journey to the profession matches our biological journey through our stages of development.

It's true that each of us were focused on tiny phenomena – the articulation of the rock face, the cracks in the walls, the way the water ran through the sediment at the bottom. In the first semester of the MLA programme, we were absorbing an overwhelming amount of new information. This approach was very challenging and at first we were resistant. It took time to understand a methodology that centred non-humans and abstraction. "Unfinished" was an appropriate name for the studio because we did not arrive at a resolution by the end of the semester. However, it was extremely valuable to have to define our own work in relation to a rigorous methodology, to be uncomfortable and to not know the answer. This exercise shaped our trajectory for the rest of the programme (Figures 8.2 and 8.3).

The 2016 "Manifesto Studio" at UVA applied this methodology in an urban context. The goal of the studio was to translate the impact of the climate and the often-erased geology of three cities: Richmond, Madrid and Caracas. Creating a common language for the three cities allowed us to draw the primary matter

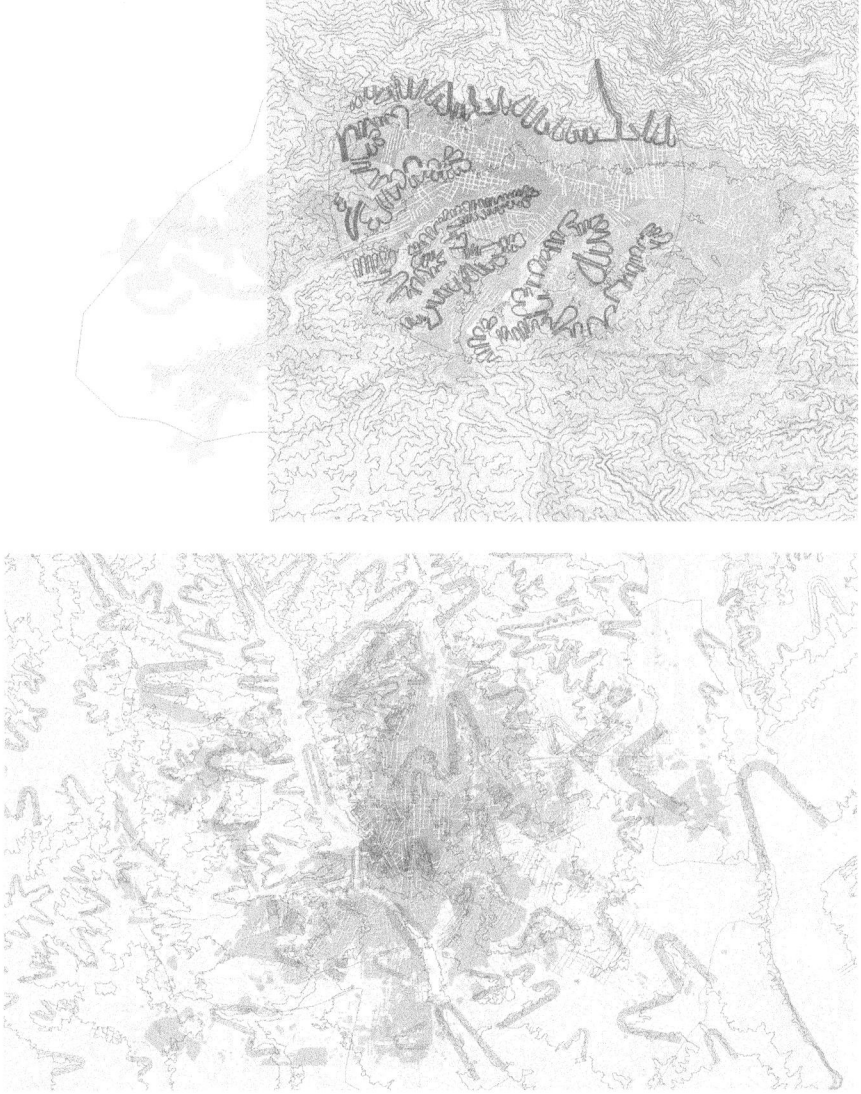

FIGURE 8.2   Geomorphology and Urban Fabric (2016). Top: Caracas. Bottom: Madrid. Collective Drawing by UVA LAR Studio 8020.

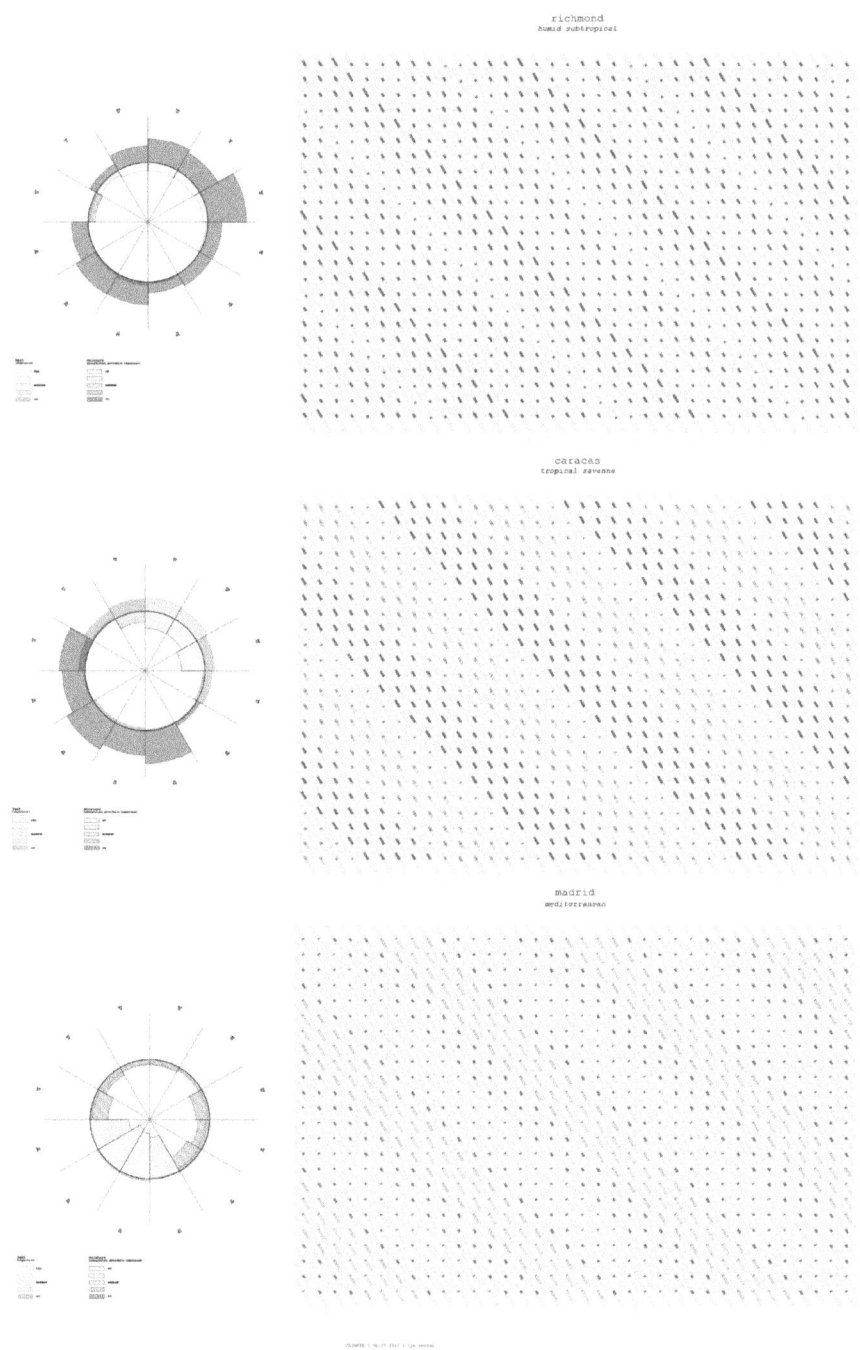

FIGURE 8.3 Caracas, Madrid and Richmond: Comparative Climate Mapping (2016). Marilyn Wenzel.

or landscape: soils, climate, rocks, geomorphology. Through a systematic over-lap of all the elements, we discovered constraints, potentials and opportunities for each site. The result of this exercise was unexpected and required a close reading The rigor of the overlap was important to us – we forced ourselves to discover what the drawing was telling us. In the overlapped drawings we discovered the isolation between communities due to the shape of the val-leys in Caracas, the dispersal pattern of rainwater flows in Madrid due to its topography, and the close presence of the bedrock in the urban landscape of Richmond.

None of us took this studio, but we saw how the students went from very specific explo-rations of a single parameter to designing systems that adapt to the conditions of each place. They really had to put aside their expectations and trust the project of the studio. Each student developed a portion of the shared drawing language, which means that everyone only has a partial vision until the languages are overlapped. Each person's success depended on the work of their colleagues. This shifted the emphasis from the individual design project to the collective production of knowledge. A strength of this method, which we have replicated in our subsequent studios, is the value of comparison. By seeing a particular landscape through its topographical, climatic and geologic context, it becomes possible to understand what makes it unique, and at the same time to position that particular place in relation to others. By developing a language for drawing climate, it becomes possible to compare tropical, temperate and dry cities. Through the comparison, the dynamics of each particular site and its limiting factors become clearer. It's impossible to understand the effect of water without going to the desert – it is helpful to understand the limits and extreme conditions (Figure 8.4).

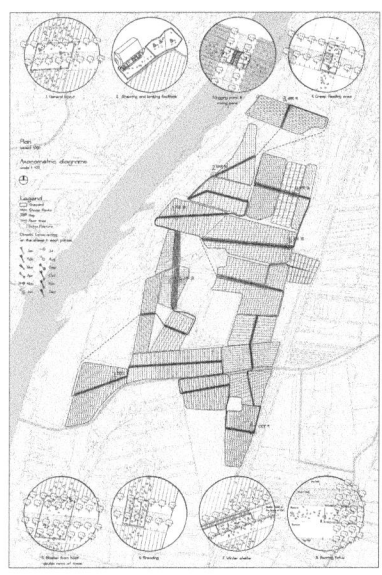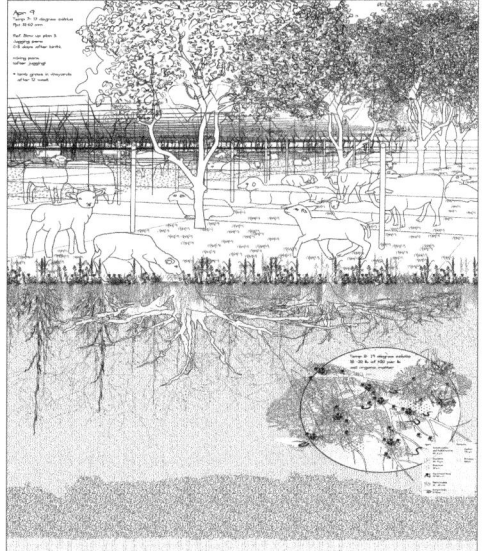

FIGURE 8.4 Lamb and Wine (2018). Oi Wai Charity Cheung.

"Rhizosphere Studio" at the Harvard Graduate School of Design built on the previous experience of the "Manifesto Studio". The students adopted the same language for geology and climate, and simultaneously spent time researching and adapting techniques and systems of regenerative agriculture. Building on the intelligence previous studios developed for understanding the site allowed us to move our research forward and focus on the translation of scientific articles and fieldwork in Arles, France. Students proposed new assemblages of actors: human and non-human. Designs for new forms of production that were more complex, integrative and transverse appeared easily because of the common drawing languages. New potentials and synergies between associations of living systems became possible.

We visited as critics for the course, and it was impressive to see how the students had been able to build on the work from previous studios and to integrate the logics of animals and practices from regenerative agriculture into their design proposals. Beginning with the soil and geomorphology of La Camargue, the studio aimed to enrich the rhizosphere, the living system of the soil, through applied research about animals and their management. These management methods, and the structure put in place by Teresa, provided the students with clear logics and parameters. This framework enabled them to explore how to design in collaboration with non-human agents. The work of this studio contributed to an emergent vision for landscape architecture as a system of interventions that happen over time in relationship with non-human actors – an idea that Teresa has since called The Garden of the 21st Century. This approach integrates living systems and non-human actors into design, which requires that the project take seriously its relationship with time, cycles and relationships (Figure 8.5).

The "Foundation Studio" in 2020 – the first semester of the Master of Science in Landscape Architecture at ETH – began with a series of modules taught by scientists on the primary materials of landscape: climate, water, soil, plants and ecology. For the soil module, a common language was defined by the instructors as a starting point, developed from years of research and exploration. We then asked the students to jump in with proposals that built on that existing knowledge. When preparing the methodology of the studios, it is important for the instructor to define the starting point for research in studio work because two levels of expectations conflict: students are usually at the beginning of their journey, while instructors have a critical vision for the field. At master's level, we think the best studios come from an honest and critical dialogue between these two positions.

As former students who are now co-teachers of this studio, we understood the aims of the assignment and the amount of time it takes to invent an abstract language. Since this module lasted only one week with first-semester students, we decided to develop a drawing language in advance and share it with the students. The language served as a structure for synthesising dense lectures from the scientists, and it facilitated the translation from science to design. By having a system that required students to think about soil texture, pH and organic matter, they were able to interpret the scientists' lectures in an active way, through drawing. Each of the students was given soil data from a location in a different part of the

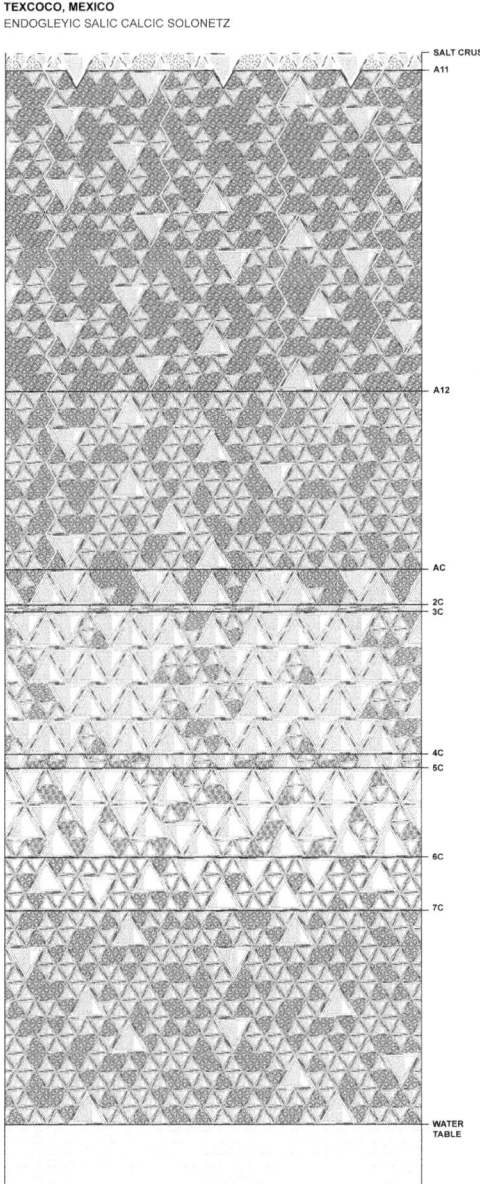

FIGURE 8.5   Soil of Texcoco, Mexico (2020). Angela Stadelmann.

world. The drawing language provided a triangular base as a structure for continuing to add information, and each drawing grew in complexity as the lectures progressed. The students' drawings were a test of the validity of the language, and over the course of the week they challenged it and proposed changes.

Throughout this process we were reflecting on the best place to begin. Providing a coherent language based on years of development raised the level of the conversation in the studio, but, returning to our experience as students in the "Unfinished Studio", we wonder if something was lost by the students not having to struggle through the process of abstraction, of getting lost and needing to find their own path? What is the best starting point to facilitate the collective production of knowledge?

The work of these studios has been to translate the hidden potential of places and pursue a new relationship between living creatures, humans among them. With the consistent methodology of defining languages for living systems, the studios have explored many iterations for drawing water, soil, trees, climate and geology. These investigations collectively build a more robust way of articulating the material of landscape architecture, and we are continuing to develop and apply this methodology in our research now at the Chair of Being Alive at ETH. Translation seeks to understand climate and geology – their performance, constraints and gifts – as the foundation for learning from the primary matter of landscapes over time. By developing a new language in collaboration with the students, we hope to give voice to the others, and redefine the relationship between humans and non-humans where their multiple individual milieus overlap in a shared environment.

We have been fortunate to be able to engage deeply and critically with this methodology as students, practitioners and educators. It has shaped how we approach the field, and we hope to continue to evolve, develop and test these approaches through our own work. The question of language is important for us – it allows us to work across cultures, to have a historical grounding in the discipline and to push the field forward from a solid foundation. A studio is a very hopeful act, because by crossing difference (generations, cultural moments, trainings) and focusing closely on a specific situation through a clear methodology, it is possible to discover new models of landscape architecture that are adapted to our time.

9

CONVERSATION WITH NINA-MARIE LISTER

What is your definition of ecology?

As someone trained first in the natural sciences, I work from a definition of ecology familiar to science, in that I study living creatures and their relationship to the physical environment. More specifically, I study the connection between, and the relationship between, living organisms and their physical abiotic habitat that supports them with and in their communities. In this context, ecology is relational, by definition, but it is also a set of observable empirical phenomena within an evidence-based framework that guides how we see and what we assess, measure and include in our designs. Relative to those trained in design, I suppose this definition may seem confining and literal. Others will interpret ecology as inspiration for or as a metaphor in design, and while I can appreciate that this has value, it is not my position. This is a fundamental principle for me because we are in a time of climate and biodiversity crises and it's clear that we need as much credible, accessible information as possible to inform and support sustainable equitable designs for living well within planetary boundaries. Ecology is not metaphorical for me at all; it is the foundation of my work with observable, traceable, phenomena made visible through repeatable, systematic science, including traditional ways of knowing and living on the land.

Could you discuss how your thinking and pedagogical approach has evolved in response to the urgency of the climate crisis?

The twin crises of anthropogenically induced climate change and biodiversity loss have long been the motivation for and subject of my work in adaptive complex systems, and this underpins the perspectives I have brought to design over the last 20 years. In many ways, the value and impact of this work within design has only intensified since, which is at

DOI: 10.4324/9781003145905-12

once uncomfortably affirming and terribly disconcerting. Climate change and biodiversity loss (in particular) were at the forefront of my research when I first began to engage with landscape architects, and more specifically, in a conversation I had with Jim Corner in 1999. Jim had read a paper I had written[1] for a scientific journal on biodiversity conservation and the urgency around biodiversity following the first Earth Summit and the establishment of the global Convention on Biological Diversity (CBD)[2] and he asked me to join the team for the Downsview Park Competition.[3] My research into complex adaptive ecosystems showed that human systems of governance (including planning and design) were neither accounting for nor understanding functionally the role of biological diversity within ecosystems, nor in climate. No tangible understanding of how ecosystems function was being captured and reflected in the way we made decisions on land – including the ways in which ecosystems sustain human life, in addition to the millions of (known and unknown) other life forms around us. This lack of understanding (or "seeing") has of course contributed to further loss and decline of habitats and biodiversity, and it is now exacerbated by climate change. At the same time, the recognition of biodiversity and its decline has and continues to have remarkable, if largely untapped, opportunities for design that are at the centre of my studio-based teaching and design-based research.

In the two decades since the Downsview Park Competition,[4] the recognition and appreciation of landscapes as complex adaptive systems has become more mainstream and has certainly become embedded into a more nuanced understanding and representation of landscape processes. What's different now is a much more widespread acknowledgement of the extent and urgency of the climate crisis – which is not one crisis, of course, but two. The twin crises of climate change and biodiversity loss are now understood to be intertwined, as was widely reported by the Intergovernmental Science-Policy Platform on Biodiversity and Ecosystem Services (IPBES Report 2019).[5] I think many of us working with landscapes and ecology have understood that for a long time; but collectively, and perhaps institutionally, it's only more recently that we acknowledge this collectively and specifically. Again, in this context, the material structures, functions, processes and relational qualities of ecology are critical, in fact, central to the work that landscape designers do, and fundamental to our teaching and practice. Ecology today is elemental.

Could you please discuss your approach to teaching; how you get students to engage with these ideas and knowledge?

When I was first asked to speak with and engage with landscape architects about the role of ecology in design, it certainly wasn't expected or even appreciated to do so as a kind of moral obligation, or perhaps even in terms of form or material quality. Initially I think designers were more interested in what the systems and processes of ecology could offer as metaphors for different approaches to landscape and, in particular, to urbanism. Studio-based teaching about landscape processes and systems as drivers for a new type of (for example, sustainable, resilient or even post-human) urbanism offered huge insights into what it means to embrace uncertainty, adaptation and complexity in a rapidly or even violently changing world. Projective and studio-based teaching centred on experiential learning, in situ or in real places (even remotely) offers a way to ground ecological complexity; it can be made approachable and tangible through imagining future scenarios, rooted in the flora, fauna and lived experiences of people in place, while also being freed from the illusion (and tyranny) of machine-like predictability and control of the "masterplan".

The honest answer is that it's not possible to teach creatively about the role of complex ecological and adaptive systems in design, unless you find ways to capture the imagination of students – and with it, their creativity and their humanity. Otherwise, the subject of this work can quickly become a soul-destroying exercise in counting what's been lost. The truth of the matter is that biodiversity conservation is necessarily and often about counting what we're losing as the genetic blueprint of life is bleeding away in plain sight. But you can only say that so many times before people stop listening and feel powerless, or worse, hopeless. My emphasis in teaching has been to foreground the crises, but to focus with compassion and humility (and ok, with joy that comes from empowerment) on the possible. As teachers and research-designers we need to take the stark reality and data-based knowledge and project it into creative thinking about design that's responsible and inspired, and that is frankly hopeful.

Working with designers, for me, is also a kind of ecology: an emergent synergy that comes from building relationships across difference. Whether I am working with a landscape architect, architect, urban designer or engineer, we are working across qualitative and quantitative data – with words, numbers and images "that count" – to visualise complex problems as much as their solutions. Much of my recent design-based research is focused on green infrastructure and remediating connectivity across landscapes for both people and wildlife. This collaborative work in applied ecological design is, in part, about making visible what we can't see (for example, asking why are so many species of wildlife at risk on our roads? Can we design a better road that provides safe passage for both humans and non-humans?). This work is not intended to be revelatory for its own sake, but rather, it makes both a problem and its solution *legible*, so that people understand how something works, why it solves a problem, and that it can do so in a way that captures the human imagination.

Design students are curious, and in their creativity they are often open to embracing uncertainty, and with it, the need for legibility. Design education cultivates and amplifies the ability to reveal what research and data tell us beyond numbers and helps us to see aspects of a system that aren't visible, or how seemingly unrelated system elements can come together to create something new. Many contemporary landscape studios and projects have focused on visualising complex systems to imagine a diversity of potential futures, from (e.g.,) reimagining dredge and sediment, to how bird habitats shift and change with migration, and making visible those processes so that people have better language for expressing their relationship to a particular species or place. This is why I emphasise to the reciprocal and relational nature of ecology: it is about building relationships and making visible those relationships whether between species or places (or both).

Your work provokes a need for radical shifts and new imaginaries, and it means addressing the future resilience and adaptation to the climate crisis. How do you imagine these shifts in terms of our region and urban environments, and what radical shifts do we actually need?

There is (at last) a growing recognition that we need to manage *ourselves* (rather than the environment) radically differently if we want a liveable planet. Unless we shift our institutionalised systems of governance away from those dominated by colonial (but illusory) ideas of endless extraction from and control over the environment, we're unlikely to be able to stop the current destructive trajectory. That seems intractable to most people, and revolutionary to others. While radical shifts like the Green New Deal are necessary and gaining traction, there are concomitant opportunities for locally grounded place-keeping strategies that empower people and sustain ecosystems. There are many things we can do in small ways that may act in synergy, across landscapes as a kind of acupuncture. Projects that engage relational ecology, between people and place can inspire hope and optimism within an individual and their community, and help citizens activate a small connection to land and life, which in turn may lead to and amplify other positive changes, from understanding to stewardship (e.g., pollinator pathway projects that connect equity-seeking neighbourhoods spatially and empower citizens to learn about locally resilient flora and declining fauna). These small acts are really incremental radicalism.

Social science shows that perception and behaviour are not easily correlated: people can readily change their attitude, but changing behaviours is more difficult. The power of design lies in the ability to project different futures, contingent on some change in behaviour (e.g., post-carbon futures). The connection between attitude and behaviour can be facilitated by good design, or by democratising design through working with communities in place through practice. More recently, in mid-career I have engaged intentionally in design-based activism, or design activism, meaning that I undertake projects to reveal, critique or deliver specific public goods or civic outcomes. I see this as central to the responsibility I have as a public educator and a civil servant, to use my platform within the university to lead by example, from a normative stance, and to help other communities to develop capacity, uncover agency and find voice, or frankly, in the service of other species.

Recently, I've been collaborating with Dr Tim Beatley and other colleagues internationally in the Biophilic Cities Network.[6] Biophilic cities are those that are biodiverse, and they demonstrate through policies and design of green infrastructure, the power of connection to nature. Cities are now the only landscapes that many children will ever know, and so it's more important than ever that urbanised places protect and enhance biodiversity, i.e., the non-human world of species that sustains us through the ecosystem services that healthy ecosystems provide (e.g., food systems, pollination, stormwater infiltration, soil health, urban heat island reduction, stress reduction and so on). Building relationships between people and nature in cities can be focused on green infrastructure design that includes (e.g.,) parkland and natural heritage systems to green roofs, green streets and bioswales. These can be vernacular community projects from rain gardens to pollinator beds, but also radical when "stacked" or scaled as citywide natural heritage systems or a ravine park system. When we activate strategies by which communities can demand and create access to healthy green places, "scaling across" private and public spaces and connecting city regions becomes very important. We have seen this clearly and emphatically during the COVID-19 pandemic during which equitable access to both local and regional green spaces and to connected trail systems of parks, ravines, waterfronts, etc. has become a human right for both physical and mental health and wellness. Under the emerging Green New Deal movement and in urban planning and design broadly, the public policy discussions about post pandemic recovery are increasingly focused on health and wellbeing through equity of access to the outdoors, and to nature more broadly.

My colleagues in public health (for example, Dr Jenny Roe of UVA's Healthy Cities Lab)[7] are at the forefront of research that is providing empirical evidence that exposure to urban nature is related to human health and wellbeing outcomes. Again, there is an emerging and significant role here for landscape design – one that calls for engaged community work at all scales, from the smallest pocket park and bioswale to extensive regional park networks. Working with communities in this context requires designers and planners to understand that equitable access is essential, that it is a human right for health. This also means that we ought to be humble and compassionate in understanding communities' needs while building relationships to urban nature that is less about grand design and more about engaging the mundane ecologies of the everyday[8] (or what Richard Weller has called "mongrel" ecologies).[9] Local ecological relationships are significant to understand and to sustain, and the cultivation or nurturing of these relationships in design might frankly be the kind of low-hanging fruit, but important attitudinal shifts that get us to a place of radicalism.

It reminds me of the protest you made in terms of your front lawn. It demonstrates the notion of the acupuncture that radically shifts both societal expectations around the lawn, but also issues of governance.

I love that you've asked about this – it's my unintentional design activist project! Really, it is such a weird and rewarding project for me, because it's personal and I never intended for it to be a research project, or even a public statement. It's the front yard of my family's home. We are fortunate to have a house on a beautiful corner lot in mid-west Toronto with a steep slope on which no reasonable person would attempt to grow a lawn. A more practical approach to the site and its ecology was to design and plant a pollinator garden in a natural meadow, where layers of herbaceous perennials, native grasses and forbs intermix with shrubs in a no-mow landscape that slows and infiltrates stormwater really effectively. The meadow also flowers non-stop for three seasons, so there is plenty of fragrance and colour, and visual interest, including a diversity of textures and forms in our long and dreary winter. For me, it is really important that it also provides great habitat for migrating and breeding birds, many pollinating insects (including moths, butterflies and ground nesting bees) and even some amphibians (toads and spring frogs). Certainly, my own values (and my family's) in biodiversity conservation and enhancement are reflected in how we chose to approach the design. We focused on locally adapted native species, many of which are roadside and farm rescues, along with some cultivars we inherited and which survived a few seasons of neglect, and even a few non-native plants that I let stay because they are great pollinators. In short, you could say that my garden reflects who I am and what I do: it's me "walking my talk" as a landscape ecologist, a researcher and a professor in an urban planning pro-gramme. But the garden became political in August of 2020 when a bylaw enforcement officer showed up at our home.

We were informed that our yard was in violation of the Toronto Municipal Code, Chapter 489 or the Long Grass and Weeds bylaw; we had to mow the yard or the City would do it for us and charge us. My first thought was that this was a prank; I could not believe that the City would still have such an archaic and vague bylaw, much less enforce it during a pandemic. What specific kind of grass was in violation of the bylaw? Which weeds were offensive? Alas, there were no answers to these questions, which rapidly became the source of my pro-test. Apparently, many neighbours had complained that our yard was "messy", "out of control" and "weedy" and so the City took action to issue a Notice of Violation. I had never been concerned about my neighbours, and I never imag-ined that I would start a controversy by installing a biodiverse garden in the form of a low-maintenance no-mow meadow. But when the City decided to enforce the bylaw, I knew I had to respond politically and publicly, principally as an act of design activism to bring attention to the problems of the bylaw as much as to the absurdity of enforcement. As a white, educated and privileged person, my voice as a university professor affords me a significant platform and obligates me to the public I serve. I saw this as my responsibility to take the matter public and to show how design can be act of insurgence.

I also decided to fight the bylaw and its enforcement publicly for reasons embedded in your question: I wanted to expose, first and foremost, the myriad problems with the dominant ideal of the lawn – a deeply entrenched symbol of colonialism that is rooted in slavery and is a monoculture of turf-grass, devoid of biodiversity and dependent on chemical and nutrient resources. In this way, the lawn is offensive to and at odds with our ethnoculturally diverse city, our indigenous population and our public policies that support and encourage biodiversity. It is an affront to citizens that our property standards uphold an outdated and colonial image of the lawn rather than healthy, diverse and even edible yards that sustain and nourish us, spiritually, mentally and physically.

On a practical basis, I also wanted to draw attention to the vague and arbitrary nature of the bylaw itself. Neither grass nor weeds are defined: there are 12,000 species of graminoids, and there is no scientific definition of a weed. (As the popular saying goes "a weed is just a plant growing where it's not wanted".) Without a scientific basis for defining grass or weeds, the bylaw was effectively attempting to regulate aesthetics. I also knew that there had been previous court decisions that upheld citizens' rights to grow "natural gardens" as a freedom of expression. I enlisted the help of environmental lawyer David Donnelly, who determined that, for these reasons, the City's bylaw was unconstitutional and indefensible. After several media articles and lots of news coverage, I was suddenly informed that no Notice of Violation had been issued, and that I could seek an exemption if I wanted to register my yard as a "natural garden". I was publicly emphatic that I neither sought nor wanted an exemption, for two reasons: first, I found it both preposterous and offensive that growing a healthy biodiverse garden was not already as-of-right in this time of climate and biodiversity crises, but rather an aberration for which one needed to apply and be granted an exemption to the lawn norm, and second, the application for exemption required a payment and navigation of a licensing system, during which your address becomes publicly posted and for which you must reapply annually. It was clear that the exemption system privileges people with education, capacity and money, and certainly does nothing to alleviate the arbitrary and vague nature of enforcement, which at the best of times is unfair and inequitable, and during a pandemic especially, is a waste of public resources.

Even more ironic in this matter was that I had spent several years of community service volunteering as an adviser to the City's otherwise progressive public policy on biodiversity, including the Toronto Biodiversity Strategy, the Ravine Strategy and the Pollinator Protection Strategy. My efforts to highlight the problems with the municipal bylaw were also intended to emphasise the disjuncture in city policy, and that policies and practices for public and private lands alike have to be aligned and work together for effective biodiversity protection. As the matter became more public (including a challenge we issued to the mayor to come and tour the meadow), the City backed down, eventually saying that our yard was exempt and that there would be no enforcement. This too was both problematic and ridiculous: by giving me the exemption whether I wanted it or not, meant I had no legal basis on which to pursue the matter in court. It also reinforced my position that I had simply been putting into practice in my yard the same commitment to biodiversity protection through ecological design as I have in 20 years researching, teaching and practising publicly. Importantly, this also allowed me to emphasise that as a white, educated person of privilege, of course, I can get an exemption. But what happens to more vulnerable people, newcomers to the city who want to put their hands in to the soil in a rented garden? Or to black people who are growing food in their front yard? What protection is afforded to indigenous people who are struggling through the effects of intergenerational trauma and want to grow a medicinal healing garden? By asking these questions, my yard became a small act of resistance to the lawn and all that it symbolises, and in the process, to call for much bigger change in the way we relate to nature in the city. As design activism, the project illustrates that a small act of resistance, even symbolically as much as literally, can be deeply radical, revealing that what we accept as a norm may be deeply flawed and part of the problem. It not only exposes but sets in motion the fall of a "house of cards", revealing that in many ways, our systems of governance are not working, and they're deeply flawed given the challenges we face for sustainability and resilience.

I'd add that this project found me at a time when I'm more inclined to embrace activism again. Certainly, I am at a career stage where I can afford to be radical, but I have always understood my work as normative action, even as informed by science. It's also distressing that 25 years into my career, the problem of biodiversity loss and climate change (which I started out to address) has gotten worse. Although I've essentially been building on the same work since 1999, I am asserting it more actively, and collaboratively across disciplines and divides.

It exposes this tendency to say, well, in Australia, particularly, "The effects of colonialism are gone. We've dealt with it." But it's embedded in every microregulation at every level. So, I think that's also a really important thing that's revealed.

Yes, I think that's true. I did a sabbatical in Australia, and I felt some solidarity there, coming from a Commonwealth nation. Certainly, we share many tensions (and ties) with the colonial legacy that is still alive and well in both nations. Together with my colleague, Pierre Belanger, I had a chance to explore the legacy of colonialism on landscape through *Extraction*, the Canadian exhibit he curated and directed at the Venice Biennale 2016. Our countries have of course become very good at the twin processes of territorial control and resource extraction which we learned painfully from our colonial masters. Extractive practices and the institutions that enable and govern them are deeply rooted and not going away anytime soon. So even small acts of resistance and pushing back on one aspect of colonial control are important. Design activism broadly can help reveal institutional structures that continue to degrade, denude and destroy ecological health in the face of a growing climate crisis and accelerating biodiversity loss. I'd say this activism is necessary now more than ever.

About wildness and coexistence: could you expand on the ways in which coexistence could occur in the potential of hybrid or, as you noted, a "mongrel" form of action?

I wanted to write more about wildness and wilderness over the last few years because it was integral to the origins of my work in ecology and yet I felt it had been missing for a while, perhaps in part because design applications of landscape and ecological urbanism had become the focus of my work between 2004 and 2010. During this time, I made a conscious decision to "urbanise" my work, exploring and developing the relationship between ecology and landscape architecture offered itself through the lens of urbanism, along with many insightful and valued collaborations with designers. Most of my early training and field experience was outside of cities, in wild landscapes, yet cities are full of "wildness", of course, of all different types. Humans long for wildness, across cultures, and our own humanity is defined in part by what it is to be wild. A broad and rich literature across the environmental humanities, psychology and biophilia attest to the importance of wildness as a concept that is embedded deep in our collective psyche. We fear it, we love it and we need it; it's the antithesis of civilisation, but it's also embedded in what it is to be human. When we lose the wild, and the diversity of known and unknown species that define it, we lose a piece of ourselves.

Wilderness and wild things appeal to us in part by engaging our curiosity and wonder, by frightening us, and by holding our gaze: we don't know what it is to be alive without those feelings. It is a really important component to consider in design, which by its direction and intention usually subsumes and controls the wild, "smoothing" out or even smothering wildness. If my garden is any example (and it's a pretty benign one) it is very difficult to emulate the wild in a place of domesticity and taming. Our relentless pursuit of wildness is evident across a spectrum of activities from children's play to extreme sports to radical ecotourism and in hobbies from bird watching to forest bathing to swimming with sharks. Many of us seek out places and experiences where we feel on the edge of community or alone in some way that helps us to discover who we are. Such places are increasingly rare in high density cities, in particular, but we seek them nevertheless – even if through something as simple as allowing children to become filthy, dirty and wet as they play. There's an element of wildness there, and it's becoming more important to foster as the natural world disappears around us.

Coexistence is a more difficult concept for us. If we consider urban culture, we can see that Toronto and Melbourne, for example, are very similar in terms of ethnocultural diversity (e.g., demographics, immigration, linguistic background and ethnocultural identity). These two cities have progressive precedents, institutionally and in policies and governance, for how to embrace human cultural diversity. We're not as good at embracing the diversity of other species. We tend to like the creatures that call to our romantic sense of a disappearing wilderness or those that are aesthetically appealing to us for some reason. It's interesting that we often dislike the species with which we are most likely to coexist in cities: that is, the species that are most like us in their adaptability, such as rats, racoons, possums, cockroaches and so on.

Of course, there are relational challenges with wild diversity and the reality of co-existing with other species, but there is also tremendous joy and hope that comes from hearing birdsong downtown. Others will find this watching a fox hunting rabbits in a public park or coyotes raising their pups in a suburban golf course, while others will find it in new and colourful plants and butterflies in their gardens. People who emigrate will be unfamiliar with wild species in their new city, and there's a significant potential to tap into the joy of ethnocultural diversity to protect and enhance biodiversity – to engage with other wild creatures that belong to the landscapes we love.

In the interstitial spaces between those outlying wild places and the remnant fragments of nature in our cities, there is capacity and potential for a richer coexistence. Certainly, there are biodiverse hotspots globally where we know that urban growth pressures are likely to collide with conservation priorities. Global suburbanisation doesn't bode well for wildlife. The urban periphery offers as yet untapped potential for refugia, as well as for breeding and feeding fragments. In the more built up areas of cities, we're really only going to find hybrid systems and species, including some well-adapted invasive and non-native species that may still provide ecological assets. We can all think of lots of examples in our respective locations, and these new, hybrid or novel ecologies are likely here to stay. I think it's important to be conscious of our communities, our cultural diversity and our respective access to nature as we're talking about how we coexist and how we design for coexistence.

Hybridity is already existing. COVID is exposing that further. How do we radically shift the landscape that lends itself to support that relationship?

This is complicated and important question. What species are we "counting" as important to protect? Which species do we endeavour to nurture and protect, or to remove and eradicate – and why? Of course, with global biodiversity in sharp decline, urgent efforts are important for many rare and critical at-risk species in urbanising areas. But I'd argue that design efforts are also needed to help people engage with mundane, everyday nature and the mongrel, hybrid species and novel ecosystems emerging in cities, because for many people, these may be the only places to experience the benefits of nature.

We know that birdwatching has increased dramatically during the COVID pandemic because people were stuck at home. Deprived of vacation travel, or the regular work commute, people began paying attention to what's hidden in plain sight, right outside the window or underfoot at night, and they began seeing and hearing creatures they'd never noticed before. This is a trend that seems to be continuing, with online naturalist groups such birding clubs on the rise, and happily more often specific to equity-seeking groups who have long been left out of the environmental movement. It seems COVID has made us aware of the environment in a more healthy and inclusive way. It's also important that this rise in awareness is not focused on some rare or impossible-to-see creatures, but on life all around us. And it's not that the animals weren't there before, but rather they weren't visible or remarkable to many people who were otherwise distracted by "normal" life prior to the confinement. In cities worldwide, stores sold out of binoculars, bird guides and even a particular kind of remote motion-activated camera used by field ecologists. Suddenly, people began watching for wildlife – from watching the birds at a feeder or raccoons raid the rubbish bin or swans in the park. Even my sons got a remote camera to spy on the mink who stalked our chicken coop.

The point is not that wildlife are more abundant in a pandemic, but that people begin to slow down and look, and to really see for the first time. Familiar places are also home to a myriad diversity of creatures that aren't so much rare as they were just not noticed. When the streets are quieter, more of these creatures reveal themselves. Once again, to come back to the essential in ecology, it's relational: it's about how we relate to the other. We learn and come to know more about ourselves when we can relate to the *other* – and to other species in particular. I find it really thrilling that there's a black birders group that's emerged in Toronto during the pandemic, concomitant with the Black Lives Matter movement. There's now a recognition that birdwatching is for everyone, it should be accessible and joyful. It's not about "life lists" that demand exotic travel but simply (and importantly) about the joy of experiencing nature in your neighbourhood, whether it's learning a robin's "hop, stop and stare" style or a cardinal's call or knowing that the red finch is an urban hybrid but no less worthy or beautiful than its cross-breeding parents. On the edge of the climate crisis with biodiversity crashing, this is a critical moment for all humans to engage our biophilia. This must be inclusive, but also expressive of our basic need to relate to the other.

How do you see your work across multiple scales, the ability of the large scale to provide opportunity to the adaptation, variation and novel formations? And how might we embrace the microbial to the planetary in the future work?

> This work is fundamentally about what we choose to relate to and what we can see. If we can't see something, do we relate to it? If nothing else, COVID has brought to the forefront of our consciousness the vast and incredible life forms that are invisible to the eye and perhaps poorly aligned with our understanding of the full spectrum of biodiversity. The pandemic has also underscored the incredible capacity of life to evolve, adapt and survive, and this may get us thinking more clearly and explicitly about the

scales at which adaptability, modularity and novel emergence are critical to life itself. At least I hope for this awareness, because at the rate we are causing both phenotypic and genotypic diversity to be lost, we need adaptation and emergence across scales.

Novel ecosystems are certainly part of the future, but they can't replace the resilience afforded by billions of years of evolved biodiversity. That is, we don't want to find ourselves left only with a world of dangerous/toxic/inedible invasive species or hybrid mongrels out of a Margaret Atwood novel. We need urgently to understand the effects of hybridisation and risks and benefits of biodiversity change at (and across) multiple scales. The COVID dialogue is likely helpful in illuminating some of these risks, e.g., increasing risks to human health by breaching boundaries of remote ecosystems and increasing the risks of zoonotic disease transfer. We can think of ecosystem boundaries like the boundary that our skin or the bark of a tree provides to disease. Evidence suggests that our most biologically diverse and abundant ecosystems should probably stay some distance from intense human activity. As we learn more about the phenomenon of cross-scale emergence, along with cross-scale adaptations in ecology, it's clear that we need to be vigilant that emergence and adaptation can both benefit and harm human life.

When I think about the impact of work that I do, I'm both aware and grateful that the humble, small scale of the private yard can be as important as a masterplan for a city park system or the investment of rebuilding and rewilding coastal infrastructure. These design interventions are not unrelated: in some ways, they are a fractal pattern of connected landscapes across scales. I want to be clear that I am not describing these interventions as "scaling up", because it's not as simple as taking one thing and just making it larger. Instead, I am referring to an intervention that can be very small in scale, but which can trigger significant changes, whether it's in attitudes or behaviours, or in some other kind of ecosystemic reaction like the release of a virus. It's about being attuned to scale differences and understanding the impacts of change at one scale which might affect another. Sometimes dialling back to more humble intentions in work that is less immediately visible but powerful in its reach is a strategic way to think about scale. It's also a way to think consciously about other creatures.

There's a tension between the adaptation of material and systems thinking, and that of planning and policy and jurisdictional overlays. How do you see the role of designers as resolving this tension or conflict? And maybe it's a question of, is it in the design or is it in the person? A difference of scales?

The last quarter century has been focused on understanding and defining ecology; that is to say, the relationship between living species and their physical environment that supports them. Broadly speaking, ecological thinking has for a long time been understood from a systems perspective;

ecosystems are embedded with one another across spatial and temporal scales, they are modular to a certain extent (meaning regulated by feedback), and they're connected by energy flows. Connected ecosystems provide the conditions necessary for biodiversity, for emergence and therefore for novelty and variety – the fabric of life itself.

Most of our decision-making institutions are grounded in and reflect the opposite of complex systems: they're based on the ability to predict with control over outcomes, and they reflect deterministic, mechanistic thinking which facilitates predictability. As a result, most modern institutional systems of governance (and agencies of governments) tend to be brittle; they're inflexible, and they're not adaptive. I say they are "brittle" because they're based on a normally rigid set of operating conditions and linear thinking. They can't make space for emergence because emergence disrupts control. This disjuncture between institutional systems of governance, and living systems is, to a certain extent, intentional by design; it's needed for example, to be able to control and manipulate ecological systems for maximum economic extraction. There's a persistent friction between living systems and institutional systems that could be seen as offering creative potential for emergent or adaptive decision-making (and so too, design). The problem is that most institutional systems cannot acknowledge and would not be rewarded for their fallibility, and their brittle qualities (e.g., rigidity of hierarchical organisation) have been historically seen as assets. But there has been a growing awareness of the need to model complex systems around the so-called wicked problems of climate change, poverty and inequity, biodiversity loss, food systems transformations and so on.

Institutionally and politically, in public planning and policy there is a growing understanding of the need for more integrated decision-making that comes with an awareness of the need for resilience (under climate change, for example). Jurisdictionally, some of these changes can happen in democratic sovereign nations with a commitment to public policy. It's hard to envision a transition to sustainable, flexible and adaptive systems of governance across borders however, whether interstate or national borders. However, globally, given the recognition of the climate and biodiversity crises, we now have a language of systems change, and we have better understanding of the importance of resilience and adaptation. Designers, particularly landscape architects, both by training and by ability, have the capacity to analyse, visualise and represent complex systems, and importantly, effects of interventions across scales and where conflicts may arise. With this capacity comes the ability to show alternative scenarios and implications. When you can reveal these systems along with points of tension, you can start to explore creative solutions. Institutionally, I am not convinced governments are rewarded to undertake radical systems transformation. Our societies need visual stories of different futures to catalyse action. But we're running out of time to make changes that will keep us below 1.5 degrees, so perhaps we'll be forced into radical change. I am less optimistic than I was two decades ago, but

I am driven by the power and beauty of biodiversity, and my commitment to a diverse and flourishing planet keeps me hopeful. Much of that hope is because of my students – and is for their future.

Notes

1 See https://link.springer.com/article/10.1023/A:1005861618009.
2 See https://www.cbd.int/.
3 See https://www.vanalen.org/content/uploads/2017/08/VAR-8.pdf.
4 See https://www.gsd.harvard.edu/publication/case-downsview-park-toronto/.
5 See https://ipbes.net/news/Media-Release-Global-Assessment.
6 See https://www.biophiliccities.org/.
7 See https://www.arch.virginia.edu/cdh/healthy-cities-lab.
8 See https://www.jstor.org/stable/23460931?seq=1.
9 See https://static1.squarespace.com/static/56b5181b2eeb812c64499929/t/5a70ee7253450a286263bec2/1517350516514/World%2BP-ark%2BLA%2B.pdf.

10

EXPERIMENTAL STUDIO ECOLOGIES

A Productive Throwntogetherness

Ed Wall and Alexis Liu

Introduction

Design studios are ecologies of relations, the coming together of things through time: project sites, overlapping contexts, academic buildings, drawing boards, digital devices, students, tutors, coffee, design briefs, materials, ideas, deadlines, conversations and dreams, intermesh in the pursuit of possible futures. Every human and more-than-human entity in design studios has its role to play as associations are formed, tensions established and imbalances found. As both physical sites of meeting, sharing, learning and producing – and temporalities within academic timetables – design studios embody important spatial-temporal qualities that tend towards landscape processes and experimental design practices.

In this chapter, we explore experimental design as productive relations between design studios at the University of Greenwich's School of Design, design projects developed by students and situated (and contextualised) challenges such as the climate crisis. There are a few unique offerings university education can give to professional landscape architecture practice. In this chapter, we argue that experimental design studio, as a throwntogetherness of place for learning and inventing new approaches to design, is the most important of these. The geographer Doreen Massey uses the term "throwntogetherness" to describe place as "a constellation of processes rather than a thing" (2005: 141). At Greenwich, we consider the design studio as a throwntogetherness of place where knowledge is generated through recognising patterns, negotiating conflicts and highlighting contradictions, where project trajectories work productively from these tensions and where the speculations of new worlds come together in design proposals.

We argue in this chapter, and evidence in our Greenwich studio practices, that to confront global challenges, such as the climate crisis, we must advance two propositions. First, we need to continue to pursue and invent techniques

DOI: 10.4324/9781003145905-13

of working with design and landscape as process. Second, we must defend and expand critical design studio teaching as sites of creative practice and innovation based on rigorous research. We intersect these concerns with three conceptualisations: "throwntogetherness", "trajectories" and "hyperobjects". We borrow from Massey, who develops the term "throwntogetherness" as a way to describe "the event of place" (2005: 140), and "trajectories" as relating to the multiplicity of processes of space. We also refer to Timothy Morton's notion of "hyperobjects" (2013) to make sense of vast entities, such as air pollution, that expand beyond traditional conceptions of space and time. We recognise the overwhelming range of spatial and temporal dimensions of landscapes, always in the process of construction and reconstruction, and their potential to be transformed through critical and creative experimentation.

Studio Throwntogetherness

The university design studio is a physical place – a room, hall or building – where students develop projects, where they meet their tutors, and where they present their proposals. Design studio also exists in time: we mark design studio in the timetable, as moments during the week and rhythms across a semester. However, the initially perceived spaces and times of design studios belie conditions that become impossible to contain and even more difficult to describe. Once discussed, planned, developed and launched by design tutors, design studios rapidly become things extending beyond campus walls and enveloping all hours of student life. The coming together of briefs, sites, projects, colleagues, advice, deadlines, external collaborators, published articles, exhibited projects and influence on professional worlds beyond the university – collectively form something simply termed "design studio". Such entanglements of students, tutors and more-than-human entities highlight that design studios always exist within larger ecosystems, including economic policies, global climates and automated networks, where decisions that inform the design of new worlds are not always ours to make. From environmental events to technological forces and from design tools to other species, the extent of more-than-human influences in the formation of landscapes is core to our studio thinking. Somewhere along the journey, the vast and expanding spatial and temporal dimensions of design studios transform what we initially plan and where we expect them to go.

Through a meshwork of intersecting and expanding ecologies – relations between projects, places, people and temporalities – it can become impossible to identify a single form of design studios: "This is the event of place in the simple sense of the coming together of the previously unrelated … This is place as open and internally multiple" (Massey, 2005: 141). Once launched, design studios, and the design projects produced, can no longer be contained spatially or temporally. Design studios are the nexus from which design projects embark, projects that quickly leave the limits of the studio building, expand outside the studio timetable, and extend beyond the studio brief. Design studios become messy, negotiated moments, contested directions, where futures are waiting to be claimed. As Massey describes of the throwntogetherness of place: "There can be no assumption of pre-given coherence, or of community or collective identity. Rather the throwntogetherness of place demands negotiation" (Massey, 2005: 141). We consider design studios in ways that Massey describes events in place:

> They implicate us, perforce, in the lives of human others, and in our relations with nonhumans they ask how we shall respond to our temporary meeting-up with these particular rocks and stones and trees.
>
> *(Massey, 2005: 141)*

This pulsing, evolving and expanding nature of design studios provides an essential environment for experimental design practice. This requires openness to how design projects develop, how they form collective agency and how they engage critically with other lives and larger challenges.

No time, and no studio at Greenwich, has demonstrated repeated negotiation more than the *Glaze Unit Fire Squad Magic* design studio in 2020. Working within the context of the Greenwich school discourse of *Disruptive Ecologies: Designing for Direct Action*, design tutors Harry Bix and James Fox developed a design brief with youth education charity Global Generation. The teenagers, who call themselves "The Generators", were a client for *Glaze Unit Fire Squad Magic* and the task was to develop a landscape in London's Canada Water with a network of developers and local government stakeholders. The aim of the studio was to forge close relationships between designing and making, while also claiming the teenagers, design students and tutors as "The Generators". Camping overnight, digging clay from the site, talking together, making objects, burning things, mixing materials and polishing objects, all became ways of developing productive social spheres that formed new ecologies within and beyond the confines of the site (Figure 10.1). Projects made material and social relations of landscape projects tangible – processes that often remain invisible due to remote sites of mineral extraction and global networks of distribution.

In March 2020, when the impacts of the COVID-19 pandemic led to our school at Greenwich moving online, and the UK government requiring us all to stay in our homes, the situated ecologies of *Glaze Unit Fire Squad Magic* were hugely challenged. The messy material and working practices of the studio stalled while online relations and digital production became dominant. The resources of students' homes, and the networks they were able to maintain from bedrooms and kitchen tables, revealed extraordinary invention and determination. Also exposed were contrasting domestic circumstances, including unequal access to technology and green spaces, and varied support from friends and family. Students found themselves in places, as Massey argues, that "necessitate invention", but also "pose a challenge" (2005: 141). The productive uncertainty of experimental studio practices was challenged by new uncertainties beyond

FIGURE 10.1 Left: Glaze Unit Fire Squad Magic was challenged by the global pandemic in 2020, with design experimentations moving to kitchen tables and grandparents' gardens (2020). Right: The clay landscape of The Generators making visible material relations and social lives

our control, of when we could return to campus, how we could work together online, and who is looking after our families – not to mention how we or our loved ones would be impacted by COVID-19. Projects began to focus more on our roles as landscape architects within uncertain and changing worlds (Figure 10.2). Ecologies of material flows, social relations and future projects became dislocated, and then reassembled, as the global pandemic and digital working intersected with projects already in progress.

Design Trajectories

To engage with landscapes as processes, design projects at Greenwich are developed and planned as incremental practices. Like trajectories, design projects are entities that move forward, interact, combine, splinter, accelerate and occasionally slow down. Like design studios, the stepping-off point of design projects can be located spatially in project sites – all projects involve repeated fieldwork – and temporally in studio briefings. Also, like studios, end points become impossible to discern as projects develop their own lives through exhibitions, publications and their influence on live projects. In *For Space* (2005), Massey uses the terms "trajectories" and "stories" "to emphasise the process of change in a phenomenon" (2005: 12). Stories provide Massey with a means to describe the "history, change, movement, of things themselves": and in the coming together of trajectories "a simultaneity of stories-so-far" (2005: 12). The relations between pluralities of trajectories entangle design projects with the landscapes they set out to inform. Situated design projects become part of the processes, or stories-so-far, of landscapes, where multiple trajectories provide a means for seeking out untold narratives, otherwise excluded voices, and often silenced concerns (Figure 10.3).

Deptford Air Force Technician Youth, by MLA student Henry Wilson (2019), takes London's illegal levels of air pollution as its point of departure. In 2018, the UK's High Court ruled that the government's policy on air pollution was "unlawful". At the same time, in south east London's Deptford, 43 percent of residents reported someone in their household had developed health conditions as a result of air pollution. The largely unseen entity of air pollution, created by vehicles burning fossil fuels, was particularly acute in Deptford – a neighbourhood in south London formerly the home of John Evelyn, author of *Fumifugium*, a critique of air pollution published in 1661. *Deptford Air Force Technician Youth* forms a series of utopian and makeshift interventions. It first encloses the sources of air pollution (such as vehicle exhaust fumes and construction air particles). Second, it pumps polluted air into a large historic hanger. Then it cleans the air with the aid of elaborate planting. Finally, it distributes clean air to nearby playgrounds and local schools via giant tubes and spouts.

In the context of government inaction, *Deptford Air Force Technician Youth* recognises the common agency of environmental protesters and local residents in making change locally while also addressing a planetary climate crisis. Described

MAKERS YARD

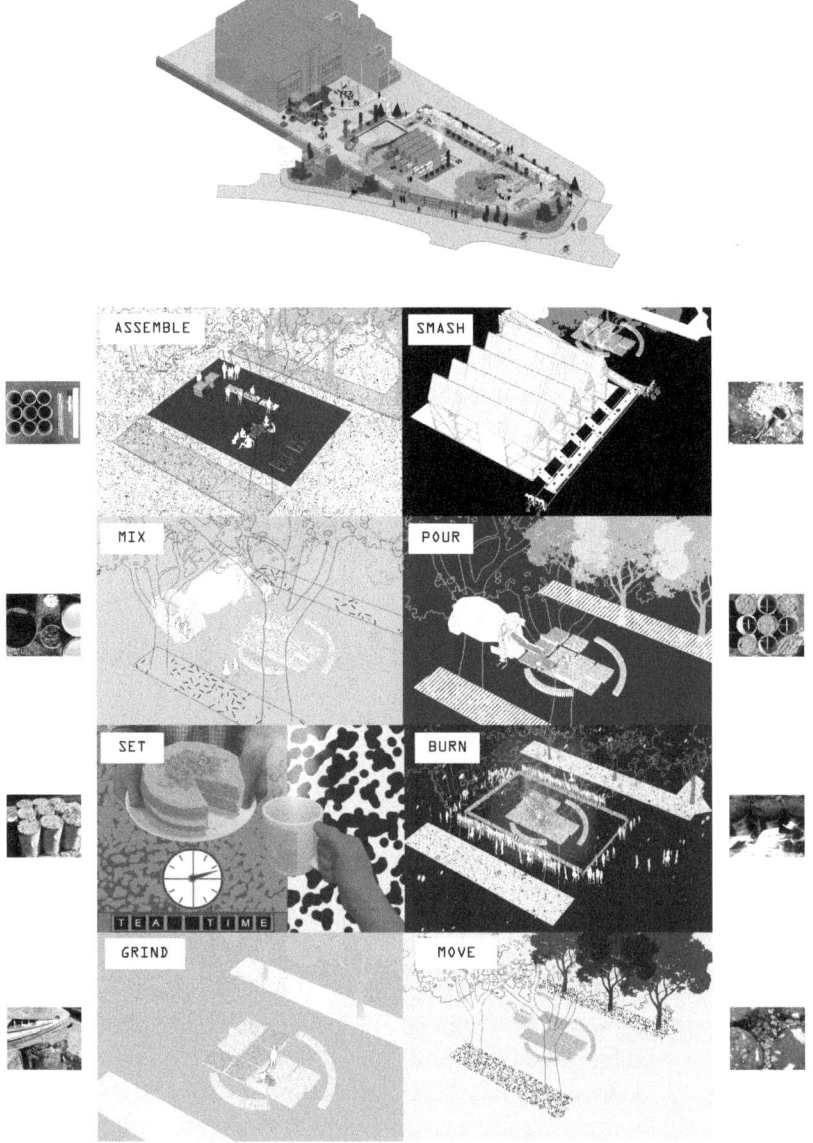

FIGURE 10.2 The landscape architect in residence, co-making a landscape based on new forms of terrazzo, constructed from found materials and local waste (2020). Ross Redman-Shaffer, MA Landscape Architecture, University of Greenwich.

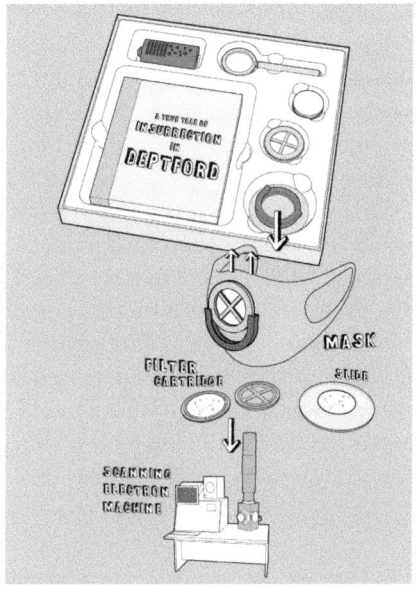
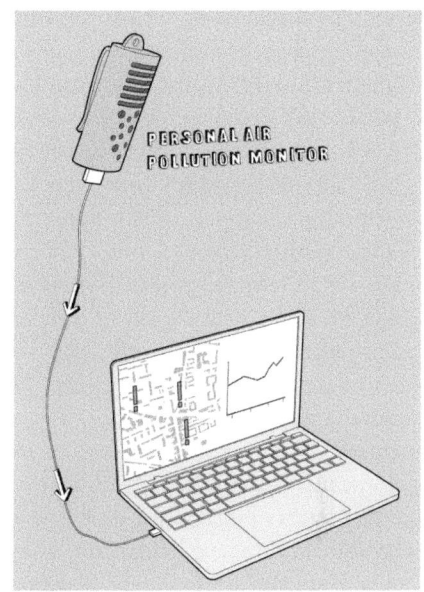
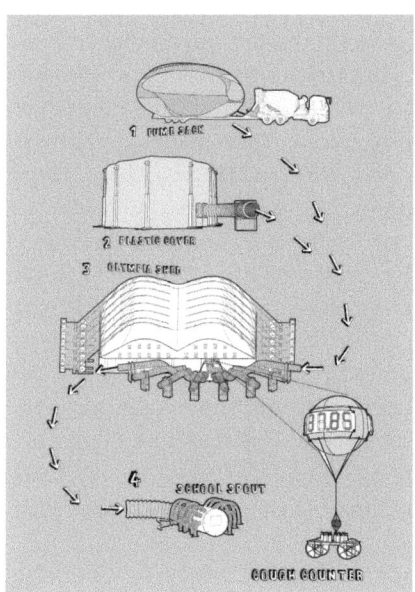
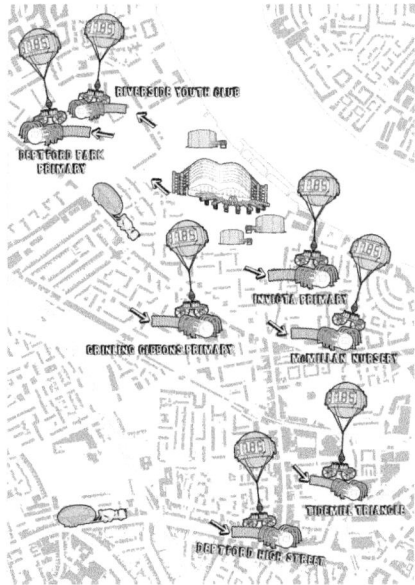

Greenwich University 2019
A D.A.F.T. Intervention

FIGURE 10.3 A kit of instruments is central to reading, measuring and protecting local activist designers from the air pollution in Deptford, South London (2019). Henry Wilson, MLA Landscape Architecture, University of Greenwich.

by Wilson as "A true tale of insurrection in Deptford", the entanglement of development practices and commuting by car comes into conflict with the day-to-day concerns of residents and an uncaring government: "The city only lives if the planet will!" Wilson writes. These interacting narratives reveal ecologies of plants and living beings considered more natural, and entities of pollution claimed as artificial. While the sciences of ecology are core to our landscape architecture teaching at Greenwich, a more deep-rooted ecological thinking and questioning is core to our design studios. Our site-specific project approaches resonate with Lorraine Code's observation that "ecological thinking relocates inquiry 'down to the ground' where knowledge is made, negotiated, circulated" (2006: 5). Ecological thinking reinforces the importance of grounded research and knowledge informing designs, while at the same time providing opportunities to design from a multiplicity of differentially scaled perspectives, across material and planetary relations. To address conditions of air pollution or global warming – entities that Morton terms "hyperobjects" (2013) – we must think ecologically: "It is not as if some abstract environmental system made us think like this; rather plutonium, global warming, and pollution, and so on, gave rise to ecological thinking" (Morton, 2013: 48).

The dynamic ecologies of design studios, that we can understand as events in time largely defined by the trajectories of design projects, provide a means to engage with vast processes such as air pollution. The unfathomable spatial and temporal dimensions of hyperobjects compel relational and process-oriented thinking. Morton defines hyperobjects through five qualities: first, they adhere to anything that they come into contact with; second, they challenge traditions of time and space as fixed; third, their vast distribution spatially and temporally cannot be understood in a specific location; fourth, the massive extent of hyperobjects makes them difficult to perceive; and fifth, they are formed from relations between more than one object. The condition of London's air pollution that *Deptford Air Force Technician Youth* aims to address resonates with Morton's definition of hyperobjects. Considering air pollution as a hyperobject reveals the challenges of site-specific practices, like landscape architecture, in effectively confronting it. However, we have found that if there are aspects of projects that have the capacity to address hyperobjects, then approaches that have the greatest potential are those that intersect disciplines, designs that have agency across multiple scales (situated and remote), and proposals that embed contrasting temporalities.

It is tempting to consider design studios as hyperobjects. Through the multiple trajectories of projects, design studios can become too big to fully comprehend, especially when working within them. But design studios differ from Morton's hyperobjects, as it is possible to gain distance from or step outside design studios. However, the extensive temporal and spatial dimensionality and relatedness to other things that define design studios reveal potential agency to confront hyperobjects of global warming and air pollution. The open-ended nature of experimental design studio ecologies, as evidenced by *Glaze Unit Fire*

Squad Magic, make design studios useful events in time from which to engage with hyperobjects. The throwntogetherness of design studio can engage with hyperobjects: "It means that the negotiations of place happen on the move, between identities which are on the move" (Massey, 2005: 158). Design studios can develop dynamic qualities that make them useful for exploring challenges such as the pollution, climate and storms. Processes of landscapes and trajectories of design projects entangle, interact and influence each other – and in so doing, design studios have potential in engaging with entities, such as global warming, that defy spatial and temporal traditions.

Structuring Processes

Tensions between material entities and entanglements of landscape processes are highlighted in landscape architecture design projects at Greenwich. In the context of the experimental and open-endedness of our design studio processes, we recognise the need to realise common milestones – tangible and consistent moments in design projects that evidence progress within what can be less predictable trajectories (see Wall, 2019). In the past two decades, academics and scientists have examined and proposed complex and dynamic ecological landscapes, but the difficulties of landscape architects working with these conceptions have been less discussed (see Wolff, 2014). Design studios, once launched, develop lives of their own, collective momentums that are increasingly difficult for tutors to steer. However, as a school we give great importance to three moments in design projects. These moments, which ground process-oriented approaches, encapsulate three aims. First, to draw conclusions from project research. Second, to precisely describe and map the dynamic actions of the proposal and site. Third, to go beyond visual representations to illustrate the construction of experiences. We achieve these outcomes, respectively, with what we term the "base drawing", the "operational drawing" and the "scene" (Figure 10.4).

Of these three compositions, the base drawing is essential for making use of research knowledge and structuring design proposals. We argue that it is necessary to make sense of research data to establish a basis for making design decisions (Figure 10.5). Research in design disciplines tends to contrast research in more academic subjects, whether in anthropology or ecology, geography or economics. At best, research by designers is briefer, being a necessary stage of design processes with the objective of developing propositions; at worst, it is a time-consuming activity undertaken with insufficient care and lacking rigour, the results of which are ignored as design proposals are composed from intuition. Greenwich design studios require students to develop clear arguments based on their site research, forming the basis from which design speculations can begin. The base drawing is a single composition constructed by bringing together, redrawing, editing and synthesising data collected from project sites and contexts (Figure 10.6). The aim of the base drawing is to make critical conclusions from

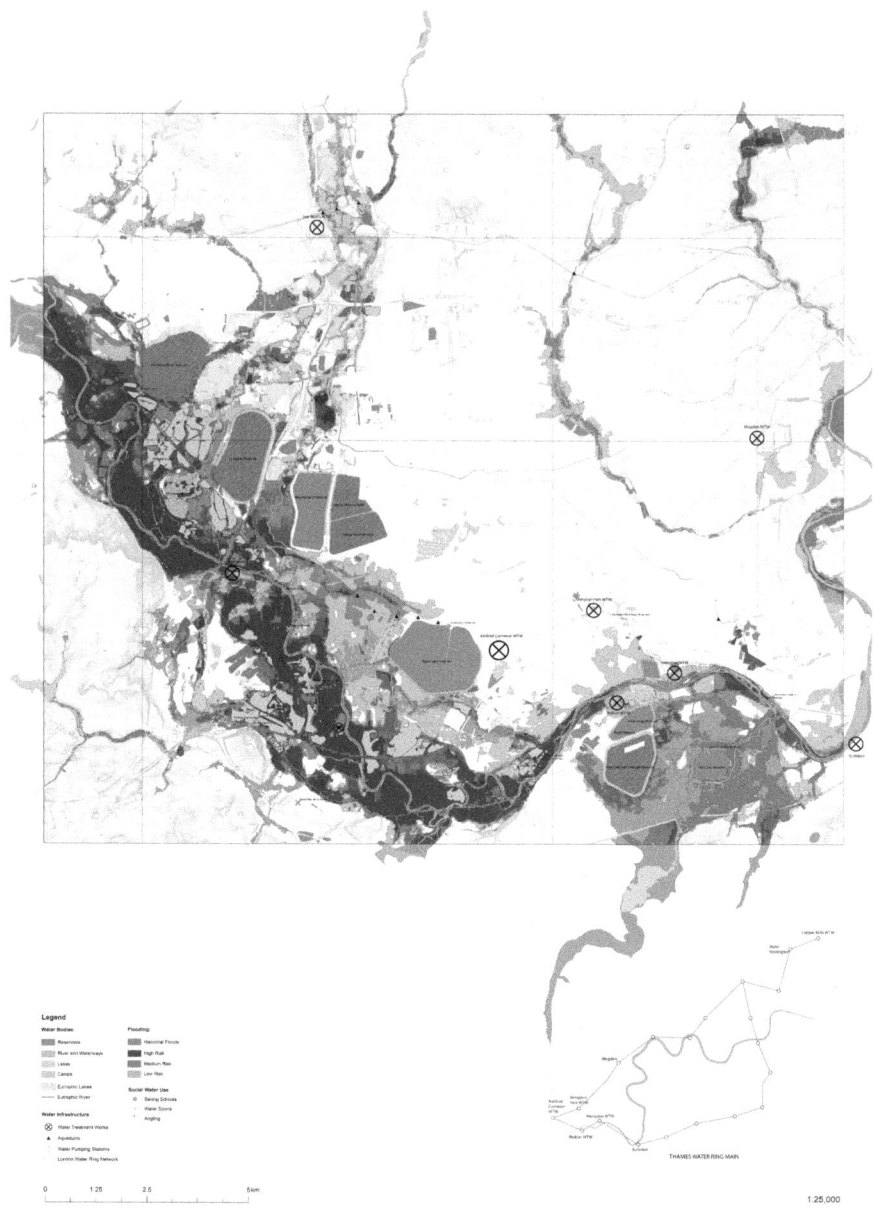

FIGURE 10.4 Reading levels and conditions of wetness of river landscapes through a composite base drawing (2019). George Armour, MLA Landscape Architecture, University of Greenwich.

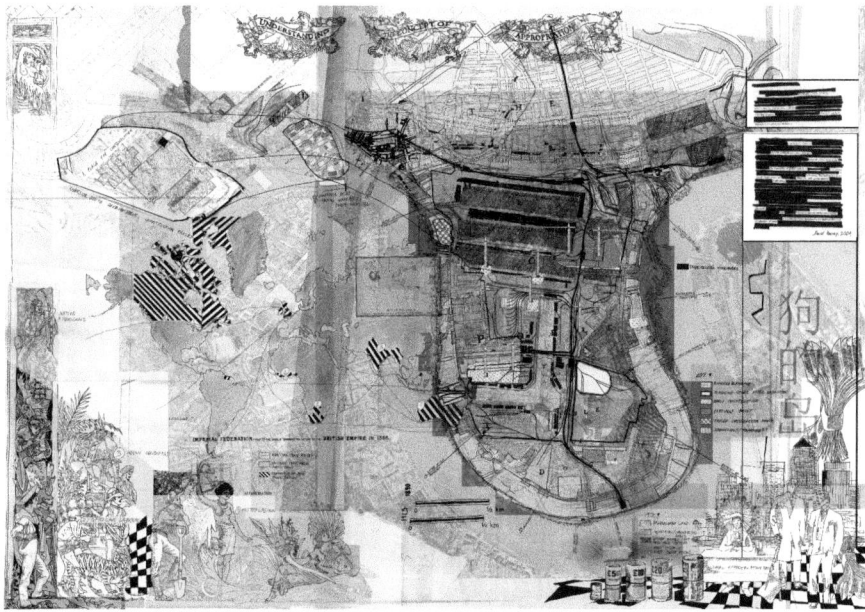

FIGURE 10.5 The climate crisis advanced by global urban development industries – still related to historic global power structures, such as British colonial trade (2017). Cesare Cardia, MA Landscape Architecture, University of Greenwich.

which inventive design proposals can be developed and tested. The base drawing requires visualising data from interviews, texts and other methods, and bringing these data into proximity, recognising relationships, with more traditional maps, plans and drawings.

The base drawing becomes an essential moment within the experimental processes of design projects at Greenwich – a point of reference within trajectories of design projects, and a milestone in the stories-so-far of the design process. Along with operational drawings and scenes, the base drawing is one of three consistencies within our design studios that are generative through their experimental approaches and diverse results. Experimentation is a core practice reliant on the essential processes of questioning, where points of departure may be located but where open-endedness needs to be embraced. Constant and simultaneous activities of unsettling and settling are required, where knowledge is continuously produced and repeatedly tested. However, amid this uncertainty, there is a need for knowledge, skill and structure. The base drawing, we argue, offers this. The other two project milestones we ask students to recognise are operational drawings and scenes. These are described further in the essay "What … is landscape?" (Wall, 2019). Building on conclusions made through producing the base drawing, operational drawings and scenes describe how the design works as an

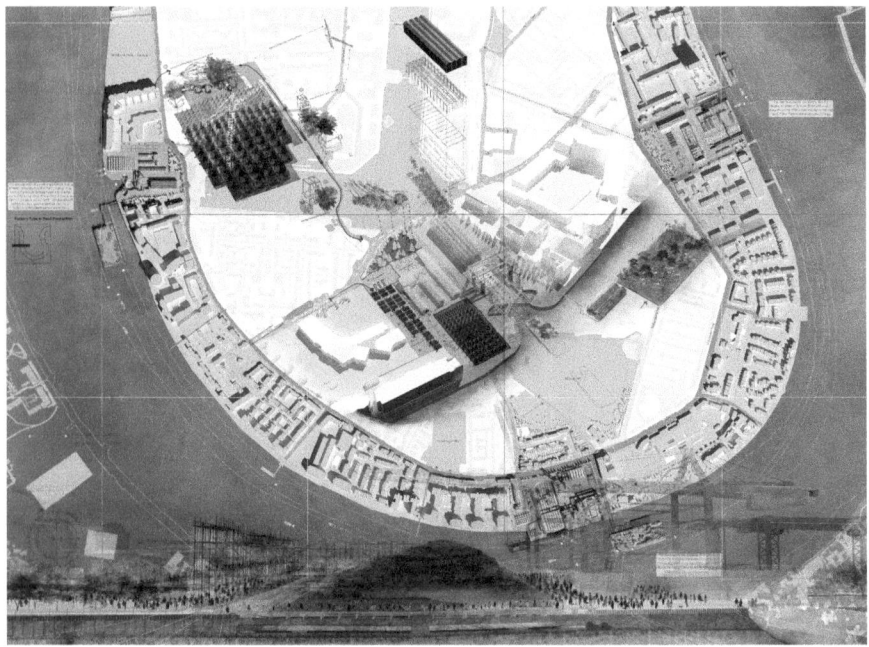

FIGURE 10.6   The Island Factories on London's Isle of Dogs reconceive a range of landscape relations, informing industrial, waterfront and planetary ecologies (2018). Mais Kalthoum, MLA Landscape Architecture, University of Greenwich.

entanglement of human and more-than-human actions, as well as the constructedness of our experiences with landscapes designed.

Closing

We claim the design studio as a throwntogetherness of places, as events in time. The creation and productive interactions of project trajectories, including accounts of conflicts, moments of tensions and contradicting stories, frame the experimental practices we have developed at Greenwich. We recognise the capacity of experimental studio practices to engage with hyperobjects of the climate crisis, such as global warming, air pollution and storms. The potential of design studios to impact outside university walls and live beyond the structure of timetables and durations of semesters, is evidenced as discourses and designs inform professional practices. The dissemination of projects through exhibitions and publications, as well as engagement with project stakeholders, grounds our experimentation and inspires those that we work with. Students also bring their concerns about the climate crisis, and their invention, to addressing such massive challenges to professional practice, impressing design studio knowledge and culture beyond the university. Through working with a throwntogetherness of design studios, and aiming to inform the future direction of hyperobjects, we embrace our roles within landscapes and test the bounds of our capacity to inform change.

References

Code, L. (2006). *Ecological thinking: The politics of epistemic location*. Oxford University Press.

Ingold, T. (2013). *Making: Anthropology, archaeology, art and architecture*. Routledge.

Massey, D. (2005). *For space*. Sage Publications.

Massey, D. (2006). Landscape as provocation. *Journal of Material Culture*, *11*(1/2), 33–48.

Morton, T. (2013). *Hyperobjects: Philosophy and ecology after the end of the world*. University of Minnesota Press.

Wall, E. (2019). What… is landscape? In Jorgenson, K., et al. (Eds.), *Teaching landscape*. Routledge.

Wilson, H. (2019). *Deptford Air Force Technician Youth* London: University of Greenwich.

Wolff, J. (2014). Cultural landscapes and dynamics ecologies: Lessons from New Orleans. In Reed, C., & Lister, M. (Eds.), *Projective ecologies*, 1st ed (pp. 184–203). Actar Publishers.

Bibliography

Tsing, A., et al. (2017). *Arts of living on a damaged planet: Ghosts and monsters of the anthropocene*. University of Minnesota Press.

11

ADAPTING PRACTICE FOR THE FUTURE OF LANDSCAPE-DRIVEN URBAN DESIGN

Anya Domlesky

As our cities and environments become more complex and face unprecedented challenges, it is no longer sufficient to design for aesthetics alone. The practice of urban design, landscape architecture and planning now demands going beyond typical design services to support deeper insights via foresight, research, experimentation and innovative advocacy.

It used to be that there was a site, and an owner who wanted to change it. They looked around for advice and found a consultant – a landscape architect who then produced design options. Today, these categories – site, client and consultant for design services – have all expanded. The site may not have a defined boundary, the client may or may not own or operate the site, it may be citywide, regional or cross multiple towns. The client may be a developer or an architect, but also could be a group of advocates, a Business Improvement District (BID), a city agency partly funded by a private foundation or a philanthropist. As funding public space gets more bizarre while US municipal budgets are diverted elsewhere, new coalitions form to fundraise, donate and partner. The landscape architect no longer only consults for requested design services. They are also an activist seeking to make projects where they discern possible benefit (see Buffalo Bayou Park). They provide strategies rather than fixed designs (see pdO, a competition entry). They have various scenarios that form the base maps, rather than a fixed plan (see Resilient Cities Project: Miami). They work with a range of partners from outside design, engineering and construction to realise multifaceted design visions (see Ningbo Eco-corridor). The landscape architect no longer only draws to communicate design intent but has a much-enlarged toolbox that includes datasets, GIS layers, thermal readings, flood risk mapping, polls, remote sensing, performance simulation, fluid modelling, activation and programming schedules. This puts into the designers' hands both the tools and responsibility for greater design performance, user adoption and return on investment.

DOI: 10.4324/9781003145905-14

Practice-Based Research and Innovation

Landscape architects today increasingly work at larger scales, on more complex projects with multiple consultants, and with greater accountability for performance and sustainability goals. Landscape architecture practices must have an understanding of many issues – for instance: transportation and mobility; emerging technologies such as smart cities and IoT devices; new ways to expand community engagement around new and long-range issues such as climate adaptation; and new methods of site performance analysis and monitoring. New conditions or new ambitions in the field ask designers to understand disaster planning and mitigation, safety provisions conventionally provided by grey infrastructure like flood protection, the pressures of rapid urbanisation, re-creation or restoration of endemic or historical ecosystems and heavily degraded or polluted sites, among other things. It is also no longer adequate to have a cursory understanding of these issues, or to farm out the work to an outside consultant-expert. To develop truly innovative, novel, synthetic and holistic approaches to designing or planning with these complexities, contemporary landscape architecture practices must acquire an intimate understanding of these issues and their importance.

This is why research is so crucial in practice today. Some questions cannot be answered by a literature review, google search or an expert in "best practices". And unlike other fields granting professional degrees like medicine, law and education, academic outputs in landscape architecture today increasingly fail to generate relevant or applied research for practice. Practice-based research can ensure that information and research is responsive, timely, generalised, applicable, relevant and accessible to firm needs, clients and designers.

Answers or strategies sometimes must come from site-specific analysis or customised data-gathering that brings to light new data or information. Sometimes new technologies developed for other purposes have to be modified or "guinea-pigged" for a new use or application. For design projects that have a planned service life of 50-plus years, or projects that require flexibility for adaptation to future conditions, anticipating future scenarios is critical. Finally, novel issues confront practitioners in the field every year, from historic droughts to public health crises like pandemics, that fall outside knowledge gained in higher education, or through years of practice.

A long-held definition of research in the industry has been that the design process intrinsically employed research. "Research through design" was the 1993 coinage of Christopher Frayling, a professor of the arts.[1] In the next decade the thinking was that design did not equal research. In 2002, professors of architecture Linda Groat and David Wang asserted that design and research are neither opposite nor equivalent and that design makes episodic use of research while research is conceptually systematic.[2] A decade later, research, specifically in landscape architecture, was defined by M Elen Deming and Simon Swaffield in 2011 as multiple and defined by gatekeepers (such as well-informed editors) on a case-by-case basis. The field was too diverse to be measured against one paradigm of knowledge.[3] These latter two definitions of research in design are ones that XL Lab embraced during its formation in 2015 at SWA.[4]

XL Lab

The XL Lab[5] was launched at SWA in 2016 with the premise that design alone is no longer adequate to present, novel challenges. These are especially clear at the larger scales the firm works at – the district, city and corridor scale, as well as in multiple global contexts. Climate change is a major current focus of the Lab, along with anticipating and understanding unpredictability in technological innovation, economic shifts and tools for practice. XL Lab uses four core methodologies: foresight, research, visualisation and simulation (viz/sim), and advocacy. This means methods vary from anticipatory to analytical and may involve experimentation with new tools and experiences or an issue-driven investigation. A review of XL Lab projects to date would include research and innovation work mainly related to the six main areas of expertise within the firm: cities and urban conditions, communities, water, infrastructure, mobility and health. To build links across projects and knowledge in the field, XL Lab initiates original research projects, facilitates others' individual and group investigations, and curates and disseminates completed research, innovation and thought leadership.

Facing the Challenge of Climate Change

Climate change, one of the big drivers for establishing the Lab, has been a focus of a majority of its early projects. Climate change causes flooding, extreme heat, drought, fire, landslides, biodiversity loss and migration, which in turn stress and disrupt both natural and human systems. These shocks intersect and can be met with landscape-driven urban design. Several design and planning projects out of the firm have tackled these issues: Ningbo Eco-corridor, Santa Monica Mountains National Recreation Area, The Newport Coast, Buffalo Bayou Park, Brays Bayou Greenway Framework, Hunter's Point South Waterfront Park and California Academy of Sciences, to name a few. Apart from design projects, designers have also been advocates, working pro bono with city leaders to tackle thorny issues like urban heat in LA and sea level rise in Miami. They have also been partners in a coalition with academic researchers working on nature-based solutions for flood risk mitigation on the Gulf Coast.

In terms of structured research, XL Lab has worked on climate change in two areas: resilience and climate change mitigation. In early 2019, the Lab looked comprehensively at the meaning of *resilience* as it relates to both biotic and social systems. An inventory of disturbances, often siloed as natural disasters or discrete phenomena, was synthesised into five categories which foreground the common physical process (see Table 11.1).

TABLE 11.1 Five Types of Natural Disasters and Their Shock Type

Disturbances	Shock Type
wind (hurricane, tornado), earthquake, freeze/snow/cold	mutilation
hurricane, tsunami, landslide, mudslide, avalanche	inundation
fire, infestation, deforestation	consumption
drought, famine, lack of oxygen	reduction
heat, pollution, infection, exotics	corruption

The Lab then defined *resilience* to apply to practice:

> The power of an ecological or social system to accept disturbances while retaining the same basic structure and function is called *resilience*. This capacity to withstand shocks relies on three conditions: ability to self-organize, ability to buffer extreme loads, and ability to adapt.
>
> *XL Lab*

Generally, resilience is aided by redundancy, strong social ties, proximity and working with natural systems. Resilience is a response in reaction to disturbances but can be designed for proactively. Research can tell us more about how to help design resilience into natural and social systems at the scales of landscape architecture and planning practice. At XL Lab, recent research projects looking at resilience in the face of climate change-driven disturbances tackled various shock types. Two projects looked at flooding on the sites of two large SWA projects that had been impacted by hurricanes (Figure 11.1, top left). The Resilience Performance Case Studies, along with the pro bono Resilient Cities Project: Miami, looked at both coastal and riverine conditions. Underway now, a multipart research project on wildfire is particularly relevant to the four studios in California (Figure 11.1, top right). Urban heat and shade are site issues and citywide issues. A recent project, supported by an internal fellowship for research, tackled the larger scale in LA as an outgrowth of a public series on the subject convened by Christopher Hawthorne, LA's first Chief Design Officer (Figure 11.1, bottom left). And a project just starting up will look at the social dimension of resilience by assessing the social impact of a new linear park in Lynwood, CA, a park-poor neighbourhood in LA county (Figure 11.1, bottom right).

The Lab also addresses climate change through climate change mitigation research. These efforts are not tied to specific disturbances. Rather, they incorporate system-wide changes to reduce carbon emissions at region, city and site scales, using material choice, density, multimodal transportation, energy saving techniques, water use reduction and carbon sequestration. Recent climate change mitigation projects in the Lab focused on materials, transportation, energy reduction and density. The project "Rethinking High Performance: The Case for Dumb Materials" foregrounded materials that perform well in terms of reducing carbon emissions over long time spans (Figure 11.2, top left). A

FIGURE 11.1 Resilience. Top left: XL Lab "Resilience Performance Case Studies" that studied real world tests of two SWA designed sites during and after hurricanes. This riverine site in Houston revealed lessons about the value of certain flood mitigation measures over the long term and multiple flood events, as well as the need to redesign an oxbow area. The inset shows Buffalo Bayou before Hurricane Harvey, and after, in the larger image. Top right: Newport Coast project, SWA proposed fuel modification zones to increase turgidity in the chaparral plant palette, an irrigated wet zone, and a thinning easement in 1988. The designers used blowtorches while working with the county to determine resistance to fire. The project was built (inset), and a few years later, tested by the Laguna Beach fire. The interventions proved successful in suppressing fire in a 10,000-acre area. XL Lab is now mining the past 60 years of projects for applied techniques for fire risk mitigation, to fill the crucial missing gap in guidance between the scale of the individual house site and the scale of a national forest. Bottom left: A fellowship project explored disparities in urban heat and shade in Los Angeles. In council district 5, with just over double the median household income of council district 9, there was five times as much tree canopy coverage. The fellows used thermal imaging (inset) to look at the inequities on the ground. Bottom right: A current XL Lab study looking at the social impacts of Ricardo Lara Linear Park, a narrow park serving a park-poor neighborhood. Social ties, community building, and extended networks reinforced through public space can increase social resilience in the face of climate disturbances. Top left: Jonnu Singleton, SWA and Geoffrey Lyon. Top right: SWA. Bottom left: Han Fu and Qiaoqi Dai. Bottom right: Jonnu Singleton.

short project on micromobility reimagined streets around a range of transportation methods between walking and driving (Figure 11.2, top right). A part of "Urban Sensorium" looked at the citywide effects of energy reduction technologies (Figure 11.2, bottom left). An early project, "MEGA MARGIN", looked at a region facing sea level rise and responded with a plan for increasing density at the same time (Figure 11.2, bottom right).

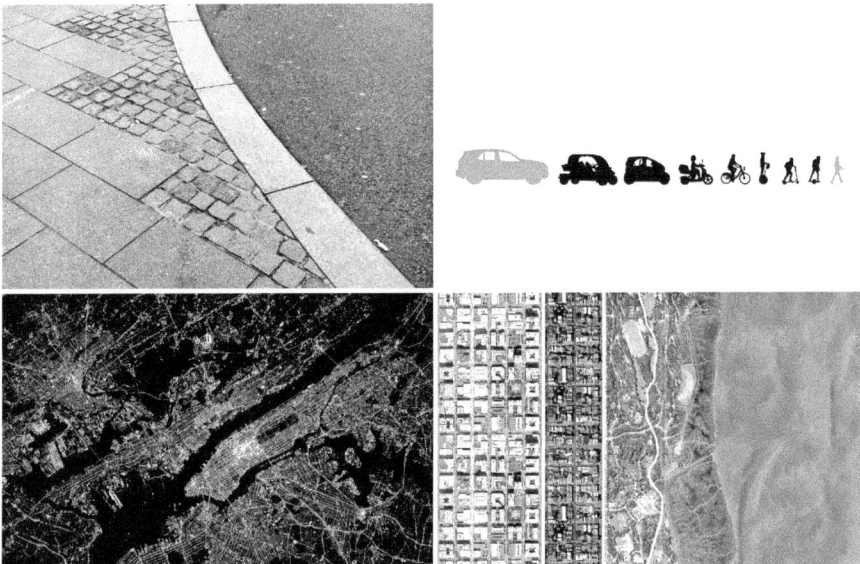

FIGURE 11.2 Mitigation. Top left: The ongoing XL project "Rethinking High Performance: The Case for Dumb Materials" makes an argument for redefining material performance to apply beyond design project lifetimes, where attributes such as convertibility, responsiveness, and deconstructability define "high performance". For example, natural stone is durable, has a high life cycle expectancy, and can be reprocessed. Igneous rock like granite is superior to cast concrete in strength, abrasion resistance, and cost of maintenance. Granite curbs and cobble like these in Copenhagen, are some of the oldest road building materials, and due to their fairly standard sizes, reusable without additional processing. Top right: The imperative to incorporate system-wide changes to reduce carbon emissions at the city scale prompted this research on micro-mobility. A component of the annual summer studio for students, the group explored the newly exploding range of passenger micro-transit in urban areas that falls between the car and the pedestrian. Bottom left: XL Lab project "Urban Sensorium" anticipated potential energy, ecology, climate, transit, and food scenarios for five major cities. In developing the scenarios, unsurprisingly, the majority of the drivers of change city-wide related to climate. In this image showing New York City, the phased rollout of a city initiative to increase energy efficiency shows up from space as old high pressure sodium and metal halide streetlights are changed out for brighter LEDs. The perception of the retrofit's effect on public space has been largely negative, maligned as "mass civic vandalism." Bottom right: Climate change mitigation challenges us to think about increasing density to reduce the energy required to serve spread out agglomerations. In MEGA MARGIN, an early XL Lab project, the slow crisis of sea level rise and the indefinite housing crisis in the San Francisco Bay area were twinned in hopes that one intractable problem could actually help solve the other. Densities were adjusted upward locally in belts (far left) and unified around the Bay, supplanting the current regional spoke-and hub model. The speculative project advocates for a way not just to rebuild after a crisis, but to prebuild in advance of one. Top left: Anya Domlesky. Top right: SWA. Bottom left: NASA. Bottom right: SWA.

Designing for a Volatile Future

Looking at the near future can appear prismatic – multiple scenarios, unpredictable events and disasters, slow stressors and sudden ones, unknown effects of collective mitigation efforts and so on. To meet this uncertainty, practice needs two things. The first requirement is a commitment to practice-based research that evolves over time. The second is the foresight to initiate design experimentation, and the follow-through to complete post-occupancy follow-up. Too often the scripts we write about project intent become the narrative. This works fine when we think about landscape design as an art practice, but not when looking for positive outcomes related to climate shocks or mitigation. The real story starts after the construction fences come down, and sometimes the reading only gets good a few years after that. As a large firm, SWA currently works on about 700 projects a year, some ongoing, some big, some small. If each team were to test one thing in each project related to resilience or climate change mitigation, then follow up and share the results, an extraordinary amount of data and insight could feedback into subsequent projects. At the moment, this formalised process happens with a fraction of total projects.

When it does happen, design experimentation allows research to inform design and design to inform research. In 1988, SWA was commissioned to do a large-scale, landscape-driven masterplan in a fire-prone area of Orange County, CA, adjacent to a state park. According to principal Sean O'Malley, at the time there were limited guidelines in place for fire mitigation related to landscape. For the project, The Newport Coast, SWA proposed fuel modification zones to reduce available fuel while increasing turgidity in the native chaparral plant palette, an irrigated wet zone and a thinning easement. Led by the Orange County Fire Authority, blowtorches were applied to various plant selections to determine resistance to fire. The project was built, and a few years later in 1991, was tested by the Laguna Beach fire (Figure 11.2). The design interventions proved successful in suppressing fire in a 10,000-acre area. Today, as the science and fire conditions evolve, SWA is currently working on several post-burn projects, redesigns and new construction in the increasingly wider swathes of California at risk of burning. XL Lab is now mining the past 60 years of projects for applied techniques, with the purpose of filling the crucial missing gap in guidance between the scale of the individual house site and the scale of a national forest – the range in which SWA commonly operates. Along with a partnered research project, these design experiments will inform future fire-prone project sites. The intention is that SWA will be able to implement the guidelines in designed sites, test them after construction, learn from the outcomes and iterate.

Post-occupancy assessment, especially when there is a clear performance intent to the design project, is the necessary pair to design experimentation. How else can we know at the site scale what worked in situ rather than in concept? XL Lab has completed post-occupancy assessments to assess a project's flood risk mitigation, social impact and economic revitalisation of adjacent areas.

Partners have been clients, academics, non-profits and technology specialists. On one project, "Plaza Life Revisited", XL Lab partnered with a data scientist to do an assessment, using machine learning, of how people behave in ten New York City plazas. On two others the Lab partnered with the Landscape Architecture Foundation to assess the efficacy of the design in flood mitigation. One can always wish for longer monitoring times or finer-grained historical data. However, following up on projects as climate conditions, users and urban context change is key to making sure new iterations of similar projects are informed and adapted to changing conditions.

To address increasingly complex environmental, technical and cultural issues, new forms of practice in landscape have emerged, and established practices are evolving. In the face of issues such as extreme weather events, rapid technological advancements and increased urbanisation, it is no longer sufficient to make landscapes and urban environments that are solely beautiful and well built. We must reconsider how to design intelligently with uncertainty and indeterminacy. In many ways that means finding ways to research better, develop collaborative transdisciplinary networks and be at the forefront of changes in the field, the market and the science. We all must adapt to increasingly complex conditions and be proactive in shaping it. The hope is that XL Lab can help to further a conversation on how we as designers can be more informed, more able to build, and evolve for this demanding design environment.

Notes

SWA research:
XL Lab: Plaza Life Revisited, Urban Sensorium, Resilient Cities Project: Miami, Resilience Performance Case Studies
https://www.swagroup.com/ideas/research-innovation/
XL Lab: MEGA MARGIN
https://www.fastcompany.com/3063519/could-rising-seas-force-the-bay-area -to-solve-its-housing-crisis
Flood Risk Mitigation on the Gulf Coast
https://www.asla.org/2012awards/264.html
Select SWA design projects addressing climate change:
Buffalo Bayou Park
https://www.dwell.com/article/landscape-architect-kevin-shanley-wants-to -reconnect-cities-with-their-waterways-dd31886c
Buffalo Bayou Promenade: https://www.asla.org/2009awards/104.html
Ningbo Eco-corridor
https://www.asla.org/2013awards/253.html
https://www.asla.org/2016awards/173014.html
Hunter's Point South Waterfront Park
https://www.asla.org/2014awards/467.html
California Academy of Sciences

https://www.asla.org/2009awards/111.html
https://www.asla.org/sustainablelandscapes/cas.html

Other

pdO competition entry for Portland
https://yeoncenter.uoregon.edu/finalist-bio-swa-group/

Notes

1 Christopher Frayling, "Research in Art and Design", *Royal College of Art Research Papers* 1, no. 1 (1993): 1–5.
2 Linda Groat and David Wang, *Architectural Research Methods* (Hoboken, NJ: John Wiley, 2002).
3 "knowledge in a diverse practice-oriented field such as landscape architecture must be consensually produced within an intellectual and professional community. The questions that are asked and the significance reported depend on the needs of the field itself, not upon some externally referenced school of thought or normative paradigm of knowledge". Margaret Elen Deming and Simon Swaffield, *Landscape Architecture Research: Inquiry, Strategy, Design* (Hoboken, NJ: Wiley, 2011).
4 SWA is a landscape architecture, urban design and planning firm with eight international offices.
5 SWA is one example of a design firm addressing emerging complexities through a platform for structured research and innovation projects. XL Lab has learned from dedicated research teams in allied fields such as architecture and engineering, where research entities that inform practice have operated for longer than in the field of landscape architecture. The launch of XL Lab was a way to formalise research direction and methods, as well as improve dissemination and adoption of research, innovations and information throughout the eight decentralised SWA studios, even though the firm has conducted research outside of design projects since 2007. Although many AEC firms do ad hoc research projects based on intermittent grant money, partnerships, or client requests, SWA has benefitted from having a dedicated research team. Today XL Lab operates with one director, two full-time staff, a small committee of principals, eight rotating members in each design studio and project assistants and consultants as needed.

12

FRAMES AND FICTIONS

Designing a Green New Deal Studio Sequence

Billy Fleming

In the time of upheaval and crisis, what is the point of design education? In nearly every school of architecture or design, there is a central, unspoken rejoinder to this question: the point of design education is to condition each successive generation of students for a lifetime of exploited labour that is detached from any critical relationship to the role designers play in aestheticising and instrumentalising global capitalism. This goal is not written in mission statements or strategic plans – to do so would threaten the machinery of student recruitment and major gift fundraising. But it is there, plain to see, in the tendency of design institutions to reproduce their most typical traits: valorising individuality and competition, endless work and infinite production, and service to the elites that fund much of the field's work. Put another way, design education exists to reproduce the social and racial order of capitalism. And the core of its pedagogy is rooted in the studio.

This is hardly a novel analysis. Peggy Deamer deftly recognises this relationship in *Architecture and Capitalism: 1845 to the Present*, writing that "while the construction industry participates energetically in the economic engine that is the base [of design practice], architecture operates in the realm of culture, allowing capital to do its work without its effects being scrutinized".[1] While reputation laundering has burst into the public's consciousness through scandals related to the Sackler Family and opioids, or Jeffrey Epstein and MIT's Media Lab, design has not yet reckoned with its role in laundering capital across the world. Douglas Spencer gestures in this direction in *The Architecture of Neoliberalism*, writing that

> while architects and architectural theorists have generally been less brazen [than Schumacher] about their enthusiasms for the subsumption of the urban and architectural orders to those of the market, they have tended,

DOI: 10.4324/9781003145905-15

since the mid-1990s in particular, to push those same truths of the way of the world as have served [the logic] of neoliberalism.[2]

This, borrowing from Fisher's theory of *capitalist realism*, situates the design fields alongside the rest of the liberal moral order in finding it "easier to imagine an end to civilization than an end to capitalism".[3]

Neither Deamer nor Spencer directly relate their analysis to the central role that studios play in design education. But their formulations of the post-critical, projective design fields are inexorable from the ways in which the studio shapes design education, and in which the limits of professional practice are often reproduced in studio settings. The post-critical turn in architecture and landscape architecture is embodied in various ways by Somol and Whiting,[4] among others, within the former, and Corner, Meyer and Waldheim[5] among others within the latter. All of part of the broader rise of post-critical theory that coincided with the rise of neoliberalism and the late 20th century infatuation with "the end of history".[6] This orientation has produced a form of epistemic hegemony in schools of design that treats markets and capitalism as natural. At the same time, it treats criticism, theory and left thought as, at best, unproductive, and, at worst, irrelevant to practice.[7] Design is best when it builds, not questions, in these understandings of the fields. That said, it is acknowledged that individual faculty can and occasionally does work to subvert this tendency.

Yet, the studio is rarely imagined in such nefarious terms. Quite the contrary – many, if not most, studio critics are working in earnest to train their students to build a variety of skills, whether organised around technical expertise or critical thinking and analysis. While their efforts are surely important, they are no match for the power of an institution. Though I won't purport to speak for the other design disciplines, I can say with confidence that landscape architectural education remains mired in this mode of toxic reproduction. It tends to do so through the same narrow band of issues that have dominated the field for more than half a century: aesthetic and formalist experimentation aimed inward and towards elite audiences; techno-futurist design fictions aimed at prefiguring the whims of Silicon Valley; and ecological pseudoscience aimed at producing verdant, non-human imagery for a mostly white, elite-led environmental movement (often with Malthusian undertones through concepts like the ecological footprint). For all of landscape architecture's self-referential talk about imagination and creativity, much of its praxis is organised around these themes and filtered through capitalist realism that masquerades as a kind of realpolitik – despite the total lack of political education in landscape architecture beyond vague, non-partisan calls for voting and electioneering.

So, I ask again: what is the point of design education? For those who fancy themselves capitalist to the bone, then perhaps what I have described requires little, if any, remedy. However, for those of us who do not so position ourselves, formulating alternatives to this institutional exploitation is urgent, necessary work. It is in this search for alternatives that I began developing the Designing a Green New Deal (DGND) studio sequence in 2018.

Framing Climate Justice in a Design Studio

Though some of my colleagues may refute the framing above, it has been ubiquitous – sometimes acknowledged, often implied – in nearly every studio review across a range of institutions over my four years directing the McHarg Center. It is also there, on the tip of most design students' tongues, whenever they are asked and given space to honestly answer questions about their experience navigating schools of design – something that's become clearer to me as I conduct exit interviews each year with former studio students. The machinery of design education exists to reproduce exploitation, and it has gotten very, very good at doing so.

As I began formulating the DGND studio sequence in the summer of 2018, I began with two simple, linked premises. First, one must look for ways to break this cycle of reproduction. Second, we cannot look at the world around us, creaking and groaning under the weight of multiple overlapping crises, and reasonably conclude anything other than how designers are trained is failing to accomplish anything outside narrow aesthetic, technological or faux-ecological aims. These perspectives, coupled with the luxury from the University of Pennsylvania of a three-year commitment to teach an advanced, interdisciplinary studio on whatever subject I desired, allowed me to begin mapping out a multi-year process. The elements are linked through a series of shared research questions, pedagogical experiments, and experiments with immersive, cli-fi storytelling. As of this writing, the second of three planned studios had recently concluded (21 December 2020). The DGND studios were never intended to offer a quick or easy solution to the problems of design education under capitalism. Rather, their development emerged from a recognition that any pedagogy or praxis aimed at subverting capitalist hegemony within a design school must be organised around the idea that "before landscape *problems* can be *solved*, they must be *framed*. Solutionism short-circuits this crucial step".[8]

To frame these studios, I relied on three key concepts: probing, usable speculation and platforming. Within each studio and across the entire series, the concept of *probing* is derived from Lutsky and Burkholder's "Curious Methods" essay. In it, they write that

> probing is a mode of exploration that informs but does not limit. It is a creative process that involves asking and enacting questions … and is a non-linear operation … involving three components: inquiry, the process of asking and enacting questions; insight, which is generated through that process; and impression, or the representation of those activities.[9]

Within each DGND studio, probing is structured into the work through a hybrid seminar-studio model of pedagogy. Over the first 4–6 weeks of each semester, students are engaged in close reading and discussion of critical historical, theoretical and sociological texts tied to issues of climate justice,[10] the energy transition,[11] statecraft[12] and particular places,[13] before any drawing, mapping or design work is allowed to commence. Each seminar day involves a student-moderated discussion with at least one of the texts' authors, and a preparatory "seminar slack chat". The chat is an hour-long exercise on slack in which a small group of students and I talk about key concepts and takeaways from the day's readings. This is a way to prime the conversational pump for the entire class.

As these seminars conclude, students then work in teams focused on a specific region (Appalachia, the Mississippi Delta and the Corn Belt) to develop two key deliverables informed by their seminar discussions. The first is a manifesto, which serves as both a conceptual framework for the students to understand their region's political, economic and cultural history and present, *and* an argument for how and why to propose futures for them. The second key deliverable is an atlas of the region blending into a coherent package of images, fieldwork (oral histories and interviews with local activists), counter-cartographies and other spatial phenomena. These assignments are directly linked to Judith Schalansky's conceptual framework in *Atlas of Remote Islands* that states,

> I didn't realize then that my atlas—like every other—was committed to an *ideology*. Its ideology was clear from its map of the world, carefully positioned on a double-page spread to that the Federal Republic of Germany and the German Democratic Republic fell on two separate pages… geographical maps are abstract and concrete at the same time; for all the objectivity of their measurements, they cannot represent reality, merely one interpretation of it.[14]

Each of these assignments – the seminars, the manifestos and the atlases – are tethered to the principles of inquiry, insight and impression that comprise probing. This work forms the first half of each studio, with the manifesto and atlas forming the core of the mid-review.

For the rest of the semester, students are challenged to engage in what I've termed *usable speculation*. The modifier "usable" is doing significant work here. It draws on the concept of "usable pasts" from public historians, in which "past national experiences can be placed in the service of the future",[15] and that "we must learn how to make a better world out of usable pasts rather than dreaming of infinite futures".[16] This, I argue, is distinct from most other forms of design fiction and speculation, nearly all of which draw on Dunne and Raby's *Speculative Everything*.[17] It is a delightful (and frivolous) book that treats speculation as a mostly technological and aesthetic proposition, eliding past the blinders of capitalist realism, the constraints of contemporary politics and crucially, the demands of movements. In fact, usable speculation in these DGND studios is framed explicitly as a rebuke to Dunn and Raby – an entire seminar day is dedicated to deprogramming the kind of cultish individuality and frivolity that it engenders in designers. It has come to mean centring the demands of the climate justice movement in producing climate and design fictions. Over the entire series of studios, it means investing, with those groups, the time and resources into long-term collaboration and trust-building that's necessary if a design school ever intends to live up to its stated goals.

More instrumentally, the concept of usable speculation also engages with the theory of change that no one will ever understand or rally around an energy transition through molecules (carbon in the atmosphere) or electrons (electricity in their circuits). Rather, people will only understand it through material investments in their lives and livelihoods – through the buildings, commutes, offices, parks, public works and civic infrastructure that stitch together everyday life. So, in co-producing cli-fi with leaders from the climate justice movement, these studios aim both to *illustrate* their demands and, in doing so, to *advance* them by giving form, aesthetic and visual culture to their demands. This is about reframing conversations about climate justice and the Green New Deal from ones of scarcity (for example, a ban on airplanes and hamburgers) to ones of dignity and plenty. Within this framework of usable speculation, students are then charged with proposing and developing their own storytelling vehicles – things that have ranged from graphic novels and zines to cookbooks and farmers' almanacs to children's books, among many others (more on them below).

Finally, and most simply, these studios are dedicated to the principle of platforming. Here, I rely on Edward Herman and Noam Chomsky's analysis of institutions and their role in reproducing a particular, self-serving ideology in *Manufactured Consent*. They write that "the beauty of the system is that dissent and inconvenient information are kept within bounds and at the margins, so that while their presence shows that the system is not monolithic, they are not large enough to interfere unduly with the domination of the official agenda".[18] Though they are writing about the media ecosystem and elite capture within it, their analysis holds for much of design education. Indeed, nearly all elite schools of design are funded in part by the same class of global elites at the core of Herman and Chomsky's analysis. In these studios, platforming holds dual meaning. Materially, it relates to the ways in which the studio serves as a vehicle for elevating the demands of the climate justice movement – demands that, at best, are placed on the fringe of most design programmes and, at worst, are banished altogether as unrealistic and divisive. Instrumentally, it relates to the ways in which this studio serves as a platform for students in their final year of study to begin building alternative career pathways for themselves. The use the exposure and relationships these studios have provided to find modes of practice outside the private, client-driven practice of contemporary design.

This latter point has been among the most generative throughout the DGND studios. They have not served as a way for me to reproduce the exploitative model of a "studio" or "firm" book, in which the unpaid and often un/under-credited is used by a critic or principal to "write" a book. Rather, it has resulted in students who have matriculated the studio becoming go-to experts in the climate and environmental media, developing ongoing projects with the local movement groups they met, and leading studios and policy-focused work around the Green New Deal elsewhere.

Fictions and Collective Dreaming in a Design Studio

In their visual essay on the history and future of democratic design, Liz Sanders and Pieter Jan Stappers write that "if we can use design thinking, making, and enacting to visualize and explore the future together, then we will be able to harness our collective creativity to serve our collective dreams".[19] Their notion of collective dreaming is at the core of the work produced in each of the DGND studios thus far. Because these individual studios are linked – the second builds on the first, and so on – the following is a condensed chronology of the work done in these studios to date. A third studio is planned, but its fate is a bit uncertain (more on that in the coda).

The first DGND studio, held in the autumn of 2019, revolved around two core questions of implementation priorities. First, what regions of the US must be "won" to achieve the technical aims of the Green New Deal tied to decarbonisation, justice and job creation? Second, from that more limited pool of potential regions to receive the first wave of GND investment, which belong at the front of the line? Key considerations in assessing relative priority is that a region is an historic site of public disinvestment, or that it offers an opportunity to grow the climate movement's coalition by making the material benefits of the GND real for those who do not already support it, or both.

During the first half of the studio seminars focused on the New Deal's built environment legacies, sectoral plans for decarbonisation (for example, electricity and transportation), and policy sequencing and design.[20] The studio then moved through a nation-scale, economy-wide analysis of the US, looking for critical clusters and overlaps of public housing, transportation, electricity systems, food production, water management and public landscapes – all of which formed the basis of their mid-review. Over the course of a weeklong in-person and slack-based discussion, students settled on three priority regions for rolling out the first wave of GND investments: Appalachia, the Mississippi Delta and the Corn Belt of the Midwest. I framed their work to purposefully exclude places like Manhattan, San Francisco and Boston. I did so not because they are unimportant, but because they are cities with a surfeit of financial, technical and political capital that will develop an adequate response to the climate crisis (or not), whether or not a Green New Deal ever materialises. And for me, the entire purpose of design studios like this is to operate outside market forces – to work with communities and in places that cannot and never will be served in the short-term by the firms beholden to those same market forces. So, while there is a number of other places that could have been selected through the framing, I offered, these three regions formed an excellent response to the questions above (see Figure 12.1).

FIGURE 12.1 This section illustrates how clean energy infrastructure, surface mine remediation, social housing development and public land improvements could be co-located and deployed along a range in Central Appalachia (2019). Zachary Hammaker, Tiffany Hudson, Sara Harmon, Allison Carr and Joshua Reeves.

For the final half of the studio, students were charged with curating an exhibition of ideas and producing a report for policymakers on their work. The exhibition, held on 19 December 2019, was curated by Chelsea Beroza (city planning) and Rosa Zedek (landscape architecture) – an interdisciplinary team charged with designing the entire show, from material choices to wall space allocations and so on. Students were also asked to think of their primary task as one of design communication – to create images and visual material aimed at speaking to broader publics about the kind of world that a Green New Deal could build. This led them to produce a series of propaganda posters, annotated and didactic sections showing the rollout of clean energy investments, speculative fiction zines about pre- and post-GND life in the Midwest, and a variety of other media rooted in cli-fi storytelling. I dispensed with a conventional final review in which jurors arrive in all-black clothing and students stand at their boards, sleep-deprived and largely excluded from the post-presentation conversation about their work. Instead, I charged this group with developing a public exhibition, replete with gallery tours and a series of panel discussions around the larger questions and themes that framed their work – a format I believed would allow the students to accomplish a few important professional goals. Those goals included to demonstrate their real expertise in this subject matter, to prepare them for speaking on panels and having conversations about their work, and to engender a more collegial, generous atmosphere for what is supposed to be a celebratory day. Most important, this format required them to produce and discuss their work for a large, public audience – including the movement leaders, policy experts, Congressional staffers, and other members of the Philadelphia environment and climate justice communities who were involved in their work throughout the autumn.

In addition to including invited jurors and advisers who were with the studio throughout the semester – including Julian Brave NoiseCat (Data for Progress), Randy Abreu (Rep. Ocasio-Cortez's office), Kate Wagner (architecture critic) and Peggy Deamer (Architecture Lobby), among others – the exhibition was open to the public. It drew more than 150 guests from the broader Philadelphia community, including members of the local Sunrise Movement and Philly Thrive, an environmental justice organisation. Their work was covered in *Gizmodo*. The policy report – edited by another interdisciplinary team, Tiffany Hudson (city planning) and Sara Harmon (landscape architecture) – is now being used by a large network of local climate justice groups in Appalachia and the Mississippi Delta for their organising work.[21] Their work also won the Award of Excellence for Community from the American Society of Landscape Architects in 2020 (Figures 12.2 and 12.3).

In many ways, the second DGND studio, held in the autumn of 2020, picked up where the first one left off. It began back in Appalachia, the Mississippi Delta and the Corn Belt. Rather than taking an all-sector approach to regional analysis, the studio instead focused on three key industries throughout the semester – the industrial agriculture, fossil fuel and prison systems. I framed the studio around these issues in large part because in the aftermath of George Floyd's murder, the national organising tables for the Green New Deal Network (where I often provide pro bono design and technical services), and the Movement for Black Lives, merged a number of their strategies and operations. One of the first items to come out of that merger was a focus within the climate justice movement on those three systems – and their abolition as a key step in winning a Green New Deal.[22] So, this studio revolved around how to translate the demands of abolitionist movements linked to fossil fuels, prisons and industrial agriculture into compelling, charismatic design fictions that could build up the library of images and stories about the future world a Green New Deal might build.

During the first half of the studio, seminars focused on the legacies of racial capitalism in each region, including the nexus between the plantation economy and slavery in the Mississippi Delta, settler-colonialism and absentee land ownership in Appalachia and commodity landscapes in the Corn Belt. Students then worked in three groups – one per region – develop a conceptual framework for understanding the political economy of each region. They used that framework to produce an atlas of incarceration, fossil fuels and industrial agriculture in each region, and from that proposed a series of key sites to focus their speculative fiction work throughout the rest of the course. Here, I framed their work around two key goals. The first was to develop a critical rhetorical and cartographic analysis of how the fossil fuel, prison and industrial agriculture landscapes of each region were produced and continue to be upheld. The second goal was to use that analysis to build an argument for interventions rooted in the demands of local abolitionist movements. Their conceptual frameworks and atlases formed the core of their mid-review, and they did not produce a single speculative concept or drawing until after this date (approximately week 9 of a 16-week semester) (Figure 12.4).

For the rest of the studio, students worked in their regions – sometimes on their own, often collaboratively – to develop a set of speculative futures with local movement groups identified during the first half of the course.[23] They were

FIGURE 12.2 These comics focus on the development of three characters in the Midwest, tracking their lives pre- and post-Green New Deal-related investments in the region (2019). Rosa Zedek, Chelsea Beroza, Katie Pitstick, Will Smith and Jesse Weiss.

given considerable space to conceive of and then execute their own storytelling vehicles. Several studio days were dedicated to "pitch sessions", in which students presented storytelling media precedents and assembled object lists to begin producing work for the final review. They found and developed relationships with movement groups in each region, with many of them co-producing the final work and joining the jury for final reviews. In all, the groups experimented with nearly

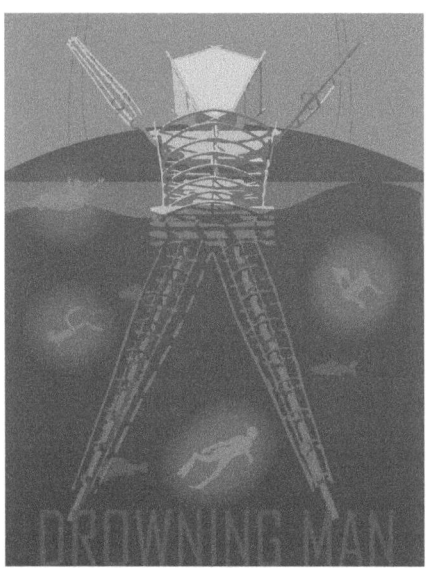
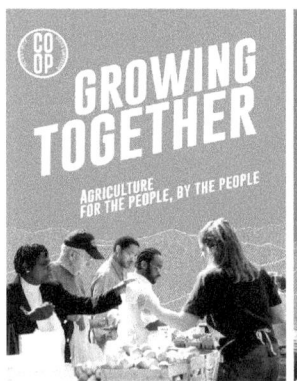
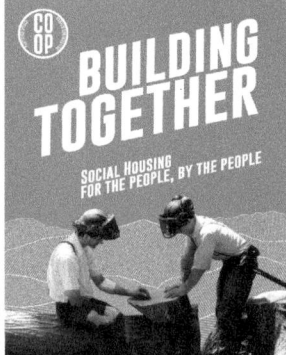
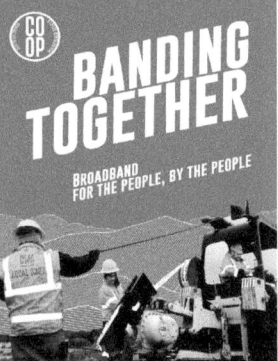

FIGURE 12.3   Top: These posters are tied to the fictional "Drowning Man" festival, designed as a way to mark and commemorate the clean-up and restoration of the Mississippi River (2019). Katie Pitstick. Bottom: These posters are linked to a fictional social housing, co-operative agriculture and rural infrastructure development programme, administered by the Appalachian Regional Commission (2019). Rosa Zedek, Chelsea Beroza, Katie Pitstick, Will Smith and Jesse Weiss.

a dozen different storytelling vehicles. There was a "Workbook for Dreaming" aimed at democratising the design process and based on the *Fumbling towards Repair* workbook, a prison-to-rural electric co-operate manual, and a climate-driven community farmer's almanac in Appalachia. There were a cookbook, children's book and a fictional National Public Radio (NPR) podcast, all linked to climate migration, agricultural co-ops and prison abolition in the Midwest. There was a

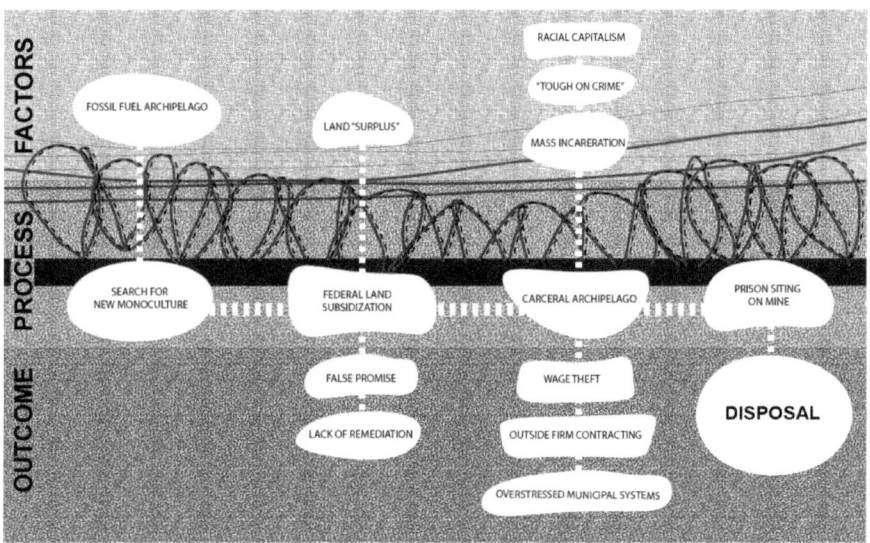

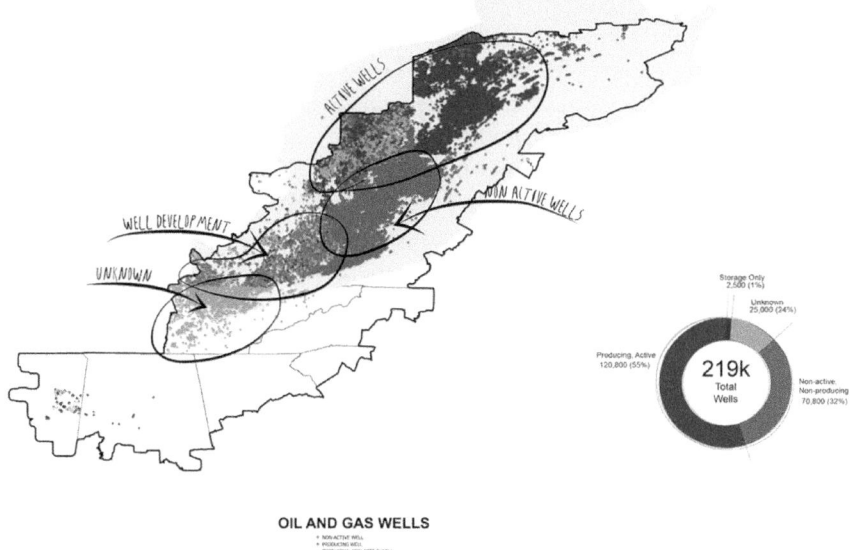

FIGURE 12.4 These images are part of a much larger Atlas of Incarceration and Fossil Fuels in Appalachia that can be found at http://dgnd.us/appalachia -atlas.html (2020). A McCullough, Chris Feinman, Diana Drogaris, Ada Rustow and Amber Hassanein.

series of zines, oral histories and character development projects in the Mississippi Delta. Rather than a final review, students were asked to create a website (dgnd. us) to house their atlases and speculative futures work, and to lead a series of panel discussions moderated by invited experts Beka Economoupoulos (Appalachia), Bryan Lee (Mississippi Delta) and Anjulie Rao (Midwest). This took the form of a zoom webinar on 21 December 2020 that included 30 invited respondents and jurors, and nearly 200 members of the broader public (Figures 12.5 and 12.6).

Throughout these studios, design is viewed as an instrument of redistributive climate justice. Speculative design in particular is framed as a medium for translating often abstract demands of the various movements for justice into compelling, charismatic images and stories about the future worlds they intend to create. This isn't merely an exercise in illustration. It is rather a way of testing those demands by giving them form and aesthetic, and by landing them in real communities and on real sites, through storytelling rather than solutionist frames that open up possibilities for the future rather than putting forward a singular, idealised future. The fight for a GND becomes at least a bit easier when the frame of conversation moves from one of scarcity to one of beautiful, communitarian, low-carbon luxury for all. In that spirit, the advisers and jurors for these studios have not been the same class of elites that are often invited into schools of design – chambers of commerce and redevelopment authorities, city councils and technocratic agencies, or mostly white-led environmental NGOs like The Nature Conservancy. Rather, I have welcomed frontline and fenceline activists into this work, as well as movement leaders at organisations like the Sunrise Movement, the Gulf Coast Center for Law and Policy, People's Action and the Democratic Socialists of America. Though this work is only just beginning, it is part of a larger project within the design professions. It is a project that involves real political education within schools of design, materialist commitments to communities and issues throughout the professions, and a more confrontational role (at least within academia) with the systems of immiseration that shape design practice around the world. If the Green New Deal is about a total restructuring of the economy, then the DGND studios have been about finding a place for the design professions within that process.

Coda

In a sense, these studios have become extensions of a larger critique of the design disciplines first outlined in an essay I wrote for *Places Journal* titled "Design and the Green New Deal". They have focused on rural communities already operating outside the markets and flows of global capitalism that bound so much of contemporary practice. They have also pushed back against the ways in which capitalist realism frames what constitutes "pragmatism" in the design professions – a term often used to describe projects that are eminently buildable, even if their construction is predicated on the preservation of a status quo that all but ensures social and ecological immiseration. Put another way, the DGND studios

FIGURE 12.5 These images are part of a series of climate fictions in the Mississippi Delta that can be found at http://delta.dgnd.us/ (2020). Christine Chung, Al-Jalil Gault, Erica Yudelman, Rachel Mulbry, Pat Connolly and Avery Harmon.

FIGURE 12.6 These images are from "A Workbook for Dreaming", part of the larger set of climate fictions developed for Appalachia and found at https://appalachia.dgnd.us/ (2020). A McCullough.

have sought to question whether it is pragmatic to uphold the very systems of exploitation and immiseration that wrought planetary climate change in the name of building out one's portfolio.

I am writing this as a coda because this sort of work is always precarious – mine as a junior and non-tenure track faculty member in particular is subject to the whims of administrators, donors and the university's reactionary politics. The DGND work is in a state of constant peril and real questions remain about how long it may be allowed to continue. But taking risks like this is the only justifiable reason to locate oneself at an institution like Penn – because it offers resources, power and a platform that can be used to point the design disciplines in new directions. These are directions that would likely not have been explored if not for the Green New Deal work I've led in the Center over the last few years. While not everyone is well-positioned to take such risks, those of us who are have a responsibility to do so – whatever the professional consequences might be.

Notes

1 Peggy Deamer, *Architecture and Capitalism: 1845 to Present*, New York, NY: Routledge (2014).
2 Douglas Spencer, *The Architecture of Neoliberalism: How Contemporary Architecture Became an Instrument of Control and Compliance*, London, UK: Bloomsbury (2016).
3 Mark Fisher, *Capitalist Realism: Is There No Alternative?*, New York, NY: Zero Books (2009).
4 Robert Somol and Sarah Whiting, "Notes around the Doppler Effect and Other Moods of Modernism", *Perspecta* (2002), 33: 72–7.
5 For more on this, see James Corner, "Critical Thinking and Landscape Architecture", *Landscape Journal* (1991), 10(2): 155–72; Elizabeth Meyer, "Landscape Architecture as a Critical Practice", *Landscape Journal* (1991), 10(2): 155–72 and Charles Waldheim, "The Landscape Architect as Urbanist of Our Age", in Landscape Architecture Foundation (Ed.), *The New Landscape Declaration: A Call to Action for the Twenty-First Century*, New York, NY: Rare Bird Books (2018).
6 For more on this, see Francis Fukuyama, *The End of History and the Last Man*, New York, NY: Free Press (1992).
7 Billy Fleming, "Design in the Time of Crisis: Landscape Architecture, Climate Politics, and the Green New Deal", in Richard Weller (Ed.), *The Landscape Project*, New York, NY: ORO Editions (Forthcoming).
8 Rob Holmes, "The Problem with Solutions", *Places Journal*, June 2020, available at: https://placesjournal.org/article/the-problem-with-solutions/.
9 Karen Lutsky and Sean Burkholder, "Curious Methods", *Places Journal*, May 2017, available at: https://placesjournal.org/article/curious-methods/.
10 Readings included but were not limited to House Resolution 109, "Recognizing the Duty of the Federal Government to Create a Green New Deal", available at: https://www.congress.gov/bill/116th-congress/house-resolution/109; Kate Aronoff, Alyssa Battistoni, Daniel Aldana Cohen and Thea Riofrancos, *A Planet to Win: Why We Need a Green New Deal*, New York, NY: Verso (2019); and Kian Goh, "Planning the Green New Deal: Climate Justice and the Politics of Sites and Scales", *Journal of the American Planning Association*, 86(2): 188–95; available at: https://www.tandfonline.com/doi/abs/10.1080/01944363.2019.1688671.
11 Readings included but were not limited to Myles Lennon, "Decolonizing Energy: Black Lives Matter and Technoscientific Expertise Amid Solar Transitions", *Energy &*

Social Science (2017), 30: 18–27.; Johanna Bozuwa, "The Case for Public Ownership of the Fossil Fuel Industry", *The Democracy Collaborative*, available at: https://thenextsystem.org/sites/default/files/2020-04/Public%20Ownership%20Briefing%20final%20v5.pdf; and Reinhold Martin, "Abolish Oil", *Places Journal*, available at: https://placesjournal.org/article/abolish-oil/.

12 Readings included but were not limited to Brent Cebul et al., *Shaped by the State: Toward a New Political History of the Twentieth Century*, Chicago, IL: University of Chicago Press (2018); Michael Katz, *The Undeserving Poor: America's Enduring Confrontation with Poverty*, Oxford, UK: Oxford University Press (2013); and Shalanda Baker, "Anti-Resilience: A Roadmap for Transformational Justice within the Energy System", *Harvard Civil Rights-Civil Liberties Law Review* (2019), 54: 1–48.

13 Readings included but were not limited to Clyde Woods, *Development Arrested: The Blues and Plantation Power in Mississippi*, New York, NY: Verso (1998); Richard Mizelle, *Backwater Blues: The Mississippi Flood of 1927 in the African American Imagination*, Minneapolis, MN: University of Minnesota Press (2014); Manu Karuka, *Empire's Tracks: Indigenous Nations, Chinese Workers, and the Transcontinental Railroad*, Berkeley, CA: University of California Press (2019); and Karida Brown, *Gone Home: Race and Roots through Appalachia*, Chapel Hill, NC: University of North Carolina Press (2018).

14 Judith Schalansky, *Atlas of Remote Islands: Fifty Islands I Have Never Set Foot On and Never Will*, New York, NY: Penguin Books (2009).

15 Van Wyck Brooks, "On Creating a Usable Past", *The Dial* 64, no. 7 (11 April 1918): 337–41.

16 Tony Judt, "The Last Interview", *The Prospect*, 173 (21 July 2010), available at: https://www.prospectmagazine.co.uk/magazine/tony-judt-interview.

17 Anthony Dunne and Fiona Raby, *Speculative Everything: Design, Fiction, and Social Dreaming*, Cambridge, MA: The MIT Press (2013).

18 Edward Herman and Noam Chomsky, *Manufactured Consent: The Political Economy of the Mass Media*, New York, NY: Pantheon (1988).

19 Liz Sanders and Pieter Jan Stappers, *Convivial Toolbox: Generative Research for the Front End of Design*, London, UK: Laurence King Publishing (2013).

20 All of my syllabi are publicly available. The 2019 DGND studio syllabus can be viewed at: https://docs.google.com/document/d/1AsvYYqJ0vFHvvuwsK0dDxUOlcWenpB1czLv5-B0ZD5Q/edit?usp=sharing. The 2020 DGND syllabus can be viewed at: https://drive.google.com/file/d/1_XpNhoxe0DlKxWUGJnUn8xvbb-ixKwoy/view?usp=sharing.

21 Throughout the autumn 2020 semester, we were surprised to learn from many of our new Appalachian interlocuters that dozens of groups were now using that 2019 studio report in their organising and advocacy work. Some of these findings are reflected in this component of the 2020 Appalachia group's project: https://appalachia.dgnd.us/A-Practitioner-s-Guide-to-Appalachian-Futures.

22 Varshini Prakash and Guido Girgenti (Eds), *Winning the Green New Deal: Why We Must, How We Can*, New York, NY: Simon & Schuster (2020).

23 These groups included but were not limited to the Gulf Coast Center for Law and Policy, the National Family Farmer's Coalition, the Federation of Southern Cooperatives, Black in Appalachia, and 100 Days in Appalachia.

13

CONVERSATION WITH KATE ORFF

Could you outline your definition of ecology in relation to your teaching?

The Columbia Urban Design (MSAUD) Program brings students from all parts of the world together, under one roof, for an intensive course of study. One year we will have students from Peru, Taiwan, New Zealand, China, and the next Jordan, Vietnam, India and Brazil. Although most students have an architecture background, there's a diversity of experience that we try to build on in the Program. Ecology in the broadest sense of the word – the study of relationships between things: organisms, people, the immediate environment – versus the things themselves, becomes useful, particularly in the studio. It helps us challenge linear thinking and move towards systems thinking. It helps us move away from conceptualising a design project as a "solution", and more towards understanding the scalar effects of an intervention over time in ways that might be disproportionate to the cause. We also ask students to draw relationships, say the geography of an object, or trying to trace water on a site, or to look at migrations of people and animals. Even though the context is exploring and embracing the complexity of the urban, we use ecological tools and concepts to draw. So, ecology informs the teaching in both a broad framing context, and specific, material and practical ways. I started teaching in an urban design programme instead of a landscape architecture programme for now because we have commitment to practices rather than disciplinary expertise. We think about urbanism, landscape, forms of social life, and integrate science and economics all together, rather than designing this park or that park.

How would you describe your teaching and research practice?

I'll focus here on the Columbia Urban Design Program and its work, not the SCAPE Landscape Architecture office. As Program Director, I see my job as setting a direction, developing a culture of inquiry and creativity, fostering a collegial environment, bringing diverse perspectives together, amplifying

DOI: 10.4324/9781003145905-16

patterns in the coursework and working as a multiplier and connector between faculty and students. This is an interpretation of ecology that you will not find in an environmental science textbook. But I do feel it is a mode of leadership based in an ecological ethos, and that you cannot separate ways of working from ways of knowing and designing. The practice itself emerges from an ecological ethos and stance of connection building. This also characterises how I lead the SCAPE office.

At Columbia I've set into motion a number of research and teaching initiatives that aim to loosely gather the interests and expertise of our faculty. The core contradiction in academia is that to progress, you have to advance a specialised research agenda, but to have impact in the world at large we all need to work together in larger collectives. So, the initiatives are a way to try to push these contradictory things forward, to foster an environment that many people can plug into and bring out the best in everyone. The initiatives also build a movement over time. For example, we started the Hudson Valley Initiative to cultivate on-site work, commitment and trust with small cities and towns in the mid-Hudson over many years, building a database of collaborators and a trust base with the river and people at the centre.

The Global Cities and Climate Change studio under the frame of "Water Urbanism" has evolved a sustained focus on issues over many years to form a compelling research agenda (13 water and global cities studios). It has also created a mesh network of cities facing many different challenges, a forum for sharing practices and ideas, and a comparative context where cities learn from each other. The Water Urbanism sequence of studios I have coordinated over many years now with some outstanding friends and colleagues, has been a source of joy and inspiration. I have so valued teaching alongside Dilip Da Cunha and Geeta Mehta, among many others, over the past years, and exploring the world and this pedagogy together. Dilip brings a piercing philosophical viewpoint of the lived experience of water's oppression by land, connecting the containment, confinement and control of water to colonialism, and power and control in society. Geeta has marshalled social capital towards urban transformation in the form of social capital credits, or "SoCCs", that incentivise community members to participate in shaping their environment. Both in different ways offer a sharp critique of capital-driven urban development, and I suppose with me, focusing on spatial design, we make a powerful team that is an ecology of insight, shared purpose and complementary perspectives.

So, it is an ongoing exploration on how water creates space for imagination. It explores the very direct ways control of water continues to marginalise the poorest of the poor. And it explores concepts that can halt and reverse that trend – we are looking at water and ecosystems as both a material, and as an imaginative and political framework. Most recently we explored how three cities along the Rift Valley – Tel Aviv-Yafo, Israel, Addis Ababa, Ethiopia and Beira, Mozambique – might forge systems and spaces to span this divide amid rapid urbanisation, and while grappling with the unique impacts of the climate crisis.

The studio's design strategies proposed new forms of urban living that embrace the complexity of water; this is critical to maintaining life along the Rift that fosters social interactions through local stewardship and empowerment models. Student design projects imagine creative alternatives to address interrelated risks faced by vulnerable populations. These include extreme heat in Tel Aviv, flash flooding due to river floodplain development in Addis Ababa and coastal inundation and offering alternatives to standard capital-driven disaster recovery practice in Beira, which was struck by Cyclone Idai in 2019.

We've also been working in India over many years and taken a deep look at the Ganges system from the Himalayas to the Bay of Bengal. We have focused on the city of Varanasi and its relationship to the Ganges, also at the cascading tank system of Madurai, the crisis in Pune happening now with concrete channelisation of the Mula Mutha, and the mangroves of the Sundarbans and their protective benefit for Kolkata. In all these studios, we've learned to focus on culture. I've learned so much from this endeavour and working alongside my incredible co-teachers, Geeta Mehta, Dilip Da Cunha, Thaddeus Pawlowski, Julia Watson and others. We've travelled to Amman and Aqaba in Jordan, Rio de Janeiro and São Paulo in Brazil, Can Tho in Vietnam and four cities in India – Kolkata, Madurai, Varanasi and Pune, among others. It has been a virtual PhD for me.

From our collaborators and students, I've learned that excellence emerges in the space between people – in open dialogue, hard work and collaboration among people with diverse and international backgrounds who have a shared purpose. A few years back, I hosted a panel on "Water and Social Life in India" at the ASLA Conference with Geeta, Dilip and Alpa Nawre. This session captured some of the big lessons for me. Over the years, we have learned water is not an abstract "issue" to be solved. To embrace a water-resilient future we have to learn from past practices, learn from small communities managing and communicating with each other. Designing with water is not just about adapting to changing conditions – it is crucially also about fostering forms of social life, maintenance and care.

What are you currently working on? Could you also discuss how your research and pedagogical focus has changed over the years?

I suppose my research and pedagogical focus over the years has just become more radical, since with each city or site in the world where we do a deep dive, there is a form of ecocide in the making that we bear witness to. In Can Tho, we learned on the ground, and saw first-hand, the effects of a massive dam building project, upstream on the Mekong in China and Laos. It's an ecocide in the making, and it sets the stage to shift from a millennium of the river nourishing civilisation to a future millennium (or not!) of political power and control based on water. So, we've gotten more recently into the basic fundamental questions on human habitation and power.

But, alongside power, control and oppression, water also opens up space to contemplate a politics of the free, and what this means for water. It also encourages bold thinking about Earth's next 100 years of habitability, and a sense that time is running out.

Your work emphasises the need to consider the act of design as occurring across social, political and environmental domains. In your article "What is Design Now? Unmaking the Landscape", you discuss the incredible amount of work required through policy, regulation and community to "unmake". What methods and techniques do you use to allow students to engage with these dimensions and different skillsets?

The mindset of "I'm the designer, here's my design and you all figure out the rest" is discouraged from the outset. I've always promulgated the notion that design can influence policy, not just react to it. In the studio every year we do an "Implementation Workshop" which foregrounds how to build coalitions and align stakeholders to the physical proposal. I think the concept of unmaking also requires a different lens – and this is where the "ecology" mindset leads to alternative design outcomes – rather than design as a constant additive process, the challenges ahead to link up fragmented water bodies, to stitch forest fragments together and to liberate rivers will require ripping out, jackhammering, exploding, protesting and many other forms of action that seem off the table now or as "not design".

Your practice is firmly embedded in the core of the landscape architecture discipline. Could you discuss what influences you draw on and how they come together in your work?

My influences came mostly from the periphery of landscape architecture. I suppose much of the task for me has been pulling these influences more to the core. As a teenager I worked as a gardener – rooting, selling and taking care of plants. It taught me a lot, and to this day I don't think you can practice landscape without a lived understanding of the hard work and labour that goes into every line, sketch and gesture we make on the page. I also initially wanted to be an artist, literally making "the thing" myself. Eventually I felt bereft that landscape architects are not "making the thing". We are making drawings that direct low bidding contractors to do the fun stuff – scoop, sculpt, pile up, etc. The book *Toward an Urban Ecology* (Monacelli, 2016) speaks to this – in response I've tried to recast landscape in light of broad participatory frameworks, stewardship and the transformative power of making landscapes together – not just gazing at them at a distance. I've tried to put forward a model of building living landscapes as a form of building community, not a "high art" that needs to be preserved like a piece of sculpture.

I wrote my undergraduate thesis at University of Virginia in 1993 on Ecofeminism and the power of combining social and environmental movements. I profiled Wangari Maathai, Vandana Shiva, Rachel Carson and others, and to this day these women working to connect the dots and push broader movements forward is a constant inspiration. My early writing and mapping of Jamaica Bay set a template for thinking about climate-changed landscapes through the lens of water and politics, and the role of activism and that has been explored in many ways since. Recently, I wrote a chapter in the new book *All We Can Save* (Penguin, 2020), and have been on Zoom calls – wine in hand! – with the amazing chorus of poets, artists and scientists in that book. After interacting with them all I feel a surge of optimism and resolve.

I also had a lot of influences inside the profession! As a student at the Harvard Graduate School Of Design (GSD), I had Ken Smith, Walter Hood, Richard Forman and Anita Berrizbeita as teachers, among many others. I did a research thesis with Rem Koolhaas pre-*S,M,L,XL* that got me writing and drawing, and I feel like I learned a lot from him just as a mode of moving through the world, also with my work at Office of Metropolitan Architects & its research, publication and branding studio (OMA-AMO). After school, I worked at SWA Group and Hargreaves Associates, and learned a lot about professional practice, and hung out with Julie Bargmann, who I just dig as a person and as a collaborator. But the work with Rem and his obsession with bookmaking planted a seed of possibility that I could do a research and drawing book like *Petrochemical America*, not be too afraid to just start organising chapters. And of course, now Richard Misrach and his way of seeing has been a huge influence. Today, I'm influenced and supported by an incredible younger cohort Principal Team at SCAPE – Gena Wirth (Design Principal), Pippa Brashear (Planning Principal), John Donnelly (Technical Principal) and Alexis Landes (Managing Principal). I cherish them so much and have learned a lot with and from them. I'm sure there are more influences out there! The influences snowball over time.

In your article "What is Design Now? Unmaking the Landscape", you wrote: "Moving forward, in light of increasing climate shocks and stressors, designing the social must be paired with new forms of architectural expression such as unmaking, undoing, subtracting, reversing, decarbonising, tearing out, ripping up, replanting, softening and connecting". Could you please describe how unmaking is an agent in your policy, regulatory and consultative work?

We can think of ecological design not as traditional "restoration" but as an outcome of systems change, as the effects of the work, not the work itself. There is a formal language that SCAPE deploys, but I do not get caught up in making a formal statement. "Unmaking" is part of that – the design gesture may be to remove a dam, or rip out a roadway, carve down a bulkhead or breach a levee. Does it expand biodiversity, set in motion new and nourishing sediment patterns, sketch a framework for sustained public engagement, rebuild wetlands, create intertidal dynamics, connect people? *Oyster-tecture* (MoMA, 2009) prototyped a series of nested ecological and social regeneration strategies. Thinking about "studio ecologies" and the time it takes to evolve new ideas into a policy context – what was once only in the white walls of a museum is now under construction! We've carried that through into a constructed reef and science-based initiative being built in Raritan Bay, now in the form of *Living Breakwaters*. The project is physically and intellectually beautiful, but its impact (a revitalised shoreline, finfish habitat, hands-on citizen science programmes, etc.) will not conform to current landscape architecture aesthetics, which can sometimes seem like variations of the Adobe Illustrator palette and spline tools. A professional practice "take" on climate might be to incorporate so-called best practices in our projects. But we have to move beyond doing less harm within the prescriptive bounds of a land-based, property-centric profession. We have to move towards our projects doing more good, working in more collective forms and on vast regional landscape ecosystems that are collapsing but that do not have "clients".

What do you see as the future adaptations and extensions of current practice, research and teaching as a means of responding to the climate crisis?

I've been trying to just push broad-front initiatives. It is so hard to break out of the frame of the "project" in professional practice. That's where the work at Columbia has been hugely important. There are missing links between the non-profit and NGO sector, federal and state governments, and private money. The piecemeal approach, the project-by-project basis, is wholly insufficient to address the systems collapse wrought by a century of carbon-driven development (itself driven by private developers), subsidised highway infrastructure, engineers controlling and killing water bodies, and planners responding to a market-based system. I can see universities, private practices and professional societies all aligning around larger efforts to hit the reset button on critically endangered regional places. That place may be the San Francisco Baylands, the New York Bight or the Mississippi River riverlands and deltas – the subject of my studio and research work now. Wherever the place, the sum of all of our actions has to add up to more good, more radical systems change.

PART 3

Generative Processes of Fieldwork

Narratives

The Australian metropolitan sewer systems, developed and implemented in the late 1800s, has changed little over time. Like the sewer system in many cities, it collects, transports, and treats human excrement, eventually disposing of it through combined sewer overflows into the ocean.

The excess fluids and detritus of human consumption often agglomerate into sewer "fatbergs". These sinewy congealed masses traverse the concrete conduits, passing through an already constrained network of sewer pipes. This extensive, institutionalised infrastructure – segregated pipes of water supply and sewer – coalesces into a vast underground network. It infiltrates the internal spaces of our homes, bathrooms and kitchens to carry various streams and states of viscous material to the ocean. This network, designed to partition and convey a waste stream, reveals a human understanding of the world, a world of human excess.

Through the Probe

A probe sent down a sewer pipe bumps into what initially appears to be a small, amorphous mass of waste. As the probe continues along the path, it becomes clear that this single mass is equivalent in size to a large school bus, a ginormous, entangled mass of food, fats, hair and wet wipes lodged in the sewer system. The probe forensically analyses the new body's surface, finding points of weakness and discontinuity. The use of this device allows interaction with the fatberg from the (relative) comfort of the control room. At this remove, the surface of the fatberg is rendered as a topographical body – a landscape in miniature. Processes of material formation most often seen on a much larger scales are replicated within the sewer. The probe reveals an anthropocentric landform that is diverse in its more-than-human temporalities and variability.

DOI: 10.4324/9781003145905-17

As the Microbe

As we peer through the microscope into the circular dish housing the living organisms of the microbes, we begin to see the congealing nature of the fat, grease and other indiscernible objects floating in the glass container. An enclosed living ecology of multiple organisms and life matter in which microbes abound, densely integrated within the fatberg in a symbiotic exchange of nutrients.

The microscope reveals a complex world that has been there, working away, all along. A finger momentarily obstructs the lens, and a planet is revealed of microbes on the skin. Microbes at home on, and within, our bodies. They are not external to us; they are inherent to and essential for our biological and logical functioning along the gut/brain axis. Equal in number to the count of cells in our body, microbes are described in relation to other entities. The mode of production and cohabitation of microbes calls into question the separability of these bodies. Microbes are the continuous and reciprocal agents between humans, infrastructure and the fatberg.

The Sewer Overflow

The miles of fully enclosed pipe transition into a large, semi-circular, brick-lined, open concrete channel. We are now at the end of the line: the sewer overflow demarcates the extents of the city and from here the fatberg becomes disconnected from its host. The once dark, subterranean sewer pipe is now exposed for the first time to the light of day. The expansive blue sky appears overhead, and the warmth of the sunlight radiating onto the surface of the water meets the fatberg. This is a moment of rapid environmental change that results in a spike in microbial activity. Nutrient levels on this new day peak with the arrival of the unfamiliar organism that is the fatberg, contributing to the manifestation of a vibrant green algae bloom. "The Antipodean Cloaca Maxima" soon disperses the sewage into the Western Port Bay. This is perceived as the end of the fatberg's journey; alas, this is not the case. New relationships emerge between the divergent material flows that coalesce to catalyse encounters between the water of the bay and the ocean beyond that encompasses expansive waters, surging currents, diverse species and abundant organisms.

Definitions

The climate change predicament can be linked to the systems of measure, land demarcation, classification, and objectification. The act of mapping the land predicates the linearization of inherently live systems, and the rationalization of connecting isolated objects. Historically, this was achieved through various instruments, such as the sextant defining a projection of angles, or the projection of longitude and latitude emphasised by the development of satellite technologies. These instruments give agency and exude power for territorial dominance, or for specific agendas. These representational systems of colonisation, patriarchy, and capitalism have shaped landscape architectural practices and pedagogies.

'Processes of Fieldwork' is defined as a suite of interactions that occur between the observer and ecological processes. These conditions change over time and capture the transformations of the systems at play. The act and techniques of observation identify new relationships and measures that are inherently ingrained in the materials of the observational systems, and in the constraints that issue from the design of the devices of translation.

'Processes of Fieldwork' explores a fundamental aspect of landscape architecture which is concerned with being out in the field, working in the field to construct meaning and knowing through visiting the site. In response to expanding complexity due to the climate crisis, the scope, scale, and location of the landscape architectural project initiates modes of fieldwork that critically question, frame, and re-present the landscape. In this situation, ecological thinking allows for the positioning of the designer as an agent in the field, one who interacts with other processes, species, actors, and agents in time. This section examines several types of fieldwork that interrogate both the extents of the field, and how the landscape architect operates within it. The intention of examining the work of academic practitioners actively engaged in constructed conditions, both virtual and/or physical, is to generate processes and systems that respond to the changeable contexts and extents of the discipline.

This section considers the ecological framework as a series of processes that require alternative forms of representation through various techniques that capture multiple perspectives. The contributors to this section have both academic and professional practices that use fieldwork to identify site conditions which are not explicit. They utilise technologies such as GIS systems, drone technology, and social media networks to draw out unforeseen networks, processes, and systems at play at planetary and territorial scales.

Pedagogies

The design studio pedagogy historically mimics the processes of undertaking a project in professional practice. Site in the landscape architecture discipline is often regarded as a bounded entity distinct from its surroundings. Professionally commissioned projects often have a predefined client and a designated area or property allotment, which is determined through property ownership, designated overlays, and historical subdivision that occurs through transforming land use and development agendas. Historically, pedagogical agendas have reinforced this approach, consequently limiting the designer's role in determining site and its extents, the design brief, and who should be involved in its deployment – all of which reinforces current development-dominated practice paradigms.

'Processes of Fieldwork' supplies a means of shifting this form of practice. It argues for the site to be considered as a 'relational construct' (Khan & Burns, 2020) that reveals ecological processes of known and unknown conditions, describes transformations, and registers novel combinations of unseen amalgamation between and within systems. These characteristics consequently generate new systems of measure that imbue complex systems with matter, time, and order. Together, these new systems advocate for a formation of practice that is both collective and specific to the endeavour at hand.

'Processes of Fieldwork' examines various social, environmental, political, and economic ecologies through virtual and physical interactions and observations. It utilises the notion of the *expedition*[1] as the approach by which various interactions between ecological systems are observed and translated. A range of roles and positions constituting different observers and stakeholders is utilized to define a variety of tools and techniques of observation that purposefully operationalize intrinsic parameters and constraints. This approach generates multiple translations of diverse, interacting ecological systems.

'Processes of Fieldwork' explores techniques and technologies for examining and modelling generative forms of analysis, discovery, and building design briefs. The way is made open for alternative, unpredictable assemblies of heterogeneous relations, in which the landscape and human interaction are considered inseparable. It utilizes various tools and techniques that transcribe information between the physical and the virtual, thus developing processes that can respond to an ever-changing, unpredictable landscape to shift away from understandings formulated around colonial ideologies of land ownership, and associated systems of measure that prioritize parcelisation. The shift is to different types of spatial knowledge and site formations. Knowledge systems and forms of classification are developed to become more attuned to biological, chemical, and ecological processes, where the demarcation of land is much more fluid and dynamic and prescribes a multiplicity of cohabitation in which the observer and the observed intermingle.

What I call an ecology of practice is a tool for thinking through what is happening, and a tool is never neutral. A tool can be passed from hand to hand, but each time the gesture of taking it in hand will be a particular one—the tool is not a general means, defined as adequate for a set of particular aims, potentially including the one of the person who is taking it, and it does not entail a judgement on the situation as justifying its use. Borrowing Alfred North Whitehead's word, I would speak of a decision, more precisely a decision without a decision-maker which is making the maker. Here the gesture of taking in hand is not justified by, but both producing and produced by, the relationship of relevance between the situation and the tool.[2]

Notes

1 In this context, *expedition* shifts from historical endeavours of discovery and projected classifications to a trek where the observer and the observed are seen through a lens of subjectivity, where relationships are understood, and where novel ecologies are revealed.
2 Stengers, Isabelle. (2013). Introductory Notes on an Ecology of Practices. *Cultural Studies Review, 11*(1), p185. When we deal with practices, recognition would lead to the question—why should we take practices seriously as we know very well that they are in the process of being destroyed by Capitalism? This is their 'sameness', indeed, the only difference being between the already destroyed one, and the still-surviving ones. The ecology of practice is a non-neutral tool as it entails the decision never to accept Capitalist destruction as freeing the ground for anything but Capitalism itself.

Bibliography

Bohm, D. (1981). *Wholeness and the implicate order.* Routledge & Kegan Paul.

Chakrabortty (2008).

Corner, J. (1999). The agency of mapping. In Denis E. Cosgrove, *Mappings.* Reaktion Books.

Cosgrove, D. (1999). *Mappings (critical views).* Reaktion Books.

Del Tredici, Peter. (2014). The flora of the future. *Places Journal.*

Ellis, Erle C. *Anthropocene: A very short introduction (very short introductions).* Oxford University Press.

Hobbs, R. J., Higgs, E., & Harris, J. A. (2009). Novel ecosystems: Implications for conservation and restoration. *Trends in Ecology & Evolution, 24*(11).

Kahn, A., & Burns, C. (2021). *Site matters: Strategies for uncertainty through planning and design.* 2nd ed. Routledge.

Latour, B. (1987). *Science in action: How to follow scientists and engineers through society,* Open University Press. pp. 64–65.

Stengers, I. (2013). Introductory notes on an ecology of practices. *Cultural Studies Review, 11*(1).

Tsing, A. L. (2015). *The mushroom at the end of the world: On the possibility of life in capitalist ruins.* Princeton: Princeton University Press.

United Nations. (2011). *The global social crisis: Report on the world social situation 2011.*

14

TALES FROM THE DARK SIDE OF THE CITY

Unknown Fields (Kate Davies and Liam Young)

Unknown Fields is a nomadic design studio that ventures out on expeditions into the shadows cast by the contemporary city, to uncover the alternative worlds, alien landscapes, industrial ecologies and precarious wilderness set in motion by the powerful push and pull of the world's desires.

The dislocated landscapes we survey – the iconic and the ignored, the excavated, irradiated and the pristine – are connected to our everyday lives in surprising and complicated ways. They are embedded in global systems that form a vast network of elusive tendrils, twisting threadlike over everything around us, criss-crossing the planet, connecting the mundane to the extraordinary. Unknown Fields makes provocative objects and films from this expedition work, exploring the dispersed narratives that coalesce to form a contemporary city.

Our material things set in motion a vast, planetary-scale infrastructure. They carve holes like canyons, they move mountains, they remake our world from the scale of the pixel to the scale of the planet. Our city casts shadows that stretch far and wide.

We tell tales from the Dark Side of the City. Stories form a notional road trip through a reimagined city that stretches across the earth. We create portraits of a place that sits between documentary and fiction, a city of fragments, of drone footage and hidden camera investigations, of interviews and speculative narratives, of toxic objects, reimagined landscapes and distributed matter from distant sites. We tell stories from the constellation of elsewheres conjured into being by the city's wants and needs, fears and dreams.

Across the last nine years, Unknown Fields has undertaken 14 expeditions through the distributed territories that lie behind the scenes of the contemporary city.

Aboard a cargo ship, following the trail of our electronics through the South China Sea, the shipping ports, factory floors and rare earth excavations of

DOI: 10.4324/9781003145905-18

China; through the irradiated wilderness of the Chernobyl exclusion zone to the precarious wilderness and the wild west gem stone stones of Madagascar; through the lithium fields and charged landscapes of Bolivia and Chile, the climate-change landscapes of far north Alaska; to all the sites on the edges of the world where our technologies begin and end their lives.

A World Adrift: South China Sea to Inner Mongolia

(Figure 14.1)

Our cities are extraordinary constellations of products, goods and technologies. From the smallest and most inconsequential of objects to the most intricate and complex, these material things set in motion a vast, planetary-scale infrastructure. Our cities cast shadows that stretch far and wide.

In a world of bytes and bitcoins, cyberspace and clouds, 90 per cent of the world's cargo still travels by sea. It is not beamed or teleported or conjured into

FIGURE 14.1 Rare earthenware. Unknown Fields/Toby Smith.

existence along strings of fibre optics. Rather, it is dragged across the planet in heaving steel megaships, gizzards full with glistening gadgets and gizmos from distant lands.

The secret lives of objects span across a notional factory floor that reaches from the high street pound shop all the way to the resource fields of the Far East. We travelled to China and beyond, tracing the shadows of the world's desires across the China Seas and along supply chains and cargo routes, to explore the dispersed choreographies and atomised geographies that global sea trade brings into being. These are the contours of our distributed city, stretched around the earth from the hole in the ground to the high street shelf.

Our journey through East Asia marked a cross-section of this supply chain. From source to sea, we followed the routes of this and that, of bits and bobs and things.

It's been just over 45 years since the Apollo moon landings, and some would have it that we are failing to build big anymore. Stand on the bridge of a container ship docked in a megaport in Korea, however, and it's clear that's just not true.

5,000 ships make up the global containership fleet.

3.6 million containers are in motion worldwide.

The surfaces of our planet's oceans – for centuries a space of mystery and myth, of expanse and desolation – have been rationalised. Once an enigmatic, awe-inspiring place, the sea has become a zone of efficiency, little more than another channel for the automated supply chain network.

The captain of our ship tells us that years ago the sea used to be filled with phosphorescent algae that would glow when the waters were disturbed.

As the captain says, "We would leave this luminous green trail behind us in the water as the motors churned up the algae. Toilets are flushed with seawater, and you could turn the lights off in the bathroom and flush the loo and the whole room would glow neon green".

Nobody on the ship knows what lies inside the containers the ship carries. The ship's captain, and the dockside crane operators who have also been made obsolete, are now just passengers in the machine, their bodies repurposed as a component in the landscape-scaled robot that stacks the containers ready for transport, bringing our goods all the way home.

The shipyard worker says, "We use GPS-guided cranes to move the large sections of ships in the assembly process. We then weld the sections together by hand. These are some of the largest handcrafted objects in human history".

"Every welder signs his welds so that if there is a flaw, we can trace it back to who did it. Every component of the ship has the signatures of those who made it hidden under the paint".

Before objects set sail for our stores they are bought, sold and traded in the vast halls of Yiwu International Trade City, a wholesale market the size of a city. The Wholesale City consists of 80,000 shops, all identically sized 2.5 metre by 2.5 metre cubes containing 10 million products stretched across 10 square kilometres.

"Sorry we can't do it", a market trader says, "my minimum order is 100,000". Every one of us owns something that has passed through the Wholesale City. It's the little items that fill your desk drawers; the free pens that salesmen give you, and the toys your children break or forget. It's the hundreds of disposable products that fill 99 cent stores and gas stations. It's the stuff you buy on impulse, or because it's momentarily funny. And all because it's cheap.

Christmas is made in Yiwu. More than 60 per cent of the world's Christmas decorations are here. That tree lighting up your lounge. Those decorations hanging from the ceiling. That novelty stocking filler you bought for your child. They're all made here.

Shenzhen makes 90 per cent of the world's electronics. These are the human machines of the production line, all choreographed by efficiency algorithms, their bodies matched in speed to the conveyor in front of them. These are the real robots of our cities of technology.

We understand who we are through the trail of objects we leave behind. As we follow the technology, we arrive at a village organised around metals and hardware components. The inhabitants of Guiyu collect e-waste in their houses, surrounding their living, sleeping and eating spaces. They mine their domestic landscapes for lead, germanium, gallium, tin, nickel and copper. Next to the pot of noodles simmers the acid bath, dissolving circuit wafers and separating metals, and flavouring soup.

And finally, we arrive at the shores of the ten-square-kilometre mine tailings waste lake filled with a cocktail of acids, heavy metals, carcinogens, and radioactive material roughly three times background radiation. China produces more than 95 per cent of the world's rare earths, and two-thirds of this is in Baotou. This is some of the first footage of the toxic waste that sits beside the world's largest rare earth mineral refinery. We take a selfie with our phones and see our reflection in the mirrored screen. The material to polish its glass and run its software produces this very lake. Collapsed together in a single luminous surface, we see ourselves and this black, black earth.

We brand our technologies with terms like cloud, air and featherweight, but in reality they are violently wrenched from the earth. As our personal electronics tend towards the invisible, they conjure in their shadows an undeniably visible grey mountain, a one-kilometre-deep pit and a ten-square-kilometre radioactive tailings lake – a counterweight to the apparent immateriality of computing, communications and electric energy.

From this black sludge we have a set of vases made from the amount of waste created in producing objects – an iPhone, a MacBook and cell of a Tesla electric battery. A new material aesthetic for technologies born of the earth. In silhouette the three vases echo highly valuable Ming dynasty porcelain vases. Ming vases are particularly iconic objects of high value as well as being artefacts of international trade. The Ming dynasty – a one-family global superpower – presided over an international network of connections, trade and diplomacy. These three "rare earthenware" vessels are the physical embodiment of a contemporary global supply network that displaces earth and weaves matter across the planet. They represent the undesirable consequences of our material desires.

We have followed the unmaking of these objects of technology – reversing their journeys from container ships and ports, through wholesalers and factory floors, all the way back to the banks of the barely-liquid radioactive lake in Inner Mongolia that is continually pumped with tailings from the rare earth refining process. The unmaking of our technologies is the making of these vases, carefully crafted from their toxic by-products.

Our cities cast shadows that stretch far and wide.

Treasured Island: Madagascar

(Figure 14.2)

In "Treasured Island", Unknown Fields travels through Madagascar to catalogue the push and pull of economy and ecology and meets the illegal traders of the world's luxury brands. One of the planet's most precious ecological treasures is home to one of its poorest nations, and it raises difficult and complex questions

FIGURE 14.2 All up in my grill. Unknown Fields/Toby Smith.

about the relationship between necessity and luxury. Hidden amid political uncertainty, the island's fragile and unique ecology is being smuggled out illegally, boat by boat, gem by gem. Rare tortoises leave in rucksacks, forests are carved into one-million-dollar rosewood beds to be sold in China, and precious stones are shovelled from the earth and smuggled onto the stage in popstar bling.

As the beat drops and the stage lights strobe, a pop star flashes their designer bling for the camera in flurry of choreographed dance moves. Another world away, in a hole in the ground in the wild west mining town of Ilakaka, Madagascar, another choreography of bodies moves in rhythm to dig dirt by hand out of a gemstone mine. The majority of the world's sapphires are pulled out of the ground by the human conveyor belts of Madagascar's gem fields. It contains an extraordinary amount of high value resources. Precious gems were deposited here by an ancient river that once flowed across Africa, before a tectonic shift ripped it from the mainland to form the island of Madagascar. The stones collected in a pocket along the twists and turns of the riverbed, resting patiently, beneath 20 metres of sand, and the future boom town of Ilakaka.

"We dig holes to try and find the river again". These are the words of Mark Nouvera, one of the major players in present-day Ilakaka. One of the only Europeans in town, he moved here 18 years ago, chasing the stones, hunting his fortune. He is the anonymous face reflected in every piece of jewellery you own.

> "I like stone better than humans, and I like stone better than money".

In 1999 he drove into Ilakaka to live in a tent and watch the place explode. When he first arrived, there was only one building here, and now the landscape is overrun with almost 100,000 miners. There is only one road in or out, and it is lined with gem shops and sweaty men, with guns on their hips.

> "I finance the pit. Here in Ilakaka the land is free, all you need to mine is a local arrangement with the chiefs and elders. There is no government here. There is nothing. There are thousands of people here, but it is the most lonely place in Madagascar".

This is the landscape produced from unregulated desires. If you want to mine with machines, you need a formal contract and money for fuel and maintenance.

> "To dig with people you don't need anything, just a bag of rice".

Each worker gets US$2 a day to work in the mines, and 50 grams of rice. Here it is much cheaper to pay workers in rice than it is to buy and maintain mechanical mining equipment. The 20 men of his "Swiss Bank" mine shovel dirt in perfect synchronisation, each paid with rice, their bodies repurposed as machines.

"This mine took 12 years to dig by hand".

Here their movements are traced like the early photographic studies of Frank Gilbreth. He was mapping the choreographies of the production line, looking to optimise every movement, constrain every motion, with the elegance of tuned engines. The digger is a robot, just one component in the gemstone conveyor belt: 2,880 shovels per day, 5.76 tonnes per day.

"This is my hole", says one of the miners. "I started digging four years ago. Every hole here was made by our hands".

"If you are lucky, you find a good sapphire and you have a good life. With no luck you die, or you grow old digging holes".

"When we find a stone, we go all together to the Sri Lankans to sell it and split the money. We don't know what happens to them or the stones. We don't know how much money they sell them for. The money never comes back to us".

Everyone else with money in this town is Sri Lankan. The cultural relationship to sapphire runs deep in Sri Lanka. Embedded within this tradition, Sri Lankan sapphires sell at a much better price than Madagascan stones. If rough, unpolished stones can be smuggled out of the country and back to Sri Lanka they can be refined and sold on, at an extraordinary mark-up. At 6 pm the single street town comes alive as miners return from the field for a treacherous two-hour negotiation for the sale of the day's pickings. They crowd around the tiny grilled windows that line the street and watch as a Sri Lankan inside sorts through their finds. The street is washed with the focused light of a hundred tiny gem torches shining through stones looking for imperfections and inclusions, anything to drop the price.

"We find ways to send rough stones out of the country. If I find a good stone, I fly back to Switzerland to get a certificate on which I can nominate the origin of the stone. I make much more money if I say I got it from Sri Lanka. We imagine planes full of buyers, lifting off from this treasured island, their shoes, their jacket lining full of shimmering, deep blue jewels".

"We are not a city, we have no name, no mayor, no bank, no map".

A hidden black market supply chain connects these two choreographies, from the lawless mine sites to the jewellery stores, hip hop music videos and celebrity red carpets across the ocean. Material production and cultural production have never been separate. Unknown Fields has used the amount of rice the human conveyor belt consumes in a day to manufacture a precious stone that embodies the systems through which these worlds are intimately and profoundly connected. The red Madagascan rice grown endemically on this Treasured Island is a staple food of the miners. It has been collected locally and shipped to gem specialists for carbon analysis. By subjecting the rice to extreme heat and pressure in the laboratory, Unknown Fields has formed a synthetic stone encoded with the sum of the human conveyor belt's labour. After manufacture, the gemstone has been set into a gold tooth, ready for that million-dollar smile and the outrageous lyric. From kilojoules, to carats, to the nightclub. In the glare of this cheeky gold grin, we see the cost of luxury, of beauty, of a daily allowance of rice, of 20 men shovelling at the bottom of a hole.

It glistens in the light, and in the mirrored facets of the rice diamond we see ourselves.

Snowing in the Supercomputer: Far North Alaska

(Figure 14.3)

To locate the environmental forecasts and data landscapes of the Unknown Fields city, the studio travels to Alaska's far north, to visit a territory that sits in the collective imagination as one of the last remaining wildernesses. Here, climate change is not a condition that is going to play out in a possible future; it is a phenomenon unfolding in real time. Whale migration patterns are shifting, coastlines are disappearing, the ice is melting. Unknown Fields spends the winter solstice with climate scientists from around the world who are camped out in the most northern cities on the planet to collect data that is fed into the climate modelling supercomputers and environmental policies further south.

FIGURE 14.3 Snowing in the supercomputer. Unknown Fields.

This is the sound of the supercomputer. The white noise of millions of calculations per second. From the oracles and augurs of the ancients to the predictive modelling of modern digital prophets, through the ages both wise men and charlatans have claimed to see into the future. Cultures differ in their concept of time and their attitudes towards the future, which are central to these acts of prediction. Alaskan Inuit, informed by ancestral memories of their environment and its patterns, embrace the uncertainties of the future with a deep belief in their own ability to adapt. Meanwhile, the world's environmental scientists attempt to assemble their observations into climate models to predict the future as precisely as possible. Caught between improvisation and premeditation, these cultural relationships with landscape and time will define the future of the North and in turn our cities beyond.

We peer inside the supercomputer to find a set of surreal landscapes, ones that sit between tradition and technology, the real and the imagined, the present and the future.

They are landscapes narrated by native Alaskan authors and generated from the climate data and modelling software of supercomputer scientists. Against these images run panoramas of the supercomputer infrastructure that simulates them, and the doomsday statistics that are shouted at us every day, but which we do our best to ignore. Traditional data visualisations and guilt-laden headlines are no longer sufficient strategies to encourage the cultural shift required. Unknown Fields cracks open the black box to pull out the environments the supercomputer is predicting, to imbue them with new narratives. The indigenous poetry is an alternative mythology for an ever-melting landscape of fear and hope. Panoramas that are dramatisations of data, portraits of a world we may have already lost or one we are yet to find.

The following poems are by native Alaskan writer Priscilla Naungagiaq Hensley.

> *Gentle being, how do you float across the ice to me?*
> *You somehow know my heart's become overheated, that I churn, churn, churn, my own friction burning me up. I can only stand here on the edge again and look, wait, yearn. Fog rises.*
> *Strong being, you leap from pan to pan, something a child could do if we'd ever let one anymore.*
> *Somehow you know I need your courage. You burst through the mist, over and over again*
> *I try to breathe you in.*

The pink clouds floating above the ice field is a particle simulation that has been produced from a global carbon emissions dataset.

> *In the morning there came a new bird's call.*
> *It seemed louder than all the others either because it stood out or because the other birds were shocked into near silence.*

What is it saying? Where is it from and (perhaps more importantly) are there others on the way?

Texturing on the iceberg is generated from the translation of a projected global temperature rise dataset into a three-dimensional noise map.

It will be the water rising.
Not just against the measure of shorelines and house posts, but as a needed, necessary thing.
Fear fluttering in parched mouths. The journey will be long.
It will seem novel, risky.
It may be.

The waveforms of a soundscape generated by running a Fairbanks supercomputer data tape through a reel-to-reel player have been used to form a layer of cirrus clouds above the churning ocean.

And so much is gone, gone.

The Breastmilk of Volcano: Bolivia and the Atacama Desert

(Figure 14.4)

Unknown Fields chronicles this electric landscape, investigating the infrastructures that serve as energy conduits. They translate matter like a luminous language – from a hole in the ground to the glow of our phones – tracing a wild journey of electrons from the radiant gizmos of our technologies deep into landscapes far, far away. This fiction is an account of a new creation story for our energy, from the Big Bang to the battery, from the birth of lithium at the beginning of the universe to the low power warning flashing on our screens. We power our future with the breast milk of a volcano.

13.8 billion years

In the beginning, the beginning of the beginning, seconds from zero, 13.8 billion years ago, the creation story of lithium begins. In a Big Bang.

It was there, at the dawn of time, alongside helium and hydrogen, just one of only three elements able to claim their ultimate origin in that hot dense primordial gas. Light is the only detectable record left of the Big Bang: it is the ghost of lithium creation.

At five kilometres above sea level on the Chajnantor Plateau is the Chilean Atacama Desert; the landscape has eyes. Sixty-six white pupils turn in unison to search the thin air of these dark skies. Here the astrophysicists at Atacama's Large Millimeter Array (ALMA) observatory are focused skyward, travelling deep through the dark interstellar clouds of the coldest, oldest parts of the universe.

FIGURE 14.4 Breastmilk of the volcano. Unknown Fields.

Another community of nomadic shepherds, the indigenous Likanantay people, used to own this land and trekked across the grounds where the antenna now stands. On the Chajnantor Plateau all eyes have always been on shadows in the sky. Silhouetted against the light of the Milky Way, the dark clouds that ALMA observe are the same shadow constellations of indigenous mythology. Dancing within a swathe of the interstellar cloud that forged lithium is Yacana the Llama, her baby and her shepherd. These creatures have trampled lithium, the lightest of metals, from the beginning of time to the crust of the earth.

4.6 billion years

4.6 billion years ago, as Yacana drank from the sky, the wreckage of an exploding supernova slowly began to condense into our planet. In this vast cloud of swirling cosmic matter, gravity and violent collapse gave shape to the sphere of the earth and embedded within it the traces of lithium. Lithium comprises seven parts per million of the planet's crust. Locked in the ground, waiting for release, it's an electric earth.

10,000 years

Ten thousand years ago, a series of lakes formed high on the Andean plateau, where Chile, Bolivia and Argentina, the three countries of the lithium triangle, now meet. Here lies the largest salt lake on the planet, containing at least 50 per cent of the world's lithium reserves.

The salt lake was once a vast plain where the Incan giants lived. Among them was the beautiful Tunupa. She chose to marry Cuzco, a strong young man, and a son called Calicatin was soon born of their union. While away on one of his trade journeys he became infatuated with a pretty young woman and they ran off together, never to return. The gods, tired of the giant's lies, secrecy and betrayal, decided to punish them all and petrified them as mountains. Tunupa began to cry, a volcano spewing ash and rock from the depths of the planet, rich in light elements like magnesium, potassium, boron and lithium. While the tears rolled down her cheeks, her breasts began to lose the milk her son had not suckled. Millennia of meltwater from the snow-capped peaks of her mountain seeped down through her rocky sides, leaching minerals into the lake below. As the giants became volcanoes, Tunupa's tears ran into subterranean brine and her milk crystalised as the crusty salt skin that now stretches endlessly across the plateau.

This charged landscape, this electric earth, remote and unforgiving, is now quantified for its energy potential. Cities, industries and infrastructures will feed at the shores of this ancient lake, playing out our electric future.

The future of green energy is made from the tears and milk of a mother mountain.

20 years

It was a little over 20 years ago that lithium erupted again in the Salar de Uyuni. A Belgian construction company found a ground too soft for construction but inadvertently had cracked open the earth to reveal the lithium-rich brine below. A hundred million tonnes of Tunupa's forgotten tears are now estimated to be trapped here.

Without the knowledge and assistance of foreign companies it has taken more than 20 years, but now Bolivia is ready and the Salar is soon to be industrialised. This natural wonder has become the most lucrative of investments and has cast Bolivia as the Saudi Arabia of the electric age.

15 months

You cannot see it on the desperately flat horizon or access it by any public road. Its mystery is protected by its isolation. Lithium development is not mining through extraction but through evaporation. A tessellated ocean of evaporation ponds where each shift in hue signals a rising concentration of lithium salts. The shores of the metal sea begin at Pond no. 15, 0.2 per cent lithium, the least concentrated, azure blue with a sodium chloride beach. Each month the ponds are drained and transferred to the next in line, and each month the colour changes and the lithium gets richer. Across 15 months the sea migrates through the holding pools of the Salar until it reaches the deep coffee waters of Pond no. 1, 6 per cent lithium sulphate.

What is left behind is massive quantities of table salt, which is piled up beside the lithium ocean, and gradually a new mother mountain grows. What will we call her, this crystal volcano? A totem for a sacrificial sea, evaporating to keep the screens glowing and the wheels turning.

2 weeks

This sea that has been slowly evolving across billions of years is now ready to leave the land of giants forever, sucked up by a convoy of thirsty 18-wheelers, and driven off the Salar to the lithium carbonate factory to be processed into batteries.

The creation story of a battery has become the creation of a nation.

14 hours

As it gently vibrates across the earth the iPhone 6 can travel 14 hours on its 1810 mAh lithium polymer battery before it comes to rest. It feels warm to the touch, and we are told stories about its lightness, and its slim lines. Reflected in its pristine polished glass is the mirrored expanse of the crystal white salt lake from which it has been wrenched. Nearly nine billion mobile phones in the world are powered by lithium-ion batteries; 5–10 grams of Tunupa's tears and breast milk are contained in each iPhone.

2.8 seconds

In ludicrous mode, the 7,000 lithium nickel cobalt aluminium oxide cells of the 90-kWh battery pack that sits in the belly of the newly born Tesla P90D delivers enough power to accelerate from 0 to 60 miles per hour in 2.8 seconds. Twenty to thirty kilograms of lithium are at the core of each electric car. As this glistening beast shrieks down the tarmac, it sends ripples across the turquoise pools of lithium brine. It is a hunter, stalking the electric future.

2 milliseconds

Three electrons orbit the lithium nucleus. Our story began with the travels of stars and ends with these tiny revolving planets just waiting for a charge.

Unknown Fields has built a glass battery – a mythic love story that trickle-charges a phone. While the world does its best to ignore that technology is forged from the earth, with marketing campaigns of ephemeral clouds and the relentless push for the smallest and lightest, this object embodies the story of the landscape in which it was made. A mass of alternating aluminium and graphite – anode and cathode – submerged in a lithium brine electrolyte, collected from Bolivia's electric Salar de Uyuni, creates a slow reaction, the drip charge of a weeping volcano. The creation myth of this landscape is told again and again as the electrons flow. The flash of the Big Bang to the flash of an electron.

Our future is powered by the breastmilk of a volcano.

Between the scale of the stitch and the planetary supply chain, in glittering gold Unknown Fields, we weave new connections and reimagine the relationships between consumption and production.

15

CLIMATE INQUIRIES FROM ARCTIC FIELDWORK

Leena Cho

In "Revisiting *Klima*", Fleming and Jankovic (2011) describe a difficult challenge placed on climate science to explain how climate relates to social and economic life. Scientific climate, which is a statistical average of weather over large spatial and temporal extents, abstracts climate from the lived experience and constructs it as a derived, distant entity. Questions close to everyday life and also critical to design and design education – such as how climate catalyses landscape, experience and imagination – therefore cannot be answered by science alone. Yet our exclusive dependence on climate science in grasping planetary uncertainties, they argue, is pervasive and needs to be situated in a broader context of myriad climate expressions and epistemologies.

Noting similar observations, Hulme (2010) problematises an overly physical rendition of climate in today's climate change discourse. Instead, he defines climate as "an idea which mediates between the human experience of ephemeral weather and the cultural ways of living which are animated by this experience" (2015: 3). Deep material and symbolic interactions that occur between weather and cultures in places are central to the idea of climate, he notes, and are inherently polyvalent, evolving and thus malleable. Instead of resorting to deterministic narratives of climate crises, Hulme (2010) suggests reframing the ideas of climate and climate change and using them as a forensic tool for reflection and ultimately for mobilisation. First, he proposes that we think much more directly about the weather so we recognise different forms of knowing climate. Second, he reminds that the function of climate change is not necessarily a physical phenomenon to be "solved"; rather it is to reveal underlying ideologies, vulnerabilities and tensions in places that shape cultural projects for futures to come.

Above anywhere else on the planet, Arctic landscapes are well positioned to take on this call. Vibrant and storied landscapes of the Arctic cannot be understood without first decoding their relationships to climate (Cho, 2020). Extreme and

DOI: 10.4324/9781003145905-19

amplified weather conditions – such as frigid weather and prolonged seasonal darkness – coupled with anxieties and prospects afforded by climate change permeate extensively through both the historical and contemporary fabrics of the Arctic landscape. Climate is something one cannot simply mask or ignore in this region on Earth; it is a matter of immediate survival, and through which unique socioecological and technological ingenuities have formed and flourished (Bocking and Martin, 2017). Understanding variegated clues and materialities of extreme climate, therefore, promises a renewed understanding of the many roles climate plays in society. That understanding prompts the question of how *else* climate can be (should be) positioned and designed within the era of new environmental regimes.

With an aim of expanding climate inquiries as a vehicle to conceive climate futures, the Arctic Design Group (ADG) at the University of Virginia has carried out annual research studios in the last seven years, with a most recent one based in Utqiaġvik, Alaska (Figure 15.1). Examining from the scales of frozen water particles to urban tundra infrastructures in a rapidly changing environment, the studio programme explores socioecological and sociotechnological landscapes of northern communities with design syntax unique to this cold region. Past research themes have ranged from landscape typologies in industrial and science towns in Svalbard, Norway, to migration strategies for small coastal communities, such as Shishmaref, Alaska. A key part of the programme that unites all the design studios, however, is an investigation of biophysical, phenomenological

FIGURE 15.1 Arctic landscapes – such as the northernmost city in the US, Utqiaġvik – have been a geographical focus for climate inquiry in ADG's research studios (2020). Arctic Design Group.

and political agencies of climate, and the unique transformations of landscape it fuels and sustains. By further exploring ideas and methods of environmental observation, experiment and manipulation across disciplines and knowledge systems – including environmental sciences, engineering and indigenous praxis – students probe the Arctic landscape as a polyvalent material and conceptual terrain, rather than the limits of climate.

In this context, fieldwork in the Arctic serves as a *designed encounter* with extreme and plural climates. It emerges directly out of a fully embodied, immersed experience of an amplified "weather-world", providing a productive contrast to the climatic "norm" familiar to students (Ingold, 2010). It further incorporates different climate perspectives of others who are an integrated part of the "bodily, social, collaborative, and inter-disciplinary engagement" in the field sites (Richards, 2011: 62). While the overall studio structure is more extensive regarding the design prompt, site context and design methodology, this chapter focuses on the processes of developing, editing and enacting fieldwork. Four operating components – Anticipating, Reorienting, Indexing and Magnifying – are discussed, based on observations made from students' fieldwork design processes. They are cumulative, but not necessarily linear, as students revisit each component contributing to fieldwork to refine climate inquiry and design.

Anticipating

Preparation for fieldwork begins "off-site" in Charlottesville, Virginia – some 3,500 miles south of Utqiaġvik – with an investigation of Arctic environmental systems generated and amplified by climate. Through composite mapping, guest lectures and literature reviews that span across environmental sciences, humanities and design, students develop a particular understanding about and position toward a climatically animated landscape, forming preliminary research in anticipation of on-site fieldwork. The prompt for a most recent studio, for instance, was Arctic ground, which is where climate "meets" and materialises with other mineral, biological and mechanical bodies subjected to radical thermal adjustments. To form an initial synthesis of research, students develop "position drawings" that straddle descriptive and interpreted narratives of climate, while identifying conceptual frameworks for design and on-site fieldwork (Figure 15.2). Parallel to this drawing-based research, they construct a series of physical models that represent select Arctic weather conditions and the environmental processes that catalyse them. Built as devices for material, spatiotemporal and conceptual experimentation, live models translate the agencies of climate into human-perceivable landscape phenomena and help students to test intervention strategies by experimentally modifying the embedded environmental processes and design parameters. An example is studying the phase change of water, from the mechanical effects of freezing and expanding (Figure 15.3, top), to the haptic qualities of vapour and condensation (Figure 15.3, bottom). The physical and experiential variations the model experiments create – albeit simplified – open

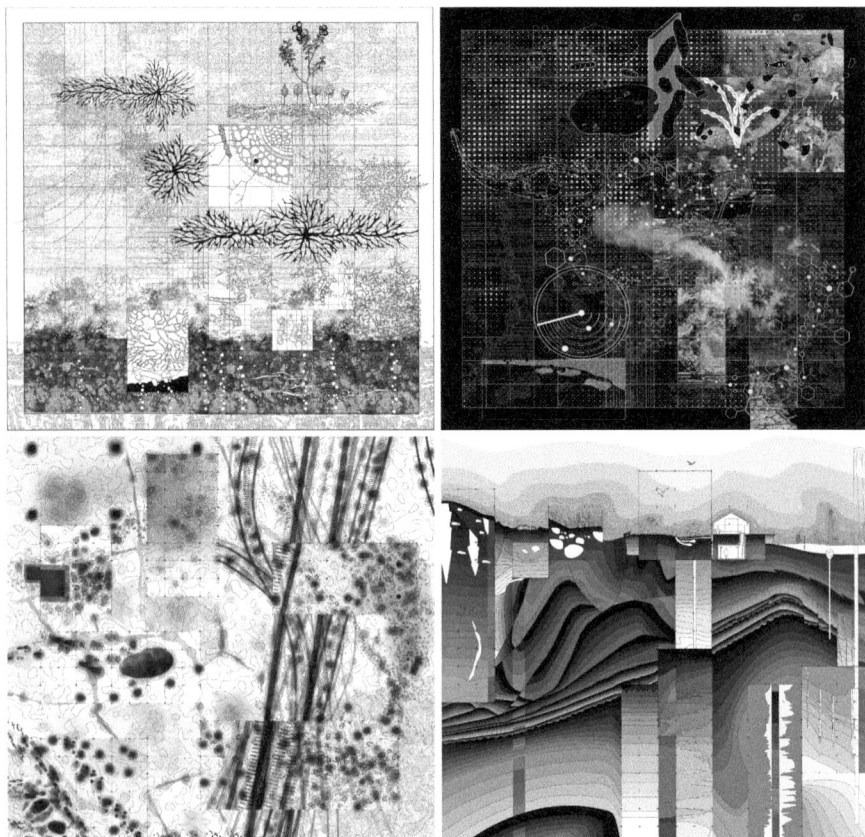

FIGURE 15.2 Top: Conceived during the ground-themed Utqiaġvik studio, the drawing pair frames ground as a site of climate entanglements seen through plant (left) and toxin (right) ecologies (2020). Yangqianqian Hu, Chenxin Sha and Qinmeng Yu. Bottom: The position drawings situate climate as a catalyst for ephemeral (left) and deep-time (right) landscapes in the Arctic (2020). Jingwei Jiang, Danni Jin and Qiuheng Xu.

up the possibilities for tapping into and designing with climate. Constructing actuated models also echoes Vannini et al.'s (2012) observation on the inseparability between weather produced and weather experienced. The processes of understanding weather behaviours and designing proxy microcosms to re-enact and edit these properties offer an anticipatory positioning and experience of the Arctic environment and imbue students' climate inquiry with agency and intentionality early on. Through this, students navigate between the physical and the conceptual, "what is" and "what if", with increasing specificity, establishing design questions to be tested during on-site fieldwork.

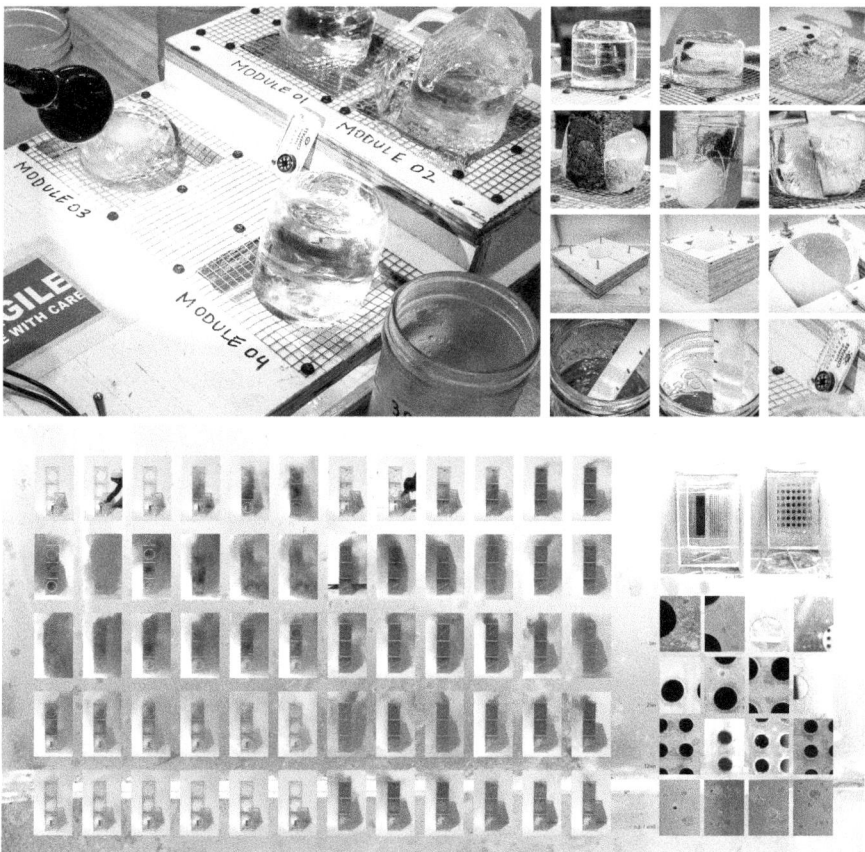

FIGURE 15.3 Top: Freeze-thaw live model investigates processes and effects of phase change of water, such as volume expansion and ice adhesion (2016). Sam Kokenge, Andrew Shea and Michael Tucker. Bottom: Vapour-flow live model explores optical and microclimatic qualities and material residues. Vaporisation and condensation are weather markers amplified in Arctic urban landscapes (2016). Kathleen Adams, Tom Bliska and Scott Shinton.

Reorienting

After the initial phase of speculative positioning, students arrive in the Arctic for what is typically a ten-day trip. They have never visited Utqiaġvik before, or anywhere close to it, especially during winter months. Surpassing all expectations, Arctic fieldwork is an encounter with extraordinary climate in action. For visitors, it is also an encounter with "counter-landscape" replete with an overwhelming sense of unfamiliarity. There are no trees, parks or paved streets in Utqiaġvik. Instead, cold mud and solid ocean stretch toward a seemingly never-ending horizon blurred with gravel dust or dim parhelia. The sun does not rise for months on end, while winter temperatures either side of a door can easily swing 120°F – from 70°F to −50°F – in a few steps. Pleasures of certainty and normalcy are thrown off. The Arctic climate and the visceral experience of its elemental landscape disrupt all sensorial devices – humans' and machines' – largely adopted to southern and temperate latitudes. Here, skins frost, pipes burst and drones shiver, even with slightest neglect, and supposedly effortless activities such as walking become part of a laborious process that requires equally extreme measures in safety and emergency training for students (Figure 15.4). The disorienting first-time encounter (regardless of how thoroughly one has prepared for it) halts much of the ready-made fieldwork plans and design ideas developed prior to the visit. Yet, as others who teach Arctic studios have observed, this disorientation is an irreplaceably generative and constructive stage of research in the design of Arctic landscapes (Larsen, 2019). Students quickly realise that climate is not only a weather or an atmosphere, but a cognitive framework for landscape experience. The mundane quietude of a city, for instance, amplifies the sounds of engine and wind from afar, distorting dimensions, distances and time. The usual syntaxes for recognising and spatialising landscape patterns such as planted form or rushing water are also seldom visible in Utqiaġvik. Likewise, the de-territoriality of Arctic landscapes,

FIGURE 15.4 Students dressed in arctic gear walking to fieldwork sites in −50°F in February. Note frosty hat, eyelashes and powerline. Utqiaġvik, Alaska (2020). Chenxin Sha.

noted by Ponte (2014), is not only real but deeply felt. The ideals of *terra* are also confronted by the provisional materiality of the Arctic, which is neither firm nor fluid, neither land nor water, further confounding familiar spatiomaterial conceptions such as boundary and simultaneity (Steinberg, 2018). Without these familiar cues for reading landscape sites, students' design of fieldwork takes a radical turn, infusing earlier discoveries made with research with the realities of intensely palpable climate.

Indexing

With the original plans mostly scrapped, students embark on revising or improvising their fieldwork, this time paying utmost attention to the Arctic climate's full, raw force. They document found material and spatial conditions of a site in detail, trace how those found dynamics might have formed, and speculate on multiple possibilities of landscape change that their interventions – both of fieldwork and design – might bring. Site information provided by the studio is intentionally limited; instead, the studio asks students to generate their own site and material data to guide their creative processes. They adopt a set of instruments – such as thermal imaging cameras, soil probes, Arduino sensors, 3D scanners – that can directly engage with the climate field (Figure 15.5, top). During this process of moving and hauling about in the landscape, they discover that generating data in the Arctic involves tremendous physical labour and astute sensitivity to environmental cues (Crate and Nuttall, 2016). As Cheyne (2015: 216) puts it, "Fieldwork fails, instruments are broken, prototypes don't fit", and these circumstances require students to develop their work dynamically, while remaining constantly attentive to external situations and contingencies. Like detectives, they follow animal footprints, tyre tracks and spotty mosses entangling single-story houses, old military shacks and sandbags along the coast. They congregate under elevated, icicled pipes and building foundations, or above buried utilidors, to measure tiny temperature changes juxtaposed to a greater surrounding (Figure 15.5, bottom). Others lie on ground, corralling blowing snow into different shapes of accumulation in an impromptu live experiment of the fieldwork (Figure 15.6, top). Digging to find more of these weather and climate traces in frozen, concrete-like ground tests the limits of bodies and tools, pushing students to invent new ways of mining and documenting them. Indexing is registering a value of things that can serve as an indication. In the Arctic, landscape indications of climate are at once both highly covert and highly conspicuous – from the warmth of a footstep that is not warm enough to melt the ice but transformative enough to turn it stickier, to the venting steams of infrastructure feeding extra moisture to lichens inconveniently hidden under freshly blanketed snow. The time spent in the field, however, brings an immeasurable insight for developing both the design tools and techniques on which students can draw. It also establishes, for the students, an awareness of what extreme places offer: the mental reorientation to sharpen their senses on how those landscapes materialise,

FIGURE 15.5   Top: Thermal index captures a temperature range one might experience on a typical winter day – from −40°F on snow to 350°F of a sheltered power generator. Moving inside-and-outside is a microclimatic adventure in itself (2020). Jinwei Jiang, Danni Jin and Qiuheng Xu. Bottom: Using 3D scanners, microtopographic data are produced of snowdrift near powerline and terrain under building foundations. Fieldwork emphasises generating students' own site and material data for design (2020). Karl-Jon Sparrman and Ziyuan Yang.

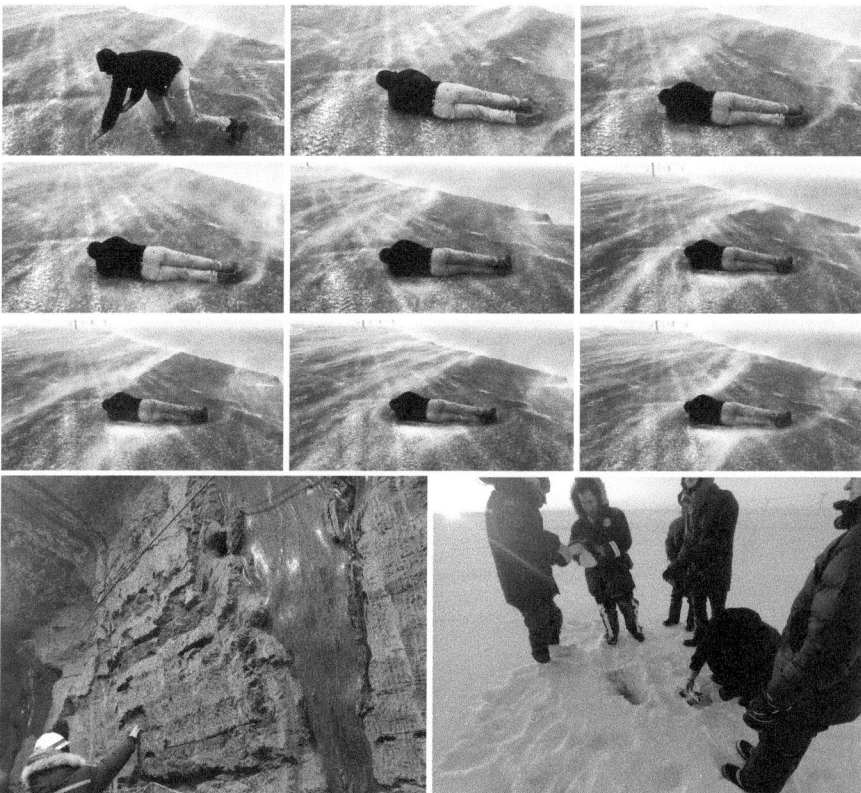

FIGURE 15.6   Top: In a live experiment, student lies on gravel road to test different configurations and temporalities of blowing-snow accumulation (2016). Sam
Kokenge, Andrew Shea and Michael Tucker. Bottom right: Before arriving
in Utqiaġvik, students visit the US Army Corps of Engineers Permafrost
Tunnel Research Facility in Fox, AK. The exposed section shows a variety
of permafrost features preserved since the Pleistocene Epoch, including
ice, roots and bones (2020). Bottom left: Iñupiaq photographer and US
Arctic Youth Ambassador, Eben Hopson, take the studio to winter foraging spots for caribou in tundra, Utqiaġvik, AK (2020). Leena Cho.

and the methodological recalibration to successfully probe, register and enact in
them (Pyne, 2010).

Magnifying

Students' evolving fieldwork strategies are grounded in the fact that climate, as
an idea, is diverse multiple and transcendental. Climate can mean many different things; it evokes seemingly unrelated but deeply connected concepts in time

and has a resonating power *in* and *beyond* its physical and material manifestations that can be far-reaching. Throughout the studio trip, students gather factual, sensorial, spatial and anecdotal data outside their fieldwork locations to examine ideas of climate multidimensionally. Descending into a permafrost tunnel, for example, students witness a gradual and palimpsestic force of cold climate (Figure 15.6, bottom left). Ancient ice, plants and bones alike, frozen in the ground come to life with a musty smell of deep time. As one student noted in her fieldwork log, climate is remembered in this subterranean world; climate is a vehicle with which the landscape of a changing environment is not only understood and projected, but also transported to other times in the history of things, even if they no longer exist in original form. Magnifying, therefore, is a conceptual broadening and interconnecting. Students seek other forms of climate narrative that the spatial bounds of a fieldwork site cannot fully convey, while concurrently reshaping the conceptual and material strategies for fieldwork with new insights. In Utqiaġvik, they follow Inupiat youths to the caribou-dotted tundra, meat-packed ice cellars, fish-drying racks and frozen waste lagoons, and assemble lived landscapes of climate discovered in a broader network of Arctic fields (Figure 15.6 bottom right). Trips to such places other than studio field sites reveal emplaced stories of climate and expand conceptual scaffolds for spatialising them. Further informed by workshops, lectures and interview and dialogues with local scholar-practitioners and residents, they construct a critical lens to develop design strategies distilled in particular climate theses, sensibilities and places. Through this experience, students define issues they must address, and identify what is at stake, and why it is important – extending fieldwork speculations beyond the individual or biophysical, and into the meanings and socialities of climates that future landscapes can forge.

Conclusion

Arctic encounters trouble classification (Barua, 2015). They challenge established taxonomies of landscape knowledge, and ways of knowing and designing. To encounter an Arctic field is thus to immerse in landscapes of amplified relations. After all, "extreme" is a relative notion which, without its opposites or associations, loses definition and potency; it is ambiguous, liminal, disruptive and transitional, as much as the kind of certitude it connotes or projects. Likewise, practising fieldwork in the Arctic creates what environmental historians Chu and Stuhl (2017) call a "curious discord", and brings students to an epistemically fruitful seam for climate inquiry. It provides opportunities to crack open preconceptions in design processes and seek alternative ways of responding to heighted climate experiences. It reveals as many limitations as possibilities – in tools, methods, perspectives, worldviews – and confronts students with an urgency to recalibrate in order to understand how the landscape of extremes is built and cared for and can be imagined. In the time of climate change and increasing environmental apprehension, landscape architecture education can benefit greatly from engaging with counter-landscapes such as the Arctic. Extreme is already "there", and is actively mutating as an *everyday* condition, not (only) as a result of exceptionally bad weather. Radical reflection and adjustment of means, assumptions and design lexicons are essential, if not inevitable, and fieldwork in this environment cultivates the founding of redirective apparatus necessary in the design of future climates.

References

Barua, M. (2015). Encounter. *Environmental Humanities*, 7(1), 265–270. https://doi.org /10.1215/22011919-3616479.

Bocking, S., & Martin, B. (2017). *Ice blink: Navigating northern environmental history.* University of Calgary Press.

Cheyne, K. (2015). Fieldwork: Uncovering cultural landscapes. *Proceedings of the Association of Architectural Educators 2nd Annual Conference: Living and Learning*, UK, 215–220.

Cho, L. (2020). Permafrost politics: Toward a relational materiality and design of Arctic ground. *Landscape Research*. https://doi.org/10.1080/01426397.2020.1831461.

Chu, P., & Stuhl, A. (2017). Reorienting world environmental history: Pedagogy and scholarship on cold places. *Environment and History*, 23(4), 601–616. https://doi.org/10 .3197/096734017X15046905071898.

Crate, S. A., & Nuttall, M. (2016). *Anthropology and climate change: From actions to transformations.* 2nd ed. Routledge. https://doi.org/10.4324/9781315530338.

Fleming, J. R., & Jankovic, V. (2011). Revisiting *Klima. Osiris*, 26(1), 1–15. https://doi .org/10.1086/661262.

Hulme, M. (2010). Four meanings of climate change. In S. Skrimshire (Ed.), *Future ethics: climate change and apocalyptic imagination* (pp. 37–58). Continuum Press.

Hulme, M. (2015). Climate and its changes: A cultural appraisal. *GEO*, 2(1), 1–11. https://doi.org/10.1002/geo2.5.

Ingold, T. (2010). Footprints through the weather-world: Walking, breathing, knowing. *Journal of the Royal Anthropological Institute*, 16(1), S121–S139. https://doi.org/10.1111 /j.1467-9655.2010.01613.x.

Larsen, J. K. (2019). Caring for Arctic and Subarctic landscapes. In K. Jørgensen, N. Karadeniz, E. Mertens, & R. Stiles (Eds.), *The Routledge handbook for teaching landscape* (pp. 148–161). Routledge. https://doi.org/10.4324/9781351212953.

Ponte, A. (2014). *The house of light and entropy, architecture Worlds 11* (B. Steele, Ed.). AA Publication.

Pyne, S. (2010). Extreme environments. *Environmental History*, 15(3), 509–513. https:// doi.org/10.1093/envhis/emq052.

Richards, K. (2011). The field. In J. A. Agnew, & D.N. Livingstone (Eds.), *The SAGE handbook of geographical knowledge* (pp. 53–63). SAGE.

Steinberg, P. (2018). Environments: Preface. In K. Peters, P. Steinberg, & E. Stratford (Eds.), *Territory beyond Terra* (pp. 87–90). Rowman & Littlefield International.

Vannini, P., Waskul, D., Gottschalk, S., & Ellis-Newstead, T. (2012). Making sense of the weather: Dwelling and weathering on Canada's rain coast. *Space and Culture*, 15(4), 361–380. https://doi.org/10.1177/1206331211412269.

16

CONVERSATION WITH PETER DEL TREDICI

Could you give us a brief overview of your teaching and research practice?

My teaching began at the Harvard Graduate School of Design (GSD) in 1990. I retired from Harvard in 2016. I then switched to MIT, the Department of Urban Studies and Planning. I've been involved with landscape architecture pedagogy for almost 30 years, but I'm retired now from teaching,

Could you please describe your approach to encouraging students to engage with ecological material?

I'm a botanist, an ecologist and a horticulturist. I worked in nurseries and greenhouses taking care of plants for 15 years at Harvard, first at the Harvard Forest in the 1970s and then for years at the Arnold Arboretum. In my late thirties, I went back to graduate school to get my PhD in plant ecology and was promoted to Director of Living Collections at the Arboretum after getting my degree. That was also when I began teaching at the GSD. I consider myself a hybrid academic – combining practical experience in horticulture with a scientific background in botany and ecology. The Arboretum is a 120-hectare landscape with over 15,000 accessioned woody plants and I had to coordinate all decisions relating to the management of the collections as well as mitigate the impacts of all the visitors who come to the Arboretum.

In my work at the GSD, I tried hard to integrate the theoretical and the practical. I used to say to my students that maintenance is about what happens on the ground after the design process is completed – about what happens when the design world meets the real world. My approach to teaching relied heavily on taking the students outside. When I first started at the GSD, I made it as much of a field course as the schedule would permit. There wasn't a lot of time to go on extended field trips, but we would walk around the campus every week and look at trees with an occasional trip to the Arboretum thrown in. Essentially, I saw

DOI: 10.4324/9781003145905-20

my role as getting the students out of the building because I felt they spent way too much time looking at their computer screens.

You describe how for many years the Arboretum provided a basis for research and teaching. Could you discuss how the students engaged with the Arboretum?

I worked at the Arboretum for 35 years, and when I became Director of Living Collections in 1991, I started teaching at the GSD. The Arboretum's plant collections are an incredible resource, but it wasn't all that easy to make a scientific collection of trees and shrubs relevant to my landscape architecture students, most of whom did not have a background in science. From the pedagogical perspective, I used the Arboretum not only as a place to look at plants and learn their names, but also as a place where we can talk about long-term landscape management issues. Most people, including my students, didn't appreciate what went into making the Arboretum look the way it does, and my job was to reveal the behind-the-scenes decision-making processes to them. In 2002, when I was no longer in charge of the living collections, my focus at the GSD shifted from teaching plant materials to teaching about urban ecology which is now a major focus of my research.

What are you currently working on?

I recently contributed an article on "The Spontaneous Ecology of Tommy Thompson Park" to the book, *Accidental Wilderness* by Walter Kehm and Robert Burley (University of Toronto Press, 2020). The chapter describes a park in Toronto, Canada, that started out as a landfill in the 1950s when the Toronto Harbour Commission started dumping construction rubble into the lake in order to expand their port facilities. The filling went on for over 20 years and extended five kilometres into Lake Ontario. Regional economic conditions changed in the late 1970s which caused the project to lose its financial viability and the Commission stopped the dumping and abandoned the project. Within a relatively short time, plants colonised the site, the birds followed after them, and eventually the landfill developed spontaneously into a wildlife refuge.

The Toronto Conservation Authority took over the project in the 1980s and, after prodding from community activists, decided that the best use of the site would be to formalise it as a wildlife refuge and to open it up to the public. They went through a design process that called for regrading the site in order to create some topographical variety as well as a pathway system to facilitate circulation. Its name was changed from Leslie Street Spit to Tommy Thompson Park and it quickly became an important ecological resource for the citizens of Toronto.

The book describes the process of going from a landfill to a popular park. I'm excited about the project because it illustrates that nature, when left to its own devices, can be a powerful creative force that can help clean up the messes that people make of the planet. We have to trust nature and give it time to do its work. When people are constantly interfering with nature, trying to direct it in a particular direction, that's not ecology, that's gardening. What I'm really interested in now is what happens when nature is allowed to do its own thing in the absence of design. It's a very different kind of process where the role of the landscape architect is not to control the design but rather to create a framework for ecology to happen. What's interesting about Tommy Thompson Park, now that it's become an official wildlife sanctuary, is that the people who are in charge of it are starting to talk about removing the invasive species such as the common reed (*Phragmites australis*) that helped create the park in the first place.

In the 1990s, thousands of pairs of double-crested cormorants began nesting on the site, eventually killing the poplar trees they nested in because of their extensive guano droppings. Park managers were upset about this, as were the local fishermen who felt the birds were competing with them for fish, so people began taking measures to control them – essentially treating an iconic native species as though it shouldn't be there. I see that as a violation of the process that created the park in the first place. The species that settled there on their own are part of the ecology, regardless of whether or not they are native to the Province of Ontario. The cosmopolitan array of plants that settled there has created what ecologists called a novel ecosystem and leaving it alone is what will sustain it, as opposed to picking and choosing which species to allow in. I'm really interested in studying what happens when ecology and design are allowed to interact without human interference. For many people this is a scary idea, but for me it's a super interesting research question.

Ecology, as I see it, is about leaving nature alone and letting the organisms work things out with each other. When people come in and insert themselves into the process, it becomes gardening, not ecology. The extent to which one can design novel ecosystems is an open question – I like the idea of landscape architects setting the table for ecology to happen – not controlling the process.

There is a clear and important distinction between maintenance and management – the former is about preserving the integrity of the design while the latter is about trying to influence the long-term trajectory of a site without knowing what the endpoint is going to be. One of the things in my experience with novel ecosystems is that things get better over time if you leave them alone. Another important question for landscape architects is whether they can incorporate time and spontaneity into their designs. This question shifts the idea of landscape maintenance from being about appearance to being about process. Future landscape architects are going to have to think hard about this question in order to deal with the ecological chaos that climate change will bring.

Unfortunately, I had to warn my students that, "You can't go into your studio and have your design strategy be doing nothing". Even though that may be the best thing to do in a particular case, you're going to fail the course if your stated strategy is doing nothing. While it's clearly not an acceptable answer in a studio, it may well be the appropriate response in some real-world situations. I think this Tommy Thompson Park is a good example of the positive things that can happen when landscapes are allowed to develop on their own.

In New Jersey, there's a park right opposite Manhattan, called Liberty State Park. The site was once a huge train depot, that had a lot of train lines ending there. Before the automobile became the dominant mode of transportation in

New York, all the commuters from New Jersey would get on ferryboats and go over to Manhattan. The central portion of this site was completely contaminated by the trains – oil, heavy metals, etc. – and the area was fenced off for 40 years because of the pollution. Over that time period, the site was colonised by an amazing assemblage of plants that are tolerant of soil contaminants and take them out of circulation by binding them with organic matter. Fencing the site off for 40 years has produced an amazing array of ecological benefits, but most people just see it as polluted landfill with invasive species. The site is still contaminated, of course, but the vegetation has helped detoxify it for its animal inhabitants.

The same thing happened in Berlin after the war, where heavily bombed sites were just left as piles of rubble scattered throughout the city. Now Berlin is hugely prosperous, and the only remnant of the once-common abandoned sites are those that have been turned into "urban nature parks". In today's world, one can still see these kind of classic "ruderal" sites in the so-called Rust Belt cities, where once-flourishing factories have been abandoned and colonised by vegetation. In cities that are economically prosperous, such as New York or Boston, land is so valuable that nobody is going to let unmanaged ecology take over vacant property. In an ironic twist of fate, a lot of the places where urban nature has been left alone to develop without interference are now being looked at as sites for dealing with problems exacerbated by climate change and urban development. Such "urban wilds" are often viewed by planners as unproductive land that would a good place for dumping storm water runoff and snow contaminated with salt. These landscapes never get any respect.

Could you describe what you mean by hybrid species, the ongoing adaptation of plants and the emergence of novel ecosystems?

The article I wrote in *Botanical City*[1] addresses the idea of **pre-adaptation** which refers to the fact that the plants that grow spontaneously in cities are not biologically adapted to urban environments, but rather pre-adapted, by which I mean they come from habitats in nature that resemble the habitats you find in the city. Pre-adaptation is an important idea because it helps explain why certain plants grow spontaneously in the urban environment while others don't, but it's only the first part of the vegetation story. The second part involves actual adaptation to urban conditions. Ecologists have documented genetic changes in numerous annual species that arrived in North America from Europe between two and three hundred years ago. The third and final phase of urban adaptation involves the process of hybridisation between once geographically isolated species that were brought together by humans. What's important about this process is that while the two parent species are adapted to specific natural conditions and the hybrid offspring isn't adapted to anything, the hybrids often out-perform their parents in disturbed habitats created by human activities. This means that urbanised environments have the potential to become hot spots for plant evolution in human-dominated landscapes. Essentially, the parent species are brought together by humans and then evolution kicks in through hybridisation to produce new species or genotypes that are adapted to the novel urban conditions.

I got to watch this process happening in real time at the Arnold Arboretum where a big part of our mission was to collect plants from around the world and grow them side by side so that their horticultural and botanical attributes could be compared. In this kind of "common garden" setting, closely related species from different parts of the world will often hybridise spontaneously. When I worked in the propagation department, I would collect seeds from the Arboretum plants and grow them in the greenhouse. I would often notice that within a flat of seedlings it was not unusual to find one plant that stood out among the others, being either bigger or smaller. I kept those seedlings separate and inevitably they would turn out to be spontaneous hybrids. Few people realise that this is the origin story for many of horticultural selections that nurseries produce.

Can you also talk about the role that urban infrastructure plays in supporting these systems, or the evolutionary process?

I would defer to my European colleagues on this, given that they've done incredible work documenting the flora of numerous European cities down to a microscale, and comparing that biodiversity with that of the adjacent non-urban areas. Remarkably, if you just looked at the total number of plant species – without making a distinction between native and non-native – the urban environments are much more biodiverse than the non-urban environments that surround them. There are a number of reasons for this, but the one that most people point to is that the habitat heterogeneity in the urban environment is much greater than what you find in rural areas and this difference allows for more species to get established. In other words, habitat heterogeneity generates biodiversity. I find it encouraging that so many plants grow spontaneously in the urban environment because it means that they understand what's going on and have adapted to the conditions that people have created.

These days, everybody is talking about climate adaptation, but urban plants have been doing it for quite a while. Whether you like these highly adaptable, often invasive, species is irrelevant, as is the question of whether they're native to your region or not. The plants clearly know what's happening to the world and the fact that they are able to track it closely is a good thing. Some people talk about restoring urban ecosystems to their "native" state, but I don't think that's either a reasonable or a relevant proposition. Going back in time is not an option – we have to move forward, and plants are showing us the way. We need to pay attention to what the plants are telling us about climate change because they know more about where we're than we do.

Note

1 M Gandy & S Jasper (Eds). (2020). *The Botanical City.*Berlin. Jovis Verlag.

17

FRAMING FUTURES

Worldbuilding in Landscape Studios

Marc Miller

Landscape architecture programmes are often mired in a desire to train young people as professional practitioners. Programme curricula are explicitly structured to reflect standardised national testing requirements and implicitly articulated to meet the expectations of employers hiring people for their offices. It is common to hear educators talk about ensuring students are "office-ready" – the implication being this is the most important outcome for students and, by association, the programme. The desire to address short-term expectations often comes at the expense of important skillsets such as speculative thinking that will enable students to adapt as designers.

Having the ability to anticipate and adapt is increasingly important for these emerging professionals. They will be challenged by a very – in some places, radically – different set of social frameworks, environmental conditions and design scenarios. It is one thing to mention these transformations as if they are a matter to be dealt with when they happen, it is another matter entirely to teach towards these uncertainties, acknowledging the changes that are already present. This includes using thinking and teaching frameworks that provide students opportunities to model behaviours related to their futures. Landscape is a temporal practice and representational mediums should reflect that.

This chapter describes examples I have used to teach beyond contemporary practices in landscape architecture to expand the discipline's reach beyond teaching for today's needs, looking towards the challenges the discipline will face in the future. The goals are to:

- Create frameworks for complex scenarios shifting from convention and optimisation to speculation and adaptation.
- Ground speculations using climate models and demographic datasets and practices.

DOI: 10.4324/9781003145905-21

- Construct narratives (not landscapes), as a means to communicate environmental futures, using media platforms that are widely accessible to the general public.
- Decentre professional practice norms in landscape architecture and introduce opportunities for new modes of practice for the discipline that will also create patrons.

Instead of finding solutions, the emphasis is placed on imagining future environmental conditions in anticipation of situations that are not yet apparent and to create empathy for people who will live then. Fundamental to this approach is the idea of worldbuilding, or the process of constructing an imagined place that serves as a narrative environment. In these examples, worldbuilding focuses on cultural and environmental change.

Worldbuilding offers alternatives to conventional disciplinary attitudes, offering students the opportunity to preprogramme the future in the present.[1] This proposition is why it is important to move to speculative investigations of the future, and beyond standard approaches and precedents that rely upon historic precedent or contemporary best practices. In the context of the classroom, multiple scenarios emerge, revealing the complexities that students will confront in the decades to come. Solutions may not emerge, but anticipation and possible trajectories for adaptation become more apparent in the process of framing futures.

As a method to project behaviours instead of solutions, worldbuilding requires emerging practitioners to decentre themselves in favour of the agent and actors who will occupy the environments they imagine. This is in keeping with activities commonly associated with worldbuilding – like science fiction, speculative fiction, fantasy literature and gaming. These activities encourage audiences and participants to displace themselves from their current situation and to imagine themselves in different settings. Like paper dolls, we can learn behaviours from the people we choose to model. Furthermore, representational formats also play an important role in the worldbuilding process. Cinematic mediums like video enable the author to create narratives that are more immersive than two-dimensional illustrative media but are also not prohibitive like augmented or virtual reality.

Framing Futures

Comprehension of disparate forms of information is fundamentally complex. These issues were effectively described by the economist and political scientist Herbert Simon, in his elaboration of "bounded rationality".[2] Bounded rationality refers to three constraints applied to any person's decision-making process:

1. Information is often limited and unreliable
2. People have limited capacities to absorb and synthesise information
3. There is only a limited amount of time to make any decision

Comprehension and synthesis are important steps towards imaging possible futures, which require that the information must be legible if not recognisable if they are to facilitate discussion and imagination. How that information is organised and applied is a challenging aspect of worldbuilding and specific scenarios. Unique or discrete datasets or histories can be difficult to process for use in understanding or managing problems. This is especially so for people unfamiliar with methods and outcomes commonly associated with design communication and diagramming.

Many decision-making processes have been operationalised through using computerised datasets. Platforms such as GIS are designed to automate the process of synthesis using a reductive approach to identify the desired outliers within a set. Other reductive methods, including computational programming, rely upon optimisation to identify the best answer. A fundamental limitation in these reductive processes is that the information is filtered through a single shared framework, limiting outcomes and perceptions. The tool becomes a convention that narrows the range of responses,[3] implicitly limiting answers to solutions.

In contrast, adaptive or emergent strategies are not reductive. Emphasis is placed on collective approaches to given conditions with multiple frames. They are formulated through decentralised, layered responses to conditions at hand. Decentralisation has the added benefit of creating a robust set of ideas and resilient frameworks.[4] Teaching that supports students to identify strategies, rather than to find solutions, prepares them as emerging practitioners who can work as facilitators for futures, instead of working as stewards who reflect on the past.

Paper Dolls: Placing Faces in Changing Communities

In the autumn of 2020, I co-instructed an urban design studio with Travis Flohr and Zannah Matson. Our selected study area was a suburban community outside Pittsburgh, Pennsylvania. Our goal was to develop a series of design ideas working with the community's office for economic growth and planning. The goal was to create a set of development scenarios illustrating a legible town centre to influence future growth. Big box commercial stores serving the immediate region with food, hardware and other retail types presently occupy the area selected for consideration in the studio.

We could not travel to the site as a class to develop an understanding of the place which was a significant complication. We had access to maps and data, along with Google Street View, but were unable to give students a sense of how current residents dwelled in the community. Additionally, we were imagining development over the long-term and had no alternative but to imagine the demographic composition of future residents. These difficult and ambiguous conditions led us to creating the Paper Dolls assignment.

Building on mapping and other computerised systems students were already familiar with, the assignment served as a surrogate for the lack of direct contact with residents. It was designed to generate discussions around implicit assumptions students were making about present residents and future residents – and who they would be designing for. Implicitly, the assignment demonstrated how aggregated data is used to construct general perceptions of communities based on remotely perceived patterns, but without the benefit of ground-truthing.

The Paper Dolls prompt asked students to create a series of imaginary individuals based on current census data, and psychographic mapping and marketing resources like ESRI Tapestry and Nielsen Claritas. Both the Tapestry and Claritas databases provide general census data but also include short blurbs of the lifestyles of people present in the community, with labels like "Soccer Parents" and "Grounded in Place". The databases provided students with gross information about spending habits and other forms of nuanced information. Most importantly, using these databases enabled us to discuss these tools their clients – namely developers – would use to analyse and create communities, based on financial models instead of environmental and spatial conditions.

Using population profiles also allowed us to discuss anticipated demographic changes in the United States. According to projections, the United States will have a "minority-majority" population by 2050.[5] This shift will impact social, political and cultural dynamics and consequently affect design decisions. In an attempt to ground our discussions about the future, students were asked to consider the community's future: how demographic changes might impact it, and how they perceived the importance of economic, social and racial diversity as they constructed scenarios for the future town centre.

Students were required to imagine four different user groups: current residents, desired residents, current residents 20–30 years in the future after ageing in place and desired residents after establishing themselves in the community. Each paper doll character followed a template with a series of prompts, requiring students to imagine and illustrate what the person looked like and how they might live in the community. Collectively, the paper dolls helped students imagine the community evolve over time. By making connections between people and activities such as shopping or recreation, students could imagine their proposals using collages that could be mined for ideas to develop their catalyst projects.

In summary, the Paper Dolls exercise supported the goals of the studio to address urban development methods from perspectives other than those of designers by introducing new resources and activities. Specific to the project, the assignment helped the students imagine who lives in the community and how the population might evolve. It also allowed us to discuss the importance of identity in constructing images that convey desired social outcomes, instead of placing people in images "just for scale". A limitation of the assignment was that it did not lead to narratives that centred around the experiences of these imagined individuals. This was due to time restrictions and the types of deliverables that were agreed upon at the beginning of the class term. A more effective use of the characters would include narrative that describes change in the community

1. Who is this person.
What is their name?
How old are they?
What do they do for a living?

2. What are their third spaces?
What is their favorite
hangout, restaurant,
or retail activities?

3. How do they get around?
Do they have other important
large possessions?

4. What is their everyday carry?
What are the items they cannot
operate without on a daily/week-
ly basis? How do these items a
reflection of their activities?

Image the paper doll using https://metahuman.unrealengine.com/

FIGURE 17.1   Paper Dolls Template. Marc Miller.

from their perspective of the character in a medium like video that is more recep-
tive to current community members (Figure 17.1).

Landscape Serials Studio – Five Futures for Braddock

The Landscape Serials Studio is part of my research interests related to worldbuilding, media and design communication. Unlike traditional studios, emphasis
is placed on problem-seeking and speculation as outcomes instead of presenting
solutions. The general studio approach also reflects a desire to communicate
ideas to the general public in a format that is more familiar to them, eliminating
the need to "educate the public" about drawing formats – and decentring the
practitioner from the problem, shifting toward the people in place. Therefore,
serial television has been adopted as the primary precedent for imaging studio
outcomes. An added consideration was that film and television are more relevant
as visual precedents than paintings or photographs to student cohorts given the
relative ease of access to media on the phone.

This particular project was located in Braddock, Pennsylvania, a small borough east of Pittsburgh on the Monongahela River. Like the rest of the region,
the steel industry played a significant role in the borough's growth, being the
location for one of the first steel mills. Now, the Braddock's urban fabric is representative of years of decreasing population, disinvestment and industrial decline.
The primary street running through the borough is faced by vacant lots, decayed

buildings and depressed properties. Concurrent with this gradual decline is a small renaissance brought about by residents including a former mayor, restaurateurs, urban farmers and artists looking for affordable studio space. In contrast, there are still residents suffering from lack of access to living wages and food. This spectrum of recovery and stagnation made Braddock a compelling place for worldbuilding, to ask questions about the economic and environmental impacts of the steel industry, privilege, community relationships.

In keeping with the premise, video was also the primary reference resource. To get a sense of the past and present of the borough, we watched documentaries by Tony Buba[6] and Rosie Hager.[7] The mockumentary *Carnage*, produced by Simon Amstell, was also presented to the students as an example of how to narrate speculative futures. These resources provided students with points of departure for research and story writing.

Serial content available on Hulu, Netflix and YouTube was also used as narrative and technical precedents. This was a significant shift for the students who were familiar with looking at photographs of designed landscapes for models. It was a necessary shift given our emphasis on telling stories about a place in the future, instead of proposing a specific solution for a site as if it were to be constructed in the next five years. Watching other forms of television was also important because students' work was being treated as something that would be broadcast, reaching a wide audience with the aim of influencing public perceptions of environmental, social and political futures.

Five student production teams developed unique approaches to the idea of using serial television as a precedent. One group looked at fugitive serials as a way to develop their story. Another group became interested in reflective documentaries to tell stories of transformation. Technically speaking, the resulting videos were not professional by any means, but illustrated five futures. More importantly, it allowed the students to investigate larger social, economic and environmental concerns without feeling the need to develop a solution.

Watching films and serials as generative mediums had the effect of flipping teaching exchanges. Instead of requiring students to look at discrete references assigned like reading, they would share the serials they were currently watching – "becoming teachers" because they were describing what was important about the serial in the context of the studio. Interests ranged from political intrigue and satire to dramas centred around serial killers. Our watchlists expanded collectively like an open reading list for a seminar – but we were still grounded in the matter of creating stories about a future Braddock (Figure 17.2).

Conclusion

These examples describe ways of teaching landscape architecture disciplines that create opportunities to create other methods for practice and making. The Paper Dolls assignment facilitated discussions about how contemporary data aggregation and mapping is a form of worldbuilding used by clients and consultants. It

FIGURE 17.2   Looking across a vacant lot towards Braddock Avenue in Braddock, Pennsylvania. Marc Miller.

also allowed students to imagine the types of social interactions they would like to promote, based on observed behaviours instead of implicit biases. Landscape Serials presented opportunities to interrogate traditional approaches to image production and precedents, making the discipline more accessible to students by using popular media.

There are still other media to be considered. Role-playing games can decentre students from assumed roles as professional designers, giving them opportunities to imagine how the built environment impacts others. Serial podcasts also have the potential to immerse audiences in other places through sound. All these practices require imagination and speculation to be employed to create landscape images, that are significantly different from representational formats that are common in design studios today. Regardless, long arcs of change and adaptation require immersive, serial media that enable storytelling to engage audiences and expand the discipline of landscape architecture.

Notes

1 Eshun, K. 1998. *More Brilliant Than the Sun: Adventures in Sonic Fiction*. London, UK: Quartet Books.
2 Simon, Herbert A. 1987. *Models of Man: Social and Rational: Mathematical Essays on Rational Human Behavior in a Social Setting*. New York, NY: Wiley.

3 Aish, R. 2011. Designing at t + n. *Archit Design*, 81: 20–7.
4 Brown, Adrienne M. 2017. *Emergent Strategy: Shaping Change, Changing Worlds.* Chico, CA: AK Press.
5 "U.S. Population Projections: 2005–2050". US Population Projections 2005–2050. The Pew Charitable Trusts. Accessed 24 November 2020. See https://www.pewtrusts.org/en/research-and-analysis/reports/2008/02/11/us-population-projections-20052050.
6 Braddock Films. See https://braddockfilms.com.
7 Braddock, PA. See https://www.topic.com/braddock-pa.

18

FINDING LANDSCAPE
THROUGH CURIOSITY

Sean Burkholder

By fostering curiosity and responsiveness in our students, and a willingness to question the fundamental structures of the world we occupy, we can assist in forming a generation of designers that can genuinely contribute to our challenging future. The work described in this chapter is from one design exercise in the autumn of 2019, representing one semester's work in the 501 core landscape architecture studio at the University of Pennsylvania. The students who contributed the work come from varied backgrounds (seldom design) and are often just learning the fundamental tools of landscape architecture. I believe it is here, at the very start of design education, that we can lay the groundwork for young designers to engage the world in productive ways, as opposed to the blind march towards rote problem-solving that so transfixes our discipline.

501 Background

I coordinate the 501 studio, the first design studio taken by many of our landscape architecture graduate students. The rapid, rigorous, immersive introduction that 501 provides has shaped it into what many students consider the most transformative class of their landscape education. Ian McHarg used to teach 501 with a strong focus on getting to know the region around the city of Philadelphia, in particular its geologic and ecological "layers". After McHarg, the course was coordinated by a range of faculty, including Anuradha Mathur and Dilip Da Cunha who led the studio for a decade. Under their instruction the studio's focus progressively shifted to a more situated and personal relationship to the landscapes of study. The subject matter moved to more exploratory and experimental methods of *grounding* and *finding*, partially influenced by material James Corner taught in the 1990s when he coordinated the 502 studio. Much of the subject matter spearheaded by Anu and Dilip has remained (at least in spirit)

DOI: 10.4324/9781003145905-22

within the pedagogy of 501. More specifically, the course remains a study of *context*. However, what context we choose to engage – and what methods we use to uncover, illuminate and build that context – continues to evolve.

Understanding the New Normal: Pressing Pause on Problem-Solving

The theoretical agenda of 501 has always been to learn how to read a landscape. This reading has come in many different forms through time, ranging from local geologic underpinnings to tracing non-human beings across a landscape. In all cases, the central assumption is that to design landscapes, you must first understand the contexts they both produce and contribute to. The course's current pedagogical framing takes this further, emphasising contexts themselves as being fabricated, not simply found; and that these fabricated contexts come with all types of baggage that must be acknowledged.

Our human-modified ecological context (primarily by way of climate change) is an ever-present concern, most often referred to as a "crisis". However, in a recent conference on climate change, indigenous scholar Kyle Whyte stated simply that we are facing a "kinship crisis", not a climate crisis (Whyte, 2020). He reminded us that labelling something a crisis tends to permit a collection of emergency actions that lean heavily on strategies that have worked in the past; yet many of these strategies have generated a long history of harmful externalities that disproportionately affected indigenous groups and people of colour.[1] "Fixing" the climate cannot be done at the expense of others. Design's coercive power has every opportunity to exacerbate this issue. That is why it is essential to teach methods of work that operate beyond the aim of myopic problem-solving.

In the United States, political rhetoric, hatred and self-interest power the growing divide between peoples of all kinds. These issues are compounded by stresses associated with the global pandemic, and the 2020 presidential election, after which I am writing this essay. Looking at our context today, it is likely quite different from what McHarg responded to in the 1970s and 1980s. Our pedagogical challenge is how to address these new contexts in ways that are fundamentally about landscape, without falling into the complexities of activism and policy now beginning to preoccupy almost everyone. This is not to say landscape architects should disengage from these topics – they absolutely should engage. However, this engagement should follow the development of a toolkit of skills associated with *design*.

When considering a subject as vast and wicked as climate change, the most important aspect for introductory design education is simply that of "change". Many incoming students see landscape architecture as a vehicle to address that change; however, the discipline's potential is undercut by tethering it to the responsibility to address these changes as problems to be solved. There are many other ways to approach these issues that do not require a landscape architecture degree. The 501 studio teaches design as a tool that can be used to achieve a wide range of outcomes, many of which could be leveraged to deal with so-called problems, such as climate change, inequality, biodiversity and water management. Designers must be facile in a way that actually finds and frames problems, as opposed to solving them – they must become curious and creative agents. Early in 2020 Rob Holmes likely made the most convincing and comprehensive case for this in *The Problems with Solutions* (Holmes, 2020). But this is by no means a new idea: Elizabeth Meyer considered a similar way of working 30 years earlier in *Landscape Architectural Design as Critical Practice* (see McAvin et. al., 1991).

Curiosity and Time

I believe, more than ever, that we need to stimulate skills in curious thinking in our students, not problem-solving. Fostering a strong sense of curiosity about the world we are enmeshed in will aid in discovering and framing new problems, as opposed to simply attempting to fix situations that are presented to us. I am strategic here in shifting towards curiosity and away from what Meyer declared as a need for "critical inquiry". Meyer's conception of the term "critique" is close to what I aspire to; however, as a modality of work the term still rests on its elitist, white, Western, colonial foundations.

Isabelle Stengers's work continually inspires me, with her adamant push towards the philosophical crafting of a science that is both productive and self-reflective (Stengers, 2018). This science acknowledges it has to learn and to listen, and to recognise it is the product of an economy of knowledge that has institutionalised scientific practices. In Stengers's opinion – one I am highly sympathetic to – science is a far-from-objective business of fact-making. As designers, we utilise these facts every day, with little concern over how impregnated they are with theories they were fabricated to support. Stengers's call for a "slow" science, one that remains sceptical of the typical classification and purported objectivity of standard science, is something that could be instructive for the practice of design. It embraces the messy realities of the world(s) we encounter. It avoids, at all costs, the instinctive desire to classify, simplify and objectify these worlds. If there is room for creativity in the process of fact-making, it reasons that this creativity extends to the process of context-making.

Essentially, this context-making process is about developing a thick, place-based knowledge. Places truly do matter to people. Bruno Latour's discussion of the "terrestrial" is helpful as a way of considering our connection to place in the face of climate change, where land is no longer the recipient or location of human actions, but an active participant with them (Latour, 2018: 42, 68). If we can accept the level of importance Latour's terrestrial ground has, then how we come to know that ground is also of considerable importance. Knowledge of landscapes and methods available for creating them can vary and has changed significantly through time and cultures. Rapid development of new instrumentation to tap into different worlds makes it more and more difficult to separate what we experience from what we passively observe. That said, our bodies are still the most complex synthetic sensing machine we have, able to process inputs from multiple sensing organs in real time, and to decipher meaning from them. Science for the most part is still interested in pulling things apart to figure them out; yet we sensing beings must put things together to experience them. This does not mean all forms of environmental sensing, remote or otherwise, have no place in establishing knowledge; it simply means that relying on them entirely, or with priority over actual experience, can lead to forms of practice devoid of multi-sensual understanding. It is exciting to think of the possibilities as our senses are expanded with the help of new sensing technologies, but we should understand them as extensions or augmentations to experience instead of replacements of it. For this reason, direct experience remains a core component of the 501 studio, both in terms of inquiry and projection.

The studio attempts to clarify two key points. The first is an understanding of landscapes as fabricated places and experiences. The second is that landscapes are enmeshed in an unending process of change. Understanding landscapes as fabricated experiences refers to the idea that what we know of places is the product of a range of inputs, including what we have been told, what we have read and what we have personally experienced. Our place-based knowledge is constructed, not simply discovered. For example, your understanding and experience of a familiar public landscape would be changed considerably if someone told you it was constructed atop a graveyard – whether the information was true or not. Students in the studio take brave liberties in (re)describing the experience of place in ways that highlight features that may otherwise go unseen (Figure 18.1). Knowledge creates places, but places also have the ability to create knowledge. David Livingstone describes this in *Putting Science in Its Place*, observing that a contextual understanding of all knowledge is deemed essential, and there is continual dialogue back and forth between place and knowledge. It is important to communicate to students the need to discover ways to be curious about the knowledge we have, and to prompt them to continually test it with methods from outside the system that created that knowledge. This is theoretically impossible, but it does provide a framework for action; one that constantly questions how we know what we know.[2] The goal is to open up possibilities for new informants to thicken our understanding of place, as opposed to simply solidifying knowledge we assume is correct. This should sound ethically suspect because it states we should constantly question science, the exact system we fight to uphold in a time of anti-science propaganda which is dismissing the reality of climate change (among other things). However, this same Western science was used to whitewash cultures of any belief systems that ran counter to it, in what stands as nothing less than intellectual genocide. There is no clean line to walk, but we must walk anyway, acknowledging the imperfections, cruelty and contradictions our pasts have created, while striving for a more just future for all.

This process of striving is also unending because cultures and ecosystems are always changing. It is impossible to ever fully understand what a place "is" because it is never that for long. New informants always have a role in reshaping

FIGURE 18.1   Top: Field Crosses. Inspired by photographs by performance artist Vito Acconci, these photographs were taken along five points of a transect through the site. For each set, photographs were taken in the same direction from four points of view: arms extended overhead, extended right arm, extended left arm and resting on hip. These photographs express a new relationship between physical body and the site. Lizzy Servito. Middle: Towers of Roxborough. This photowork was a study of views of the TV and radio towers located around the site. The site was seen as unique for the large number of such towers, as it sits at a high elevation in relation to the surrounding city. While some towers can be seen on the horizon, others are placed immediately next to the site and due to the large number of roosting birds and climbing vegetation, provide a though-provoking pairing of technology and nature. Alice Bell. Bottom: Walk. This collection of video stills documents the walking through the site's tall grasses, where sun, shadow and brambles intermix. Larger images of the grasses convey the immersion the body feels within the rough conditions of the environment. Lizzy Servito.

our understanding of place, as long as we are willing to make space for them. For this reason, the 501 studio focuses heavily on developing methods of inquiry that explore the changing nature of places, attempting to bring attention to systems that may be overlooked or marginalised by more standard methods of analysis, and make space for them to self-actualise.

As a way of responding to the conditions described here – the fabricated, fluctuating reality of landscape – Karen Lutsky and I developed what we call "Curious Methods" (Lutsky and Burkholder, 2017) as a case for enacted analytical practices guided by nothing more than curiosity. These methods probe relationships between contextual actors in ways that question our understandings of them – they populate a plurality of site-based knowledge with the objective of growing that plurality, as opposed to reducing it as science typically does. Places are thick and should be made ever thicker as a way to protect them from categorisation, classification, capitalisation and colonisation. Exploring the theory of time has served as a productive way of initiating this process in the classroom. Time, while slippery to define, provides some thematic stability when discussing issues such as place, change and other beings. The discussion of different "times" that come together in places is highly formative, both in a design sense and an intellectual sense; rock-time, tree-time, dragonfly-time, are all completely different enterprises. They crash together in physical space and are coloured by our seemingly objective human-time – an understanding of time that is *anything but* objective. Exploring the implications of these times or durations, and the changing environments they create or represent, provides a highly productive platform from which to engage other subject matter of the discipline, in addition to preconceptions design students may bring with them.

In teaching, the concept of time or "times" is used as a way of unpacking contextual actors and their agency. Students are asked to identify a series of multiple "durations"[3] on their given site and are tasked with finding ways of describing and communicating them. The most straightforward of these include the effects of site phenomena, such as weather, which students attempt to trace and track. In *The Natural Contract* Michel Serres reminds us that the French word for weather, *temps*, is the same word used to describe time that passes (Serres, 1995: 27). It should come as no surprise that many explorations students undertake focus on climactic agents (Figure 18.2). Moving deeper into the concept, students are encouraged to consider the larger range of durations in a particular place, thinking about both living and non-living beings (Figure 18.3, top). Texts to support this include *Timefulness* by Marcia Bjornerud, who pushes for a stronger understanding and application of geologic time. Texts by Valerie Trouet and Jakob von Uexküll help describe durations of trees and other animals respectively and open up opportunities for considering their experiences of time as formative to the understanding of place. Mark Rifkin's (2017) position in *Settler Time* provides a strong counterpoint to the idea there is a shared conception of time across individuals or cultures; in his critique, the assumed central focus on a Western and white idea of time has served as a powerful force of colonisation. And perhaps

FIGURE 18.2 Top: Wind Clock. For this drawing, wind was chosen to index the passage of time. The clock was constructed of a wooden bar and string and placed on a high branch. Photos were taken from below every five minutes for one hour. In the images, the length of the sting represents wind velocity, and the wooden bar shows direction. Youzi Olivia Xu. Middle: Wind Duration. Designed as an alternative wind-measuring system, photographs of a balloon were taken each second over the course of an hour along a path. Each image contains 60 still images of the balloon overlaid to represent 1 minute. As a collection they document the ranging wind conditions at very small intervals over time. Lizzy Servito. Bottom: Rainy Day. This image hints at the ephemeral states at several site locations on a rainy day. Paper and ink placed on the ground can index amount of rain, wind direction, ground surface moisture and tree canopy in a synthetic way. Keling Ni.

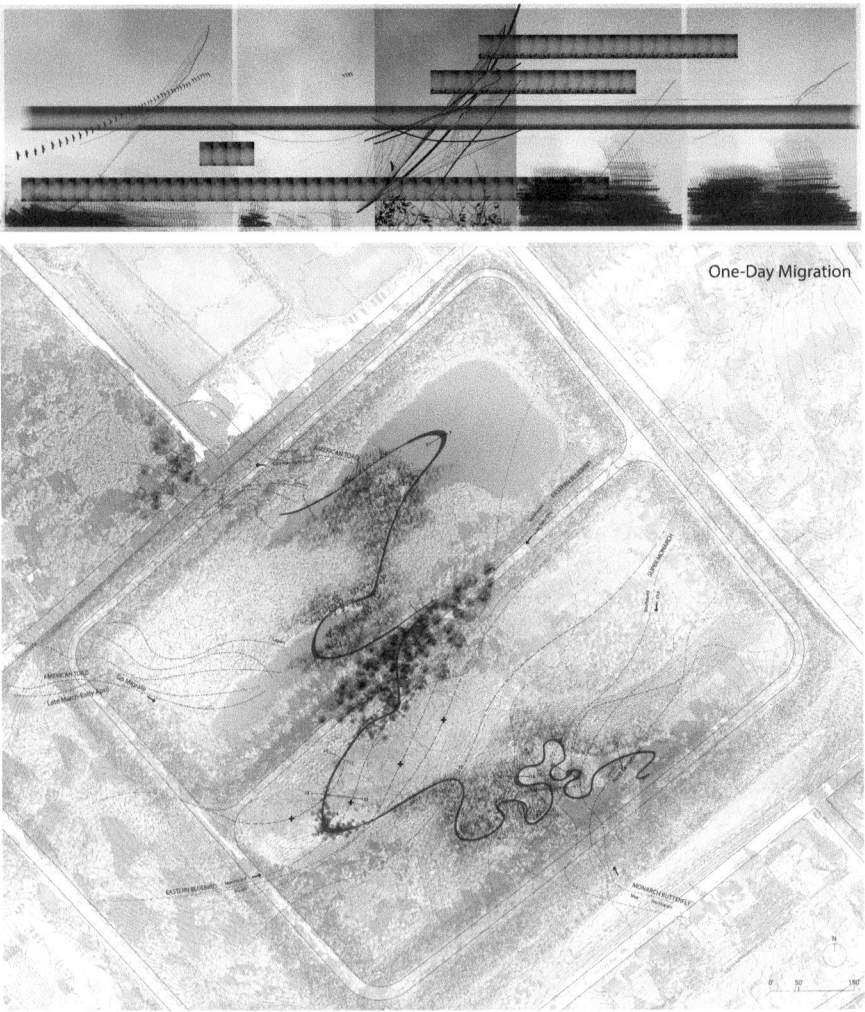

FIGURE 18.3   Top: Human and Non-Human Flights. The image is a spatial-temporal examination from multiple perspectives of the relative relationship between animals and humans, the living creature (birds) and non-living creature (the airplane), observer and the observed. It consists of a series of overlays of original photographs capturing bird and aircraft in flight, overlaid photographs sequenced by time, and abstractions of bird and aircraft flight paths. Jing Cao. Bottom: Reservoir Site Plan. This site plan for the project shows the path designed for visitors to experience the habitat of the American toad, American bluebird and monarch butterfly. Both vegetation and experiential intensity are represented on the plan, as visitors "migrate" across the site and its transitional habitats. Fan Wu.

most pointedly put, when asked about the relationship between time and labour, Philadelphia-based Afrofuturist stated, "Time is used to punish people; time is used to oppress people; time is used to create severe inequalities" (Owens, 2021). With these potentials (both good and bad) of time in mind, students advocate for a plurality of durations through the generation of a series of exploratory drawings attempting to describe these different durations in ways recognisable and processable to us as humans, with our own particular timeframe.[4]

The work included here represents a very small collection of site-based experiments tasked with uncovering or liberating a range of times present on a particular site. These experiments later inform the creation of what may be considered a more traditional site design, focused on a simple path with contemplative areas along it (Figure 18.3, bottom). Students develop an understanding of the times or durations that compose a place through collecting materials, tracking beings, indexing phenomena and so on. The work becomes an act of probing, not proving.[5]

Notes

1 The statement also recalls Naomi Klein's book *Shock Doctrine* – essentially, using crisis-induced problem-solving to justify actions that further marginalise or disenfranchise groups or beings with long histories of mistreatment.

2 By "impossible" I mean that the ability to test the validity of a system from within that system is notoriously difficult. The most straightforward example of how this is done might be Thomas Kuhn's idea of the *paradigm*, spelled out in his seminal work *The Structure of Scientific Revolutions* (1996). Imre Lakatos's (1970) conception of the *Research Program* could also be instructive here: it describes what is necessary to shift from one way of thinking to another by way of analytical or inquisitive means.

3 Here I draw from Henri Bergson's idea of *duration*, as described in *Time and Free Will*.

4 There are of course many more examples that could be added here, many of which are also presented to the students, including Robert Smithson, On Kawara, Donna Haraway and Carlo Rovelli,

5 This idea, taken from Gregory Bateson's (1988) *Mind and Nature*, was foregrounded in the "Curious Methods" article.

References

Bateson, G. (1988). *Mind and nature: A necessary unity* (6th ed.). Bantam Press.

Bjornerud, M. (2018). *Timefulness: How thinking like a geologist can help save the world.* Princeton University Press.

Holmes, R. (2020, July). The problem with solutions. *Places Journal.* https://doi.org/10.22269/200714.

Kuhn, T. (1996). *The structure of scientific revolutions* (3rd ed.). University of Chicago Press.

Lakatos, I. (1970). Falsification and the methodology of scientific research programmes. In I. Lakatos & A. Musgrave (Eds.), *Criticism and the growth of knowledge.* Cambridge University Press.

Latour, B. (2018). *Down to earth: Politics in the new climatic regime.* Polity Press.

Livingstone, D. N. (2003). *Putting science in its place: Geographies of scientific knowledge*. University of Chicago Press.

Lutsky, K. O., & Burkholder, S. L. (2017, May). Curious methods. *Places Journal*. https://doi.org/10.22269/170523.

McAvin, M., Meyer, E. K., Corner, J., Shirvani, K., Helphand, K., Riley, R. B., & Scarfo, R. (1991). Landscape architecture and critical inquiry. *Landscape Journal, 10*(2), 155–172.

Owens, C. (2021). *Are you thinking about time right now? For 'Black Futures' contributor Rasheedah Phillips, it's a lifelong pursuit*. https://www.Inquirer.com. 9, 2021, from https://www.inquirer.com/news/black-futures-book-afrofuturist-rasheedah-phillips-philadelphia-20210101.html.

Rifkin, M. (2017). *Beyond settler time: Temporal sovereignty and Indigenous self-determination*. Duke University Press.

Serres, M. (1995). *The natural contact*. University of Michigan Press.

Stengers, I. (2018). *Another science is possible: A manifesto for slow science*. Polity Press.

Whyte, K. P. (2020, September 14) [Plenary Presentations] *2020 Virtual National Tribal and Indigenous Climate Conference*.

Bibliography

Bergson, H., Ansell-Pearson, K., & McMahon, M. (2014). *Henri Bergson: Key writings*. Bloomsbury.

Klein, N. (2008). *Shock doctrine: The rise of disaster capitalism*. Picador Press.

Trouet, V. (2020). *Tree story: The history of the world written in rings*. Johns Hopkins University Press.

von Uexküll, J. (2010). *A foray into the worlds of animals and humans: With a theory of meaning*. University of Minnesota Press.

19

IN SITU/EX SITU

Geometries of Density and Spectra

James Melsom

The site-oriented research work, workshops and studio teaching discussed here borrow from the phrase "landscape as laboratory" (Fraguada and Melsom, 2018), emphasising the ability of the site to be the loci of both research activity and the generation of research questions. While not a new concept, the work also emphasises the implications of the use of the tool or apparatus on the site as a fundamental method. The relationship of a site to its abstract data sources, preparation for fieldwork and ritual onsite and manner in which the student interacts with the site, are fundamental. This relationship was developed by the author in the research office, LANDSKIP, in collaboration with Ilmar Hurkxkens. It emerged from common work, and from our diverging application of those principles (Hurkxkens, 2020), between the Alps of Switzerland and the coastal regions of Australia. The particular thread of research discussed here focuses on aspects of interaction with the phenomena of site and its representative data, in relation to specific principles from 20th-century approaches to ecology.

The 1954 book, *The Distribution and Abundance of Animals*, presented an attempt at a description of animal ecology that integrated a comprehensive understanding of all aspects of the interaction of non-vegetal organisms and their environment. The book charted the ecology of animal-based occurrence and was one of the first comprehensive studies of its kind. While the research referenced sites around the world, the authors noted they were inspired by the Australian context and environment, especially its extremes of climate and biodiversity (Andrewartha and Birch, 1954), and relatively recent incursion of anthropic disturbance, however marked.

Importantly, it also sought to identify links between the environment and reasons for fluctuations in animal numbers. An environment with diversities that stretch between alpine and coastal, desert and tropical, temperate and arid,

DOI: 10.4324/9781003145905-23

as well as its predilection to seasonal weather and climate extremes, made it the perfect research crucible; albeit one with often extreme reactions in local variations of local species. Researcher Dr J Davidson, to whom the book was dedicated, was quoted as seeing Australia as the perfect location in which to carry out this research. In this vein, it is potential for investigation in extreme environments that influenced the pedagogical framework of this particular form of site approach. Davidson's "Bioclimatic Zones of Australia" map, published posthumously in 1954 by his fellow researchers, differs greatly from current bioclimatic maps, this difference reasserting both the local tendency to extremes and the need for continuous reassessment of site.

The attempt to link ecologies to the occurrence of animals to multiple types of environmental influence and interaction challenged many previous notions of the word "ecology" as discrete, or continuous, and rather as a series of ecologies that overlap and interact. Three principles were outlined in the manner of ecological study, each of which can inform our methods of approaching a site: that some experiments can only be made in the field; that as complete a survey as possible should always be generated, with physiography, climate, soil, vegetation and living organisms included; and that the metrics, careful crafting of site experiments and modes of site understanding are conducted with awareness of the imperfect nature of data collection, avoidance of false equivalency and general distrust of the exact numbers, relying rather on tendencies and patterns (Andrewartha and Birch, 1954).

These principles were not the genus of the studio site approach outlined here. However, they help to reinforce the foci and pedagogical reference for students undertaking fieldwork. Allowing them to work both in an analytical and phenomenological mindset and differentiate local and distant approaches to site.

The complementary roles of onsite experiments and those works conducted in the remote laboratory are referred to henceforth as *in situ* and *ex situ* fieldwork. In situ fieldwork can be planned, executed and documented in the same manner as scientific research, both with carefully calibrated and ad hoc site enquiry techniques, each revealing different qualitative site readings. Ex situ site work can allow the designer to work across scales, datasets and logics, from the territorial to the chemical scale of the site. The physical limitations of in situ work mean resources usually require careful ex situ work to plan and define the scope and logistics of the fieldwork, and the relative abstraction of ex situ observations rely on the considered verification and codification, on site.

This approach to site is based on working professionally and in pedagogical modes throughout the world, and certainly influenced by the approach of Girot in Switzerland, among others. An iterative pedagogical mode of site input/output, or exchange can be installed, in which the designer can allow spatial data to inform the site work, with fieldwork as a generator of entirely new spatial data for the site. This cyclical process feeds on the dichotomy of the gaze – within the site and outside it – an iterative loop that reveals thematics and holes in the datasets and separates the phenomenology of site from the extrasensory characteristics of place (Girot, 1999). The potential exists for treating each discrete research site as a focus for scrutiny with its own data and metrics. The further scope of site interaction lies in developing site-oriented tools which will reveal its specific characteristics and potential, a term which here does not imply intervention, yet acknowledges inevitable site transformation.

The focus on Australian coastal sites for this studio research revealed contrasting ecologies and anthropic influences. The first impression of site, like an anamorphic image, is difficult to unsee once seen. It is the thesis of this site approach, however, that it is possible for an act, an object, a tool, to activate another view of site. The strength of such tools is in calibration, to allow the viewer to unsee, or re-evaluate, the site through shifting metrics and lenses. A taxonomy of tools, historical and modern, and invented for specific sites, reveals that sophisticated contemporary tools subtly distance the user from specific site phenomena and the fundamental act of measurement.

Saussure's Cyanometer from 1788 was a tool developed to determine the shades of blue of the sky from the Alps. It revealed the perceptual information particular to the site, and the water content of the air, as well as its shifts according to altitude (Fischer, 2015). Like 20th-century tools such as the soil texture triangle and soil colour indexes, they are tools that codify the data they represent; each creates a geometry or logic system for the specific data it describes. A complex combination of elements can be used to describe a wide variety of site phenomena, as demonstrated by comparing the recommended site kit of the national Swiss forest index (Düggelin et al., 2020), with the author's site kit from the August 2020 coastal site visit north of Newcastle. However, increasing complexity and sophistication of tools, from left to right, is not necessarily a positive development in pedagogic or immersive site process. Much of the work of the studio is about bringing the complex and refined measurements and site data back to the clarity and simplicity of the tools on the left of the scale.

Drones that steer themselves, cameras that expose colour/brightness automatically, videos that remove shake, microphones that remove background noise – each distances the observer from the specific site processes and conditions within which they are immersed. Each tool is simply an extension of the observer, severed. The resulting data and documents – rarefied, instantaneous and optimised – are stripped of the feedback processes of their creation.

This design research charts the pedagogic techniques, tools, and applications of fieldwork within the realities and data footprints of the realities of sites and their digital representative datasets. It focuses on the specific documents of field enquiry within a studio context and their implications, rather than the resulting design outcomes. It is possible both to harness the full technological capabilities of UAV platforms, while applying and transforming the data into a format as simple and linked to site as Saussure's Cyanometer.

Coastal Sites

Australian coastal tributaries that switch from fresh to brackish and then to saline water, or a site that experiences flux from sub-zero temperatures to drought and bushfires within months, best exemplify these extremes. Consider varied applications where a line is traced through the fieldwork within separate sites along the New South Wales coast: Tallow Beach and Rosedale, each sites of design research in 2020. Each was coordinated in close collaboration with Dr Penny Allan, and the research focused specifically on the site-based approach, data collection, the author's manipulation techniques, and how these are best exemplified in varied approaches of the resulting studio work. The specifics of each approach, and the relationship to data, were elaborated through the data and design documents created in this exchange. They began with the role of the field leader, or coordinator, deploying students onsite with specific foci, based on geographical taxonomies, rather than thematic groups. Therefore, the pragmatic landscape transect, especially in linear coastal sites, is both an ideal starting frame and neutral constant between sites.

The illusory nature of a comprehensive or complete image of site is well transferred to students through the analogy of the early exploration of Australia. Figure 19.1 (left) is the drawing of the expedition through the Western Australian desert by Dempster, Clarkson and Harper. The image shows a thin transect of the Western Australian desert, illustrating a short visual distance from the path travelled; on second glance it is entirely written in German, published 12 months after the original expedition, to great international scrutiny. The arduous journey was of great interest around the world, due in no small part to the extreme and alien landscape conditions. Such maps of desert exploration are directly analogous to modern UAV flights over a site. They give the illusion of overview, but document only a thin sliver or transect of the terrain over which they pass, with restricted operation, countless occlusions from vegetation and

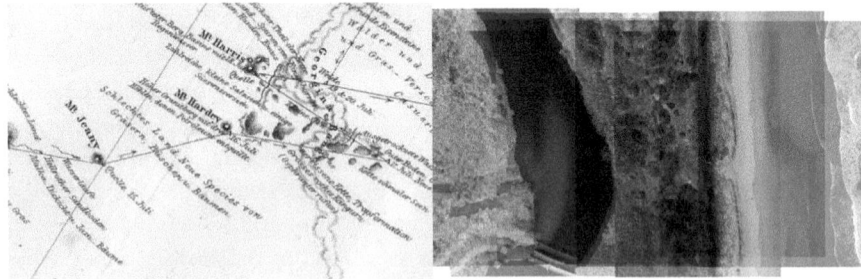

FIGURE 19.1 Tallows – Left: Historical and drone path exploration map Western Australia, 1868, original in collection of the author. Dempster, Clarkson and Harper. Right: UAV photographs over Tallow Beach, December 2019. James Melsom.

FIGURE 19.2   Tallows section analysis and tallow outputs. Top: Jeremy Chivas. Bottom: James Melsom.

structures, each constrained by the realities of fuel, weather extremes and the tyranny of the distance.

The images, taken on site on Tallow Beach, New South Wales (Figure 19.2), give the impression of overview yet have a similar, limited scope; the closer they come to overview, the lower the resolution, the paradox of scale. The transect attempts to document the transition of beach to dunes, Intermittently Closed and Open Lakes and Lagoons (ICOLL) system and dense coastal vegetation. Extensively mined for silica and other minerals in the 1940s, the dunes were entirely reconstructed. This has resulted in a complex system of care and conversion as invasive species are gradually culled and local vegetation replanted (Figure 19.3). The area is managed principally by the local Arakwal indigenous authority, in partnership with the Shire of Byron; this section illustrates several zones of care and jurisdiction, from Shire and New South Wales fisheries to Arakwal National Park.

The illustrated transect demonstrates various modes of working onsite, from conventional, yet standardised transverse photographs of the dune system, to a UAV-based photogrammetry scan, and calculation of the vegetation index (NVDI) and plant health from the infrared spectrum. The combination of these techniques can be extrapolated ex situ to determine species type, distribution, densities and health. When combined with the local bore water readings for the

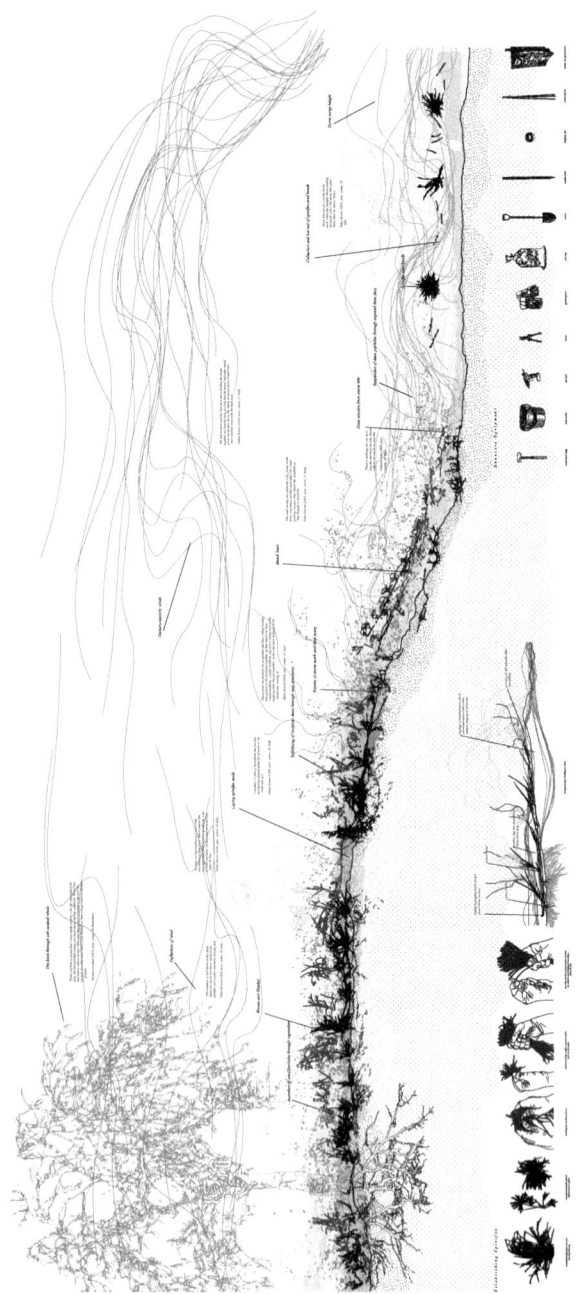

FIGURE 19.3 Rosedale dune fluctuations – geometries of wind and vegetation. Nathan Galluzzo.

depth and salinity of the water table and Tallow Creek, a detailed and nuanced image can be generated of both the ecologies and factors contributing to gradual transformation of the dune system and the interventions into its water system. The same bore water data and drilling information accessed on this site has been extrapolated by the authorities to construct one of the world's largest seamless 3D geological models, stretching over the entirety of New South Wales. The model had been reconstructed from elementary physical data sources such as historical bore water drill sites, whether monitoring, agricultural, residential or exploratory, and the Tallow site was later documented at the territorial scale (Melsom, 2020).

The resulting in situ/ex situ extrapolation of data and site phenomena, above and below the ground, facilitate a view of the site with literal "depth". The resulting design research works to represent in one document the complex ecological processes and practices of care over the site, integrating exclusive interview and photographic documentation from local dune care archives. It extrapolates the role of anthropic maintenance and intervention required for some time in the future, as the dunes gradually retreat due to storm surge events of increasing frequency and severity.

The grain of information in such documents relies on a synthesis of UAV survey data, detailed analysis and extrapolation of the species identification and distribution, and careful reconstruction of local microclimatic response and techniques of cultural intervention. Such systems, going through massive shifts through the seasons of sand movement, cycles of deposition and erosion, vegetal flux and dense human use, require iterative cycles of in situ and ex situ analysis to map, understand and project the future cultures of care necessary to balance the various factors testing these sites.

The second coastal site of focus is Rosedale, New South Wales, the site of the 2020 New Year's Eve fire event that devastated the community and the landscape. The design studio, run with Penny Allan, was a complex confluence of approaches, with detailed reconstruction of the event, and developing various approaches of living with fire (Allan and Melsom, 2020). The projects elaborated here demonstrate a specific reaction to in situ data, collected in March 2020, and painstaking ex situ reconstruction, projection and simulation.

The exceptional circumstances of the 2020 studio environment meant repeated visits to the site were not possible, increasing reliance on data collected during the first site visit. The projects took full advantage of the high-resolution aerial Light Detection and Ranging (LIDAR) point cloud data from the New South Wales government, and a UAV photogrammetry scan of the site, post-fire event.

Enquiry into UAV data need not end with directly recorded data. The log files of the UAV itself can become a mine of data, as fluctuations in the motors of the UAV can be dissected later to reveal the air currents buffeting the forces acting upon it (Figure 19.4). This reveals the region's characteristic wind speeds, directions and micro fluctuations, as well as microclimatic influences, whether obstacles,

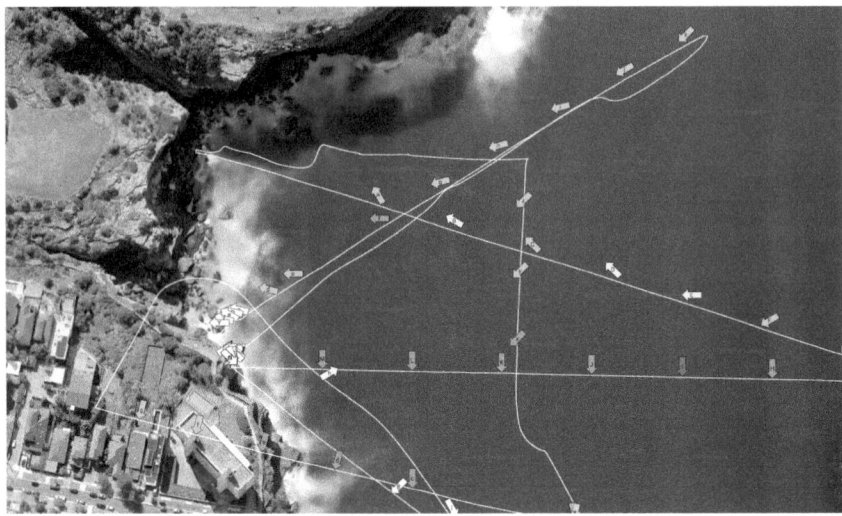

FIGURE 19.4   Wind data extracted from UAV sensor fluctuations off the New South Wales coast. James Melsom.

vegetation or shifts in air pressure. An excellent example is vertical winds. These are characteristically difficult to map, given the dominance of the European concept of cardinal directions, and the ability of UAV stabilisation systems to automatically compensate and effectively render such effects invisible. Especially noticeable in steep alpine terrains and along the stark coastal geologies of New South Wales, the dissection of the UAV's constant struggle to resist uncontrolled movement plainly maps these otherwise invisible forces. Elaborating these local effects to students has the potential to heighten their awareness of local microclimatic differences and broaden their awareness of site beyond the visual spectrum.

Drones should best be thought of as sensory rather than flying visual platforms. All the sensors designed to automate and streamline their function also act upon and log the platform's resistance to the forces affecting it. The nature of this friction or resistance within its environment is analogous to the manner in which ships charted their speed through the counting of knots on rope, the relevant physical processes need only be harnessed, and their metrics translated. It is however possible to reconnect and even amplify these lost signals and create entirely new datasets and site legibility.

The larger-scale schematic mapping of wind movements directly influenced an investigation into the relationship of fire spread rate, intensity and ember creation. Working directly with a former Rural Fire Service specialist, high-resolution canopy data was manipulated to generate alternate land management and forestry practices, a mosaic of territorial canopy textures that directly influence and incorporate the relationship of terrain slope, orientation and wind (Figure 19.5).

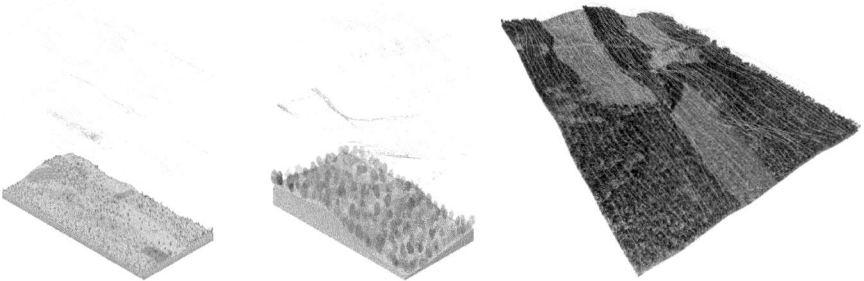

FIGURE 19.5 Rosedale wind movements and fire segmentation tests. Freya Cameron.

The point cloud colour is codified according to colour based on exposed ground, after a technique developed in collaboration with Luis Fraguada at the National University of Singapore. In this application, soil erosion and degradation are among the most substantive issues facing the landscape. As documented in the report on the 2019–20 fire season, perhaps the most worrying impact on biodiversity is the destruction and loss of topsoil due to the extent and intensity of the fires, and following erosion as identified by the Department of Planning and Environment (DPIE) New South Wales, in 2020.

Detailed reconstruction of housing sites from point cloud data lead to a project focused on the archaeology of fire, in which previous ruins become elements within the construction of communal habitation assemblages (Figure 19.6). It was only through comparing point cloud datasets that the true impact of the fires could be read in the landscape. When masked for anthropic elements, and combined with spectrum analysis techniques, the difference in biological diversity can be read in the imagery. The resulting reconfiguration of the site can be seen

FIGURE 19.6 Rosedale pointcloud fragments – destruction and reconstruction. William McRoberts.

to mediate lost canopy density within construction elements, the ground plane opened for revegetation, habitat and recontextualisation of the ruined fragments.

The current research into the extent of fire damage, and rate of recovery, brings the research full circle to geometries of site colour and texture, with 3D stratification of the colour spectrum. Photographs taken from Rosedale in areas affected, and avoiding damage, were dissected for their spectrum and histograms and turned into geometries of colour distribution and abundance.

Despite the often-intense colour apparent in epicormic shooting and regrowth, the comparative lack of diversity of colour hues is perhaps the largest indicator of unbalanced recovery in this landscape. The events are often a chance for non-endemic species to spread into the space vacated by fire. The resulting selective 3D colour profiles of the drone imagery reveal loss of spectra, limited diversity of the epicormic shoots and exposure of soil tones and open ground (Figure 19.7). The volume of 3D colour profiles, and their comparison and overlay, reveals which areas of the spectrum are missing, and can be isolated to specific species and ground cover.

Combining NDVI imagery spectrum data with the vegetal colour range, as with the Tallow site scans, allows the potential for more accurate dissection of species, vegetation health and coverage on the ground. Through this process, and when combined with extra-visual data, new metrics of health, range of variation, dryness, plant stress and diversity can be measured. Specific biotopes and plant communities can be isolated within dense vegetation with this method.

The stark contrasts of these environments lend themselves well to developing tools and methods. Even for the most banal of sites we can perhaps envision a tool to best encapsulate that landscape – the Sandometer of the Tallow Dunes, or the Carbonometer of specific stands of species affected by fire, and their underlying soils – to understand the extent and intensity of damage and recovery strategies.

Site data published in 2020 from the alpine forests of Switzerland confirm this shift, with forests recently prone to drought and explosions in insect numbers (Brändli et al., 2020). Published as part of the same study that outlines

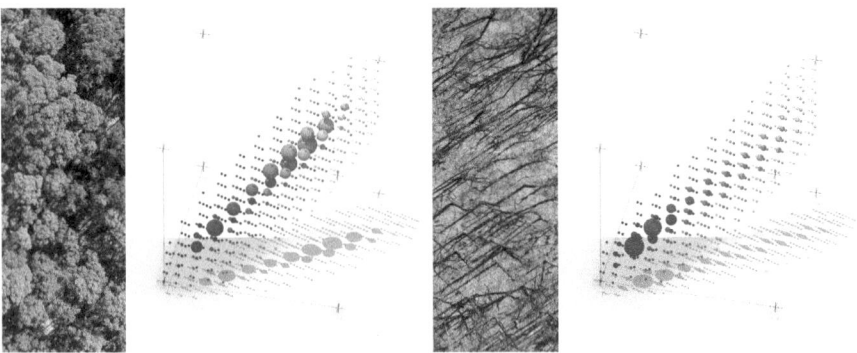

FIGURE 19.7   Pre- and post-fire spatial colour. James Melsom.

site equipment and methods applied by LANDSKIP in the field, the document reinforces the rapid change in the environment, both in empirical data and in the visible spectrum as measured both through data analysis and on foot, painstakingly measuring the distribution and abundance of Switzerland's forest substance. Similar to the data contained in Davidson's 1954 "Bioclimatic Zones of Australia", a constantly shifting image of ecology has replaced the stable models proposed with such confidence earlier in the 20th century. These accelerating shifts amplify the importance of site-based work and incremental observation.

Rather than treating this change as pathology, there is a chance to work in a constructive manner with knowledge of shifts in these landscape sites, strengthening emerging synergies of habitation and ecology. With increasing variability in climate, weather and the environment, the world is shifting to a mode of constant observation, of difference and measured abnormality. Even as ecologies shift, the abundance and distribution of all ecological material will be in a state of constant flux, and the relationship between historical data and onsite observation and data generation even more essential. Rather than empirical numerical data, perhaps the visual and textural, and shifts in densities and movement, hold the key to understanding, and working within, the shifting parameters of the site.

References

Allan, P., & Melsom, J. (2020). Living with fire. *Kerb Journal, 28*, Department of Architecture and Urban Design, RMIT University, Melbourne.

Andrewartha, H. G., & Birch, L. C. (1954). *The distribution and abundance of animals.* University of Chicago Press.

Brändli, U., Abegg, M., & Allgaier Leuch, B. (2020). *Schweizerisches Landesforstinventar Ergebnisse der vierten Erhebung 2009–2017.* Eidg. Forschungsanstalt für Wald, Schnee und Landschaft WSL.

Düggelin, C., Abegg, M., Bischof, S., Brändli, U. B., Cioldi, F., Fischer, C., & Meile, R. (2020). *Schweizerisches Landesforstinventar. Anleitung für die Feldaufnahmen der fünften Erhebung 2018–2026.* WSL Berichte: Vol. 90, Eidg. Forschungsanstalt für Wald, Schnee und Landschaft WSL.

Fischer, S. (2015). *Horace-Bénédicte de Saussure Natur- und Alpenforscher.* SJW Verlag.

Hurkxkens, I. (2020). *Robotic landscapes: Topological approaches to terrain, design, and fabrication.* Doctoral Thesis. Institute of Landscape Architecture, Department of Architecture, ETH Zurich.

Melsom, J. (2020). Multi-scalar geo-landscape models: Interfacing geological models with landscape surface data. *Journal of Digital Landscape Architecture, 5* (59–71). Wichmann Verlag, Berlin.

Fraguada, L., & Melsom, J. (2018). Code matters: Consequent digital tool making. In Cantrell, B., & Mekies, A. (Eds.), *Codify: Parametric and computational design in landscape architecture* (pp. 225–239). Routledge UK.

Bibliography

Kullmann, K. (2018). The drone's eye: Applications and implications for landscape architecture. *Landscape Research, 43*(7), 906–921.

Swanson, R. L., Potts, J. D., & Scanes, P. R. (2017). *Legacies of a century of industrial pollution and its impact on the current condition of the lower hunter river estuary.* Office of Environment and Heritage.

Fry, B. J. (2004). *Computational information design.* Doctoral Thesis, Massachusetts Institute of Technology, School of Architecture and Planning, Program in Media Arts and Sciences.

Kellogg, C. E. (1936). *Development and significance of the great soil groups of the United States.* United States Department of Agriculture, Misc. Pub 229.

PART 4

Sensing Landscapes
Narratives

The vast everchanging landscape of Lake Eyre in Australia defies customary scales of measure. Recording equipment, tools, and techniques conventionally utilised to map and scale landscapes are rendered defunct here. Alternative ways of seeing are required, ways that define a new set of systems of measure that consequently inform a geographic paradigm shift.

From 1972 to1978, research geologist Dr John Dulhunty and his wife, Roma, returned each year to observe the transformation of this prodigious landscape. Together, they progressively documented the lake's geomorphological conditions, including its phenomenal and rare transformation from a dry salt lakebed through an intense period of flood to become the largest body of water on the continent, albeit a temporary one. This was an observation of the largest flood event in approximately 500 years (Carman-Brown, 2015).

The Great Artesian Basin that feeds Kati Thanda is the largest artesian basin in the world, covering 22 per cent of the Australian continent. It is itself a registration of ancient seismic shifts and sediment accumulation that coincided with changes in the earth's magnetic field. Once the edge of the Gulf of Carpentaria, this water body divided the driest continent on earth is a remnant of an old oceanic plate. In its depths, it records the shifts in ground from the past – through fluvial, aeolian, and lacustrine sediment deposits – to the present.

This vast landscape began forming more than 200 million years ago through major climatic oscillations that occurred during the planet's Quaternary Period, the third and final phase of the Cenozoic Era. The Quaternary Period, significant for its glacial growth and retreat, and for the extinction of large mammals and bird species, saw a spread of the human species across the planet. Today, this arid and semi-arid, unregulated, dryland river system supports threatened and endemic waterbirds and aquatic species. Its ecological systems respond to starkly divergent water flow regimes throughout the Great Artesian Basin, ranging from

DOI: 10.4324/9781003145905-24

extreme to limited periods of flow, or even none at all. The cycles of boom and bust are punctuated by small and medium floods. Dulhunty's discoveries identify the lake as a 'rain gauge' (Carman-Brown, 2015) for the continent of Australia. Through its palimpsest of geological traces, this landscape captures and discloses hundreds of thousands of years of climatic history in its sedimentary layers.

As a natural meteorological device, Kati Thanda is disorienting; it throws into question usual distinctions between ground and water, human and non-human, up and down, as the infinite horizon blurs the domains of ground and sky. A shimmering of light on the lake's salt-crusted surface reveals an array of delicate, moist pink crystals slowly transforming, shifting, and gathering in the heat of the midday sun. This semi-translucent surface, with vertical undulations of only a few centimetres measured across vast horizontal distances, holds life in its material; dormant fish species and plant life, in stasis in this thin salt layer, wait to awaken. Artesian springs and aquifers bubble away from underneath, expressing their full capacity when the salt crust is upturned by spontaneous downpours flowing from north to south across this vast continent. A riot of sensation (Massumi, 2002) ensues – at the arrival of the floodwaters, multitudes of microbial transformations spawn and hatch fish, vast numbers of birds migrate to the area, and desert flowers spontaneously awake from dormancy – and these bursts of colour flow across the vibrant gradient of pinks and blues.

A fine-tuned sensing of matter, time, and the various life forms supported by the irregular emergence of the ephemeral lake is required to understand what imbues this dynamic formation and transformation. This is a landscape defined by its processes and operations of change and multitude of scales, rather than only figure and form. Here the visually striking surface of the salt plain is an expression, or sign, projecting simultaneously into the future and back into its past (Massumi, 2002). Formed by sub-surface systems of hydrology and geology – in concert with shifts in near-invisible flows of historic magnetic fields, wind and light – the landscape exemplifies the condition of the thickened ground. Mapping and constructing a figure of the ground is a process of uncovering and reading in ways that are an extension of the landscape, its structures and material processes. This landscape and subsequent earth are understood as a network of ecosystems (Margulis, 1998) in a perpetual state of service, from the microbial to the planetary, formulating the indeterminable map as the ground continues to shift.

Definitions

Sensing Landscapes describes landscape through its perpetual formations. Time is infused recursively in the material reality of the landscape through states of formation, from those that signify stability, to sequences that are predictable and observable processes of change, to those that are uncertain and instantaneous.

Massumi suggests that our own 'human' sensing of the world experienced through sensation involves a 'backward referral in time'. Therefore, a sensation is organised recursively prior to being part of our conscious chain or actions and reactions. In this process, the smoothing over of the anomaly is made to fit our conscious requirements of continuity and linear causality (Massumi 2002).

The sensed landscape is a complex system of relationships. It is composed of operational parameters and their resultant physical expression as a modulator of form that is spatial, material, and temporal. It is intrinsically difficult to fully comprehend due to its dependencies, competitions, relationships, and other complex interactions between its parts, and between a given landscape system and its environment. It expresses distinct properties that arise from these relationships as characteristics and tendencies. It also holds the potential for nonlinearity, emergence, spontaneous order, adaptation, and feedback loops, among others, that inform continual change. The landscape is formed as a result of all of these things interacting with one another to describe its 'structure'. The structure of the formation describes the motor of its matter that shape its potential future 'development' through change.

The reality of perpetual formation made visible and measurable through the Sensed Landscape describes an evolutionary model and behaviour of an ecosystem as the smallest measurable *unit able to recycle biologically important elements nested* within a network *of ecosystems* (Margulis, 1998). Parameters by which these units might be organised, defined, and subsequently evaluated typically fall under the umbrella of ecosystem services as those that purport to provide social benefits such as pollination, clean drinking water, and recreational opportunities (Ellis 2018); however, it is clear that the relationships formed between the Sensed Landscape, the information network, and the landscape architect holds the potential to shift this conventional paradigm. This includes a description of the work of nature as a hybrid labour, 'a collective, *distributed undertaking of humans and nonhumans acting to reproduce, regenerate, and renew a common world'. This brings nature into the political* economy, shifting *it from a resource to be managed and plundered, to a commons with* equal *rights through which 'it aims to call a more-than-human political collective into being, and to propose a relationship to nonhuman nature grounded in interdependence and solidarity rather than unidirectional management, ownership, or stewardship'* (*Battistoni* 2017).

The act of sensing and making the landscape is not a neutral activity, and therefore the process of representing forms a specific understanding of ecosystems and their processes. 'Actant is a term from semiotics covering both humans and nonhumans; an actor is any entity that modifies another entity in a trial; of actors it can only be said that they act; their competence is deduced from their performances; the action in turn is always recorded in the course of a trial and by an experimental protocol, elementary or not' (Latour 2004). Tools for sensing the landscapes, and the techniques by which we deploy them, have their own constraints that translate and transform information. The representations we make are constructed from a set of instruments, codes, techniques, and a lineage of conventions. Consequently, the worlds they describe and project are derived only from those aspects of reality susceptible to those techniques. These acts of sensing and seeing the landscape can formulate a view of what already exists, and set conditions for new worlds to emerge.

Pedagogies

Sensing Landscapes is focused on the landscape being defined as ecosystem services that constitute the inherent species and system and their performative capacities to function (pollinate, clean the air, etc) and serve the planet (Battistoni, 2017). It questions the standard ways and scales by which the landscape is conventionally perceived to include the urban, the territorial, and the regional. It has the capacity to identify and describe the role and work that these systems do, and those which contribute to sustaining life for communities of human and non-human species, and for the planet as a whole. The Sensed Landscape is intrinsically ingrained in processes of reproduction, regeneration, and renewal, in which its structure and performative constraints are relational and grounded through its intrinsic values.

Pedagogically, Sensing Landscapes investigates the use of specific instruments and informational systems to understand and generate a multiplicity of landscape conditions, from sensing techniques at urban and territorial scales, to registrations of environmental conditions at regional and planetary extents. The contributors of this section demonstrate a suite of performative-based approaches that show the potential of sensitive time-based registers which link the hydrological, geological, infrastructural, political, and the social.

In this section, modes of simulation go beyond ideal or 'natural' flows and shift the perceived inert view offered by technologies and informational systems. The contributors reclaim the designer as an agent acting with performance-based approaches. At the same time, they enable creative interpretations which address rapid and unpredictable changes. This is achieved by exploring sensing as a responsive informational system that has uncertain but potentially predictive outcomes, each bringing into question the relationship between designer, informational systems, and network formation.

References

Battistoni, Alyssa (2017). Bringing in the work of nature: From natural capital to hybrid labor. *Political Theory* *45*(1).

Ellis, E. C. (2018). *Anthropocene: A very short introduction (very short introductions)* (p. 119). OUP. Kindle Edition.

Carman-Brown, Kylie. (2015, April 15). A tale of extremes. The People & Environment Blog, National Museum of Australia. https://pateblog.nma.gov.au/2015/04/07/a-tale-of-extremes/.

Latour, B. (2004). *Politics of nature: How to bring the sciences into democracy.* Harvard University Press.

Margulis, L. (1998). *Symbiotic planet: A new look at evolution.* Basic Books.

Massumi, B. (2002). *Parables for the virtual: Movement, affect, sensation.* Duke University Press.

Bibliography

Haraway, D. J. (2016). *Staying with the trouble: Making kin in the chthulucene.* Duke University Press.

20

COMPUTING WITH NATURE

Digital Design Methodologies across Scales

Pia Fricker

Reflecting on the Current Notion of Digital Landscape Architecture Education

Globally, there is a recognisable trend within the field of landscape architecture education towards ecologically oriented courses. Ecosystem services, and green and blue infrastructure, are predominating terms in a multitude of course descriptions, demonstrating the importance of researching answers to the increasingly complex challenges our planet is facing. On the other hand, the topic of computational design thinking beyond the classical focus on computerisation – showcased through the predominant integration of GIS-based tools and methods – remains hardly visible (Fricker et al., 2020).

Within academia, our time is marked by pressing, but still very diverse, discussions on the future direction of digital technologies with respect to accelerating environmental, political and societal challenges. The majority of these discussions are setting a singular focus, either towards technology or towards environmentally oriented themes – hence, a holistic and future-oriented discussion of both fields in relation to design is missing.

The rapid development in the area of digitalisation, as well as the increasing necessity to deal with the effects of climate change, has had a fundamental impact on the field of landscape architecture. Landscape architecture education is greatly challenged to envision a meaningful and future-oriented integration of computational design thinking, which requires the melding of computation, design and theory as an answer to the complex challenges facing the profession of landscape architecture, today and in the future (Cantrell et al., 2018).

This discussion on situating computational design thinking as integral to curricula design took place in the field of architecture education in the 1990s. However, the field of landscape architecture is still reluctant to discuss

DOI: 10.4324/9781003145905-25

the potential of integrating this field into its course structure (Menges, 2011). We currently stand at the threshold of developing entirely new didactical and pedagogical concepts for teaching in the area of computational design thinking. These concepts go well beyond mainstream, application-oriented topics such as GIS, CAD, BIM/LIM and the mere teaching of tools and software (Fricker, 2016). Recognising the potentials for teaching and research specific to the demands of landscape architecture at the intersection of architecture, computer science and biology, requires fundamental rethinking and openness for a new area of knowledge, connecting the fields of robotics, AI, Big Data, human–machine interaction extended reality and design, to the notion of the place and its innate natural intelligence (NI) and meaning in a global context.

A Universe of Relationships

In the ongoing discussion, sustainability is often viewed as an urban problem. According to the current prognosis, the global population is expected to double by 2050 and even though urban centres occupy only 3 per cent of global land area, over half of greenhouse gas emissions are created in and by cities, where up to 80 per cent of all energy is consumed. This has led to a dramatic shift of focus and actions towards the city and the urban topic in a wider sense. The shift is evident in the advent of a huge number of novel concepts aimed at finding solutions for the problems we have caused ourselves, including a new way of designing cities enhanced through nature-based solutions (Haas et al., 2016).

This meaningful attempt to secure adaptive solutions, based on integrating nature-based strategies, can often be criticised as *greenifying* or *greenwashing* as these implemented green and/or blue infrastructures and diverse applications of ecosystem services are increasingly disconnected from the complex underlying dynamic system, and often focus on local optimisation strategies (Zari et al., 2020). This observed separate understanding of components is due to a lack of articulation, and the identification of intersection points of elements within an organisational structure.

This focused concentration on inserting performance-oriented landscape architecture strategies leads to a further division of the city and the green, and the abstract relationship of the city and the hinterland or countryside – understood as a perimeter for everything which is not the city and "where the radical changes are" (Koolhaas et al., 2020).

Today's challenges are more complex and more impactful. Cities and landscapes are designed to be more adaptive and resilient to deal with the diversity of unknown future challenges (Girot, 2017). Access to and understanding of available technology and data, has allowed us to integrate AI and Machine Learning as part of our profession, and to use them to analyse buildings and green space, user behaviour, maintenance and much more. Algorithms have been developed to uncover urban relationships and predict possible use cases. Deep-learning systems are available to generate designs with AI. We will have intelligent, self-constructing 3D printers which are able to construct onsite by adjusting the design according to real site parameters (climatic conditions, safety and so on). Robots are already integrated into project construction and maintenance at different scales, and traditional human-oriented workflows are enhanced through extended reality technologies (Halpern, 2017). The impact of local environmental hazards, like flooding or landslides, can be controlled by automated, intelligent machines. The question remains: why is the field of landscape architecture almost invisible in this field of development?

Systems Thinking and the Potential of Dynamic Patterns

To conduct a discipline-specific discussion in the area of computational education, it is essential to identify predominant challenges. In curriculum design, we are invited to develop strategies which close the numerous knowledge gaps in the area of fundamental computational tools; moreover, we are challenged to understand how to anchor the potentials in computational design thinking (Fricker, 2016). Spurred on by discussions of experts at conferences like the Digital Landscape Architecture Conference, one can recognise a positive trend towards integrative concepts which do not focus on predefined workflows and strict methods, but rather proceed along the lines of the potential of thinking in complex systems.

This approach is supported by systems theory and formulated by Murphy and Hedfors as follows: "The premise that the transdisciplinary investigation of the abstract organization of phenomena, independent of their substance, class, or spatial organization, reveals principles common to all complex phenomena and provides a basis for models to describe and manage them" (Murphy and Hedfors, 2011: 1).

When one observes natural phenomena within landscape architecture in terms of their inner logic, often the multilayered and complex structure of their patterns become clear. This inherent logic can be connected to theoretical basic principles of computer science and represents a theoretical superstructure of computational design methods (Picon, 2010). M'Closkey and VanDerSys show the potential of generative patterns for landscape architecture and refer to the potential gained in the analysis of structures to generate a new understanding for relationships and forms (M'Closkey and VanDerSys, 2017).

James Corner describes patterns in this context as "relational frameworks that simultaneously describe and project; they reveal structures, processes and relationships, as well as structure physical frameworks that give shape and form to our world" (Corner, 2014).

The logic of these connections and networks can be shown through patterns of behaviour, which manifest themselves in dynamic, active, binding, connecting and distributing attributes. Corner refers to the importance of relating this theoretical framework to dynamic processes inherent in landscape architecture, in order to "form new patterns and forms that structure new ecologies, new programs, and new modes of reception" (M'Closkey and VanDerSys, 2017).

Within our society, there is currently a strong consciousness for viewing the principles of ecosystems as potentials, in which digitalisation becomes an integrative element. In academic circles, this positive attitude, and the readiness to integrate computational potentials in teaching, is often still in its infancy. This may be seen as a chance to react to the situation creatively, since isolated observations of discrete elements does not correlate to our profession (Fricker and Munkel, 2015). In current practice, discipline-specific and isolated thinking in categories leads to undefined interfaces which weaken the potentials of landscape architecture. Badly functioning urban and landscape environments are the result of this kind of thinking and design in system components.

The Journey of Computational Design Thinking

Under the leadership of Pia Fricker, the field of Computational Methodologies for Landscape Architecture and Urbanism has been newly launched at Aalto University, Finland. One of the focus areas of the professorship is to research and define new pedagogical avenues for integrating computational design thinking at the intersection of landscape architecture and urban design.

The development of computational tools and methods is embedded into a theoretical discussion on the topic of natural intelligence, digital humanities, structuralism and systems thinking (Georgiou, 2013). An overriding focus, however, will always be the connection to the specific place: reading, recognising and simulating inner relationships as a superstructure for any further explorations. This stands in close correlation with the articulated concept of topology, defined by Christophe Girot at ETH Zurich (Girot, 2013).

In collaboration with Dr Toni Kotnik (Professor of Design of Structures, Aalto University), a computational design studio methodology has been developed, strongly based on systems thinking. Computational design is viewed as the active construction of relations or associations between entities in order to develop new hybrid forms and solutions that are as sustainable as they are versatile (Dodds, 2009). Consequently, computational design thinking is essentially topological by nature: it is not focused on form, but rather on the underlying process of formation governed by an environment of relationships. Understanding the potential of digital technology from a creative, open-ended point of view, allows us to establish new transversal connections between different fields in relation to our profession (Fricker et al., 2019). Computational design thinking in complex relationships goes beyond the mere application of digital tools and integration of diverse data packages; it defines design methods to address the changing understanding and meaning of traditional terminology like city boundaries, densification, urban sprawl, territory, countryside, landscape, natural processes and sense of place. This complex, circular way of thinking supports all stages of design, planning and construction and will open up new solutions for decoding complicated spatial schemes and tackling the social, cultural and technological issues caused by a growing population.

An Integrative Approach for Teaching Computational Design Thinking across Scale

Despite increased critique from the user perspective, academics are increasingly organised into thematic silos. However, infrastructure, urban planning and design, landscape design and planning, urban ecology and many other systems, can be understood as articulated patterns in a larger framework of systems thinking. In a next step, we understand and formulate the connection of function and form with respect to morphology, and the connection between information and communication in genetics, this is made possible through integrating computational thinking and large-scale observation of individual components (Steenson, 2017).

The research group of Professor Pia Fricker and Professor Toni Kotnik sets the focus of its exploration on the special relationship between architecture/landscape architecture and information technology. As both fields have an integrative character, the exploration lies in integrating computational and algorithmic thinking with a purpose of developing new hybrid forms and solutions that are as sustainable as they are versatile.

Understanding the Local to Impact the Global

Technology supports us in decoding our planet as a universe of relationships; acknowledging this allows us to embrace the potential of computational possibilities by being back in command to imagine new fields of application. This attitude will support our students and future practitioners to reclaim the role of key decision makers when it comes to how technology influences future practice. Integrating the potential of computational design thinking into our way of designing enables us to create a vast spectrum of informed feedback loops and iterations of our concepts. Applying the concepts of system thinking will allow us to evaluate and optimise our designs via real-time feedback of site-specific data (time, seasons including ecological aspects), design data and user-related data. This will lead to interdisciplinary, knowledge-based decision-making tools that aid high-quality design, and which assist us to respect and understand the potential of place.

Teaching Examples

Interacting Flows in Urban Systems

Within the studio, computational methods and techniques were developed to study the relation between the centre and the city across scales, and the centre's underlying flows of different data streams like, urban green, people, material and culture. This knowledge was used to speculate on proposals for new types of hawker centres in Singapore, aiming at connecting urban and green systems into a new urban typology (Figure 20.1).

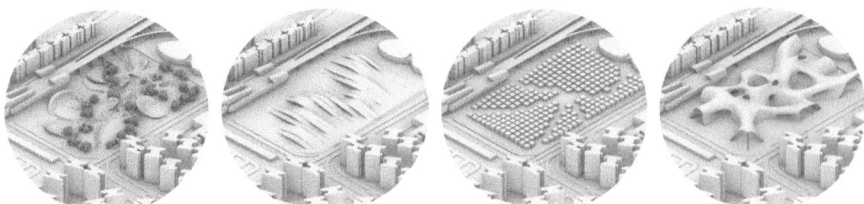

FIGURE 20.1   The overview shows a variation of computationally generated design approaches. The studio is part of a collaboration with Singapore University of Technology and Design (SUTD) teaching team: Prof. Pia Fricker and Prof. Dr Toni Kotnik (Aalto University) and Prof. Carlos Bañón (SUTD). Tina Cerpnjak and Joonas Saarinen, Nora Sønstlien and Solveig Døskeland, Saviana Theiss, Yiping Zhang and Yuyang Shi.

Computational Workflow: Computing Topography

The topography was generated by first creating a field condition with a weighted node system and using the field values to assert differences in height throughout the plot. This created a multitude of shaded spaces that blend seamlessly with the surrounding park.

The different levels encourage exploration and offer privacy and intimacy in a public space, giving it a more human scale (Figure 20.2).

Interweave

By translating the movement of people into a multilevel topography, the project, in turn, creates a water channel network that seamlessly connects the site with the new waterway park development. The channel system works not only as a visual guide, but also serves as a floodwater catchment body, filtering the water and creating a unique microclimate. The number of water bodies in the landscape creates not only a pleasing environment, but also provides water for the trees and surrounding vegetation (Figure 20.3).

Series of Coding Explorations

Students translate design concepts through the integration of the logic embedded in the area of Cellular Automata – 2D and 3D pattern system, growth – agent-based block, discrete system modelling and optimisation strategies for discrete systems, to understand the difference between computerisation and computation (Figure 20.4).

Hernesaari, a Former Industrial Area, Located in the Southernmost Part of the Downtown Area of Helsinki (Finland)

Hernesaari served as a test case for the speculative joint computational design studio of Prof. Pia Fricker and Prof. Dr Toni Kotnik. The project *Intercellular*

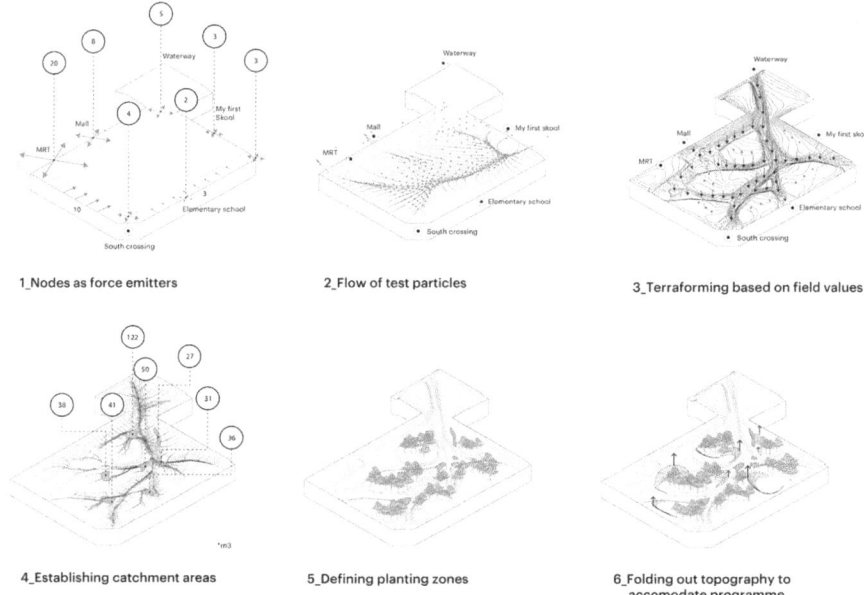

1_Nodes as force emitters

2_Flow of test particles

3_Terraforming based on field values

4_Establishing catchment areas

5_Defining planting zones

6_Folding out topography to accomodate programme

FIGURE 20.2   Left 1–3: Terraforming based on field values: By creating a field condition informed by the flow of people through the site. The topography is shaped to create a variety of multilevel spaces and ecological values. Middle: Establishing catchment areas. By analysing the water flow on site, water catchments are defined and subsequently activities and ambient in their vicinity. Right 1–2: Folding out the topography. With an operation of cutting and folding the topography, covered areas for certain programmatic functions are provided and by doing so articulating the spaces above these interventions. Joonas Saarinen and Tina Cerpnjak.

proposes a future-oriented way of designing responsive urban landscapes through the understanding and integration of natural patterns as design drivers. To create responsive urban strategies, the project avoids conventional zoning methods and applies computational design methods for mixing form and functions into a new spatial configuration (Figure 20.5).

The Project Presents a Dynamic, Evidence-Based Planning and Decision Support Tool Called CityScope Lapland

The main goal of CityScope Lapland is to use digital technologies to incorporate dynamic variables in urban and landscape spatial analysis and methodology; secondly, to improve the accessibility of the decision-making process for

FIGURE 20.3 The topography folds out to enable a covered programme with services that blend in and out of the green park-like landscape, blurring the boundary between inside and outside. They articulate the ground in a manner that accommodates different leisure activities, based on the location on site. Joonas Saarinen and Tina Cerpnjak.

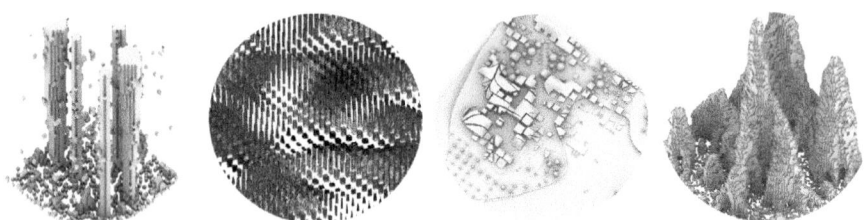

FIGURE 20.4 Coding explorations. Output generated by students of the design studio "Patterns of High Density" (Aalto University).

non-experts through a tangible user interface; and third, to help users evaluate their decisions by creating feedback through real-time visualisation of urban simulation results. This is achieved by designing an agent-based model and using different representation and abstraction features for different dynamic data packages (Figure 20.6).

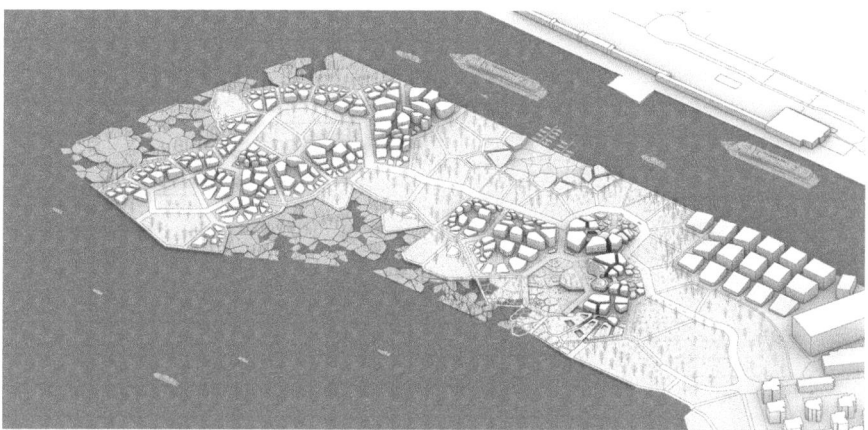

FIGURE 20.5 The zoning of the site is controlled by a cellular system, using the historical, the present and a speculative future situation as visionary input parameters. By applying this method, a highly flexible land system is generated
to realise multipurpose functions, merging environmental aspects with a
future-oriented urban strategy. The topography of the articulated ground
condition is automatically generated and influenced by environmental factors. Through this method, runoff water is purified in responsive water
catchment areas which also interact as a mediator for different seawater
levels. A systemic interplay of people, natural elements and buildings is generated as a logical consequence of the computational model, and this allows
for change and adaptation. Jiaqi Wang, Xin Ding and Pirita Meskanen.

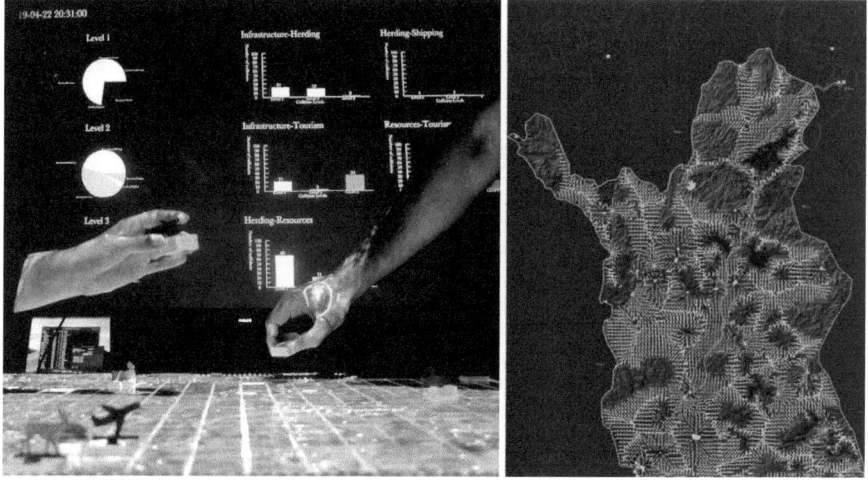

FIGURE 20.6 The project is integrated within the GAMA simulation platform and
embedded in the MIT CityScope framework – a medium for both,
analysing agents' behavioural patterns and displaying them to the
stakeholders. Ayda Grisiute.

References

Cantrell, B., Mekies, A., & Mekies, A. (2018). *Codify: Parametric and computational design in landscape architecture.* Routledge. https://doi.org/10.4324/9781315647791.

Corner, J. (2014). The agency of mapping: Speculation, critique and invention. In James Corner and Alison Bick Hirsch (Eds.), *The landscape imagination: Collected essays of James Corner 1990–2010* (pp. 197–240). Princeton Architectural Press.

Dodds, W. K. (2009). *Laws, theories, and patterns in ecology.* University of California Press. http://ebookcentral.proquest.com/lib/ethz/detail.action?docID=837312.

Fricker, P. (2016). Towards a new design and teaching methodology for large-scale landscape design in the era of digital overload. In M. Boştenaru Dan & C. Crăciun (Eds.), *Space and time visualisation* (pp. 63–75). Springer International Publishing. https://doi.org/10.1007/978-3-319-24942-1_4.

Fricker, P., Kotnik, T., & Kane, L. (2020). Computational design pedagogy for the cognitive age. In Liss C. Werner and Dietmar Koering *Anthropologic: Architecture and fabrication in the cognitive age* (pp. 695–692). eCAADe.

Fricker, P., Kotnik, T., & Piskorec, L. (2019). Structuralism: Patterns of interaction – computational design thinking across scales. *Journal of Digital Landscape Architecture, 4.* https://doi.org/10.14627/537663026.

Fricker, P., & Munkel, G. (2015). Data mapping: Explorative big data visualization in landscape architecture. In E. Buhmann (Ed.), *Peer reviewed proceedings digital landscape architecture 2015* (pp. 141–150). Wichmann.

Georgiou, I. (2013). *Thinking through systems thinking.* Routledge. https://doi.org/10.4324 /9780203607718.

Girot, C. (2013). *Topology: Topical thoughts on the contemporary landscape* (Vol. 3). Jovis.

Girot, C. (2017). *Delta dialogues* (Vol. 20). gta Verlag.

Haas, T., Olsson, K., & Olsson, K. (2016). *Emergent urbanism: Urban planning & design in times of structural and systemic change.* Routledge. https://doi.org/10.4324 /9781315579160.

Halpern, O. (2017). Architecture as machine: The smart city deconstructed. In A. Goodhouse (Ed.), *When is the digital in architecture* (pp. 123–175). Sternberg Press.

Koolhaas, R., Armstrong, R., Therrien, T. C., & AMO. (2020). *Countryside: A report: AMO, Rem Koolhaas.* Taschen.

M'Closkey, K., & VanDerSys, K. (2017). *Dynamic patterns: Visualizing landscapes in a digital age* (1st ed.). Routledge. https://doi.org/10.4324/9781315681856.

Menges, A. (2011). *Computational design thinking.* Wiley.

Murphy, M. D., & Hedfors, P. (2011). Systems theory in landscape architecture. *Conference paper: Urban nature: Council of Educators in Landsacpe Architecture CELAAt,* 1–13.

Picon, A. (2010). *Digital culture in architecture: An introduction for the design professions.* Birkhäuser.

Steenson, M. W. (2017). *Architectural intelligence: How designers and architects created the digital landscape.* The MIT Press.

Zari, M. P., Connolly, P., Southcombe, M., Connolly, P., & Southcombe, M. (2020). *Ecologies design: Transforming architecture, landscape, and urbanism.* Routledge. https://doi .org/10.4324/9780429279904.

21

ENVISIONING THE PLANETARY

Design Agency in the Climate Crisis

Clara Olóriz Sanjuán and Jose Alfredo Ramírez

Landscape Urbanism at the Architectural Association (AALU) explores the agency and roles landscape-oriented designers (architects, landscape architects, urban designers, geographers, earth scientists and others) can play in the crises produced as a consequence of planetary urbanisation processes. Policymaking at different levels often results in metropolitan areas and urban agglomerations as well as the productive, consequential and operational landscapes behind cities such as farming, agriculture and extractive infrastructures (Brenner, 2016; Hutton, 2013). Policymaking both regulates – and deregulates – planetary urbanisation and processes, and also actively creates and shapes how these worlds are and how they came to be, who they benefit and who they are leaving behind and exploiting (Purdy, 2017).

AALU recognises that from this perspective, the work of designers is but one small part of these world-making processes and usually arrives at the end, when many crucial decisions have been taken. Whether in the form of commissions or competitions, architects, landscape architects and urban planners design embellishing projects and images for which, sites, clients, users, programmes, rights, life spans and other crucial aspects have been decided, or there is little room for manoeuvre.

Within today's multiple planetary crises, many design practices acknowledge this condition and are finding alternative ways to intervene in early stages (Public Practice, 2017; Wainwright, 2017). At AALU, we are working on ways to be involved in policy design, through critically engaging the aesthetic apparatuses of landscape-oriented practices or the work that images do, envisioning alternative relationships between society and nature (Olóriz Sanjuán, 2019). Critically visualising existing policies, designing visions through policy propositions, alternative world displays and ways of seeing these planetary crises, we look at

DOI: 10.4324/9781003145905-26

one of the biggest challenges humanity is facing: climate breakdown, its wider consequences, injustices and causes.

Given the climate and ecological emergency, it is of paramount importance that designers build critical awareness of the backstage behind beautifying interventions, collaboratively working towards a socially just restructuring of the worlds we inhabit in the form of a Green New Deal (GND) (Fleming, 2019; AA Groundlab and Common Wealth, 2020). A GND is a set of policies that aims to decarbonise the economy. If implemented, this set of policies will fundamentally reshape the processes behind planetary urbanisation. This effort intrinsically depends on the health of the earth systems and social justice and requires a radical transformation of the role designers can collectively play in developing design visions, proposals, mitigation strategies, advocacy initiatives and activism (Klein, 2020: 272–9).

AALU researches, visualises and imagines how a transdisciplinary GND can be inclusively designed, prioritising the wellbeing of people and the health of the planet. To achieve this, AALU has joined forces with several institutions and thinktanks that are shaping a GND in the UK (Lawrence, 2019). AALU's annual brief and methodology have aligned with the work of these actors to explore critical design agencies, raising "the stakes for landscape" (Spencer, 2019). By involving landscape-oriented designers in GND proposals, AALU aims to open new avenues for designers to rethink their role and agency in the making of critical images, visions and displays about alternative relations between society and nature, with a view to adopting collective goals, projects and agendas with the help of many other disciplines, institutions, and communities committed to this end (Klein, 2020: 25–6).

To illustrate the programme's GND commitment alongside the methodology and pedagogy AALU is building, we are presenting a project called "Just Transition" as an example of how we imagine our involvement in policy design, and alternative design practices that are necessary to tackle climate breakdown and envision the planetary. This thesis starts by critically bringing to the fore the veiled worlds behind current decarbonising policies in the Global North and visualises the ways green policies produce landscapes – nature, material and labour relationships. Within this framework, it then moves on to a redesign of these policies, together with the local community, envisioning the reappropriation of inaccessible conservation forests through community-managed forestry. This project proposes the involvement of designers on territorial policies through alternative visions of the planetary.

Given that we live within a planetary urbanisation process, it is fundamental to develop planetary representations to unveil capitalism's global interconnectedness, while at the same time, and even more crucial, we need to avoid totalising and universal visions of it. An array of communities, people and non-humans are shaped by the planetary conditions (Gabrys, n.d.). These processes and events have various, divergent, and differentiated material and social impacts that need to become visible and clearly identifiable within their own geographies, and for which various forms of representations need to arise. From global corporate industry networks to material soil studies, the following project describes a potential alternative project to represent and envision the planetary.

Just Transition
Treherbert, Wales, UK
By Rafael Martinez, Yasmina Yehia and Elena Luciano
Through a partnership with the New Economics Foundation (NEF), AALU was given the topic of Just Transition (Powell et al., 2019) as part of an ongoing research to investigate energy transitions from fossil fuels to renewable energy, and the impacts they have on the wellbeing of communities in the UK (Heisse, Luciano and Yehia, 2019; Powell et al., 2019). Just Transition is based in the South Valleys of Wales, where draining coal in the past gave rise to an extractive system that for decades fuelled Britain. Today the valleys have been abandoned, leaving behind deprived landscapes and communities with no opportunity to achieve a just transition away from coal.

Through critical cartographic visions, AALU students explored the story of these valleys framed within three different scales: an undergoing green transition at a global scale, green policies being applied at a regional scale, and an existing green forest in a local one. This framing unveiled how – behind green curtains – these multiscalar decision-making processes impact how society relates to nature, the way landscapes are being configured as a consequence, and the images they depict.

Throughout these scales, the valleys embody the different energetic transitions the UK has seen historically: from coal extraction, when the valleys were an important centre of production during the industrial revolution up to the First World War; then the abandonment and deprivation they suffered when fossil fuel extraction moved to the North Sea and beyond for oil and gas; and now to today's appearance of windmills and other renewable landscapes subordinated to transnational and corporate capital ventures.

As a commitment to the Paris Agreement, the UK government launched the *Climate Change Act* in 2008 (UK Government, 2008) which aims at decarbonising the nation by 2050, supported by renewables such as wind and biofuels. However, the problem with this commitment is that it portrays an ideal green future for the UK because all actions in the Act are framed only within national borders: but what about the UK imports and overseas companies?

To look into this, Rafael, Elena and Yasmina produced a global atlas, a planetary projection used to show the concentration of green energy and Just Transition projects in the Global North and the gradual expansion outwards to the Global South of the fossil fuel industry. They represent recent concessions explored by companies headquartered in London, exposing the migration of UK fossil apparatuses towards the Global South, and suggesting a decarbonisation achieved upon the instrumentalisation of dirty energy extraction in the Global South: "A new green version of colonialism, waving the flag of climate emergency to justify its operations" (Martinez Caldera et al., 2019) (Figure 21.1).

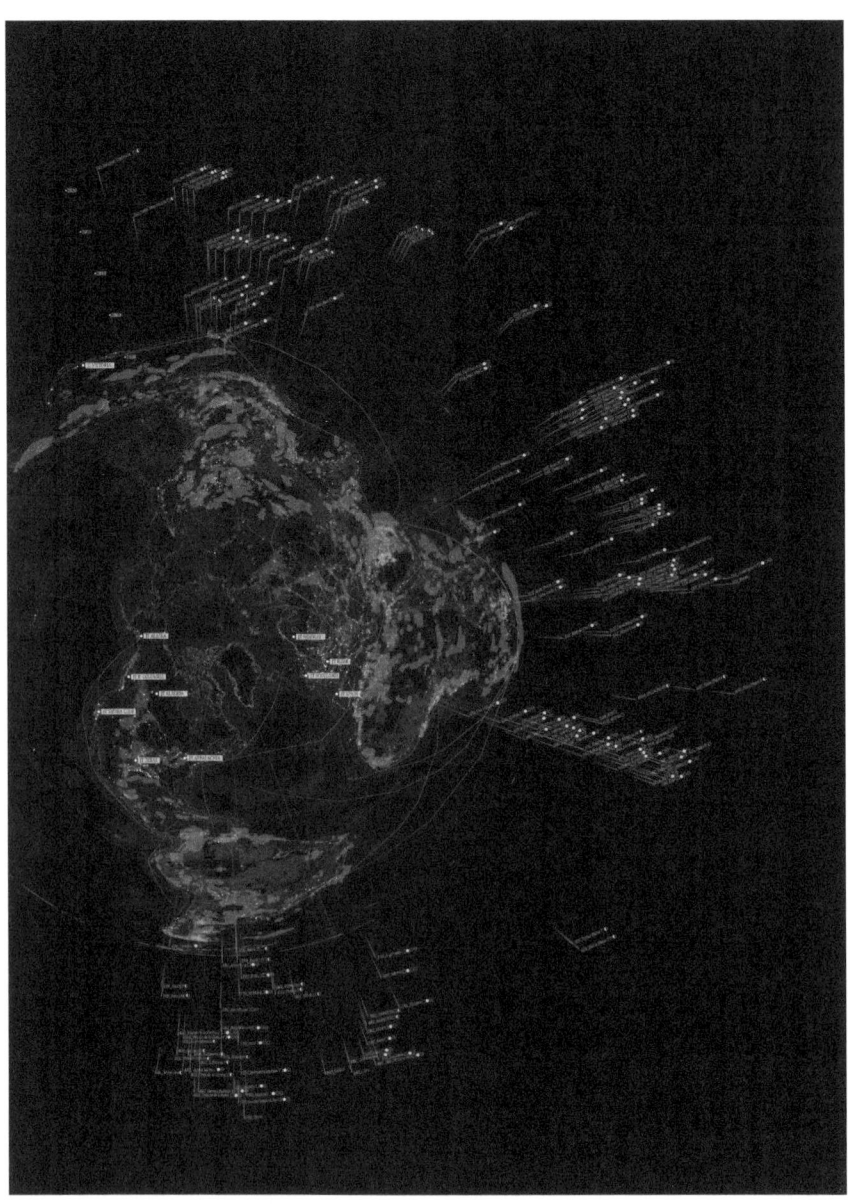

FIGURE 21.1   Global mapping oil concessions (in gold) of UK/US companies across the world and headquarter in London. Rafael Martinez, Yasmina Yehia and Elena Luciano.

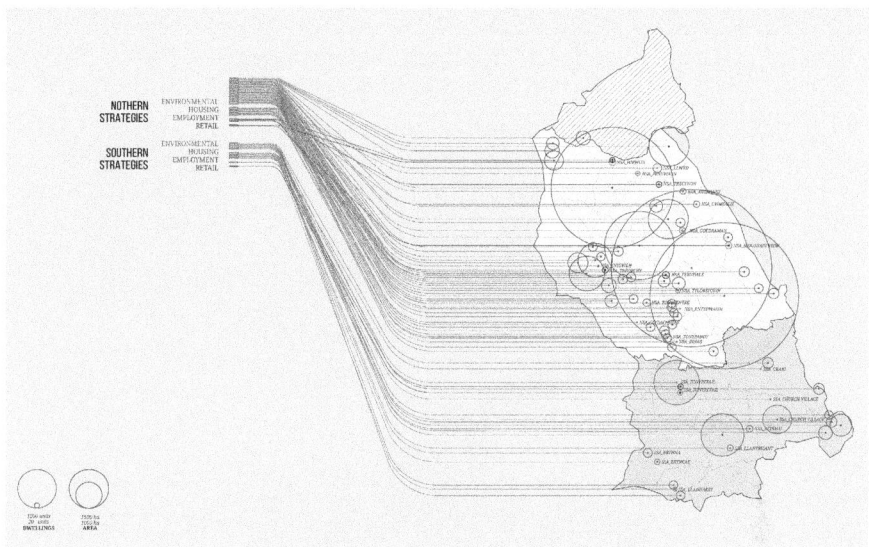

FIGURE 21.2   Map showing the collection of policies that regulates the north and south of the Rhondda Valley. Rafael Martinez, Yasmina Yehia and Elena Luciano.

What is observed on a global level is replicated in the Rhondda Valley where coal miners, contributing to the expansion of the British Empire overseas, were victims of centralised decisions. After the First World War, the coal mines were gradually shut, and people were given no choice but to transition to a different sector. Since then, the Rhondda Valleys have undergone substantial depopulation, and deprivation rates that are still the norm.

All the above was achieved through implementing policies. The project authors set out to produce a set of drawings that depict how these policies shaped the valleys. In gold are policies that refer to retail, housing and employment, concentrated in the southern area. In green, the environmental actions are overwhelmingly dominant, particularly in the northern area (Figure 21.2).

The critical spatialisation of written policies shows the dominance of green policies in the north as a veiled invitation for local people to migrate from the higher parts of the valleys to Cardiff (Martinez Caldera et al., 2019), hinting at how authorities are hiding behind a green narrative to strengthen dependencies with the so-called urban areas, and to decamp the valleys to conserve forests and allow renewable companies to appropriate land.

Treherbert, a local town within the valley, has been hit by these policies. As a consequence, two types of landscapes have flourished: a dense forest and bare land patches. They are the result of a palimpsest of land policies implemented over the 20th century, depicted in a vertical axonometric (Figure 21.3).

While most collieries were shut down in the first half of the 20th century, the UK Forestry Commission, created in 1919, acquired land to plant trees around

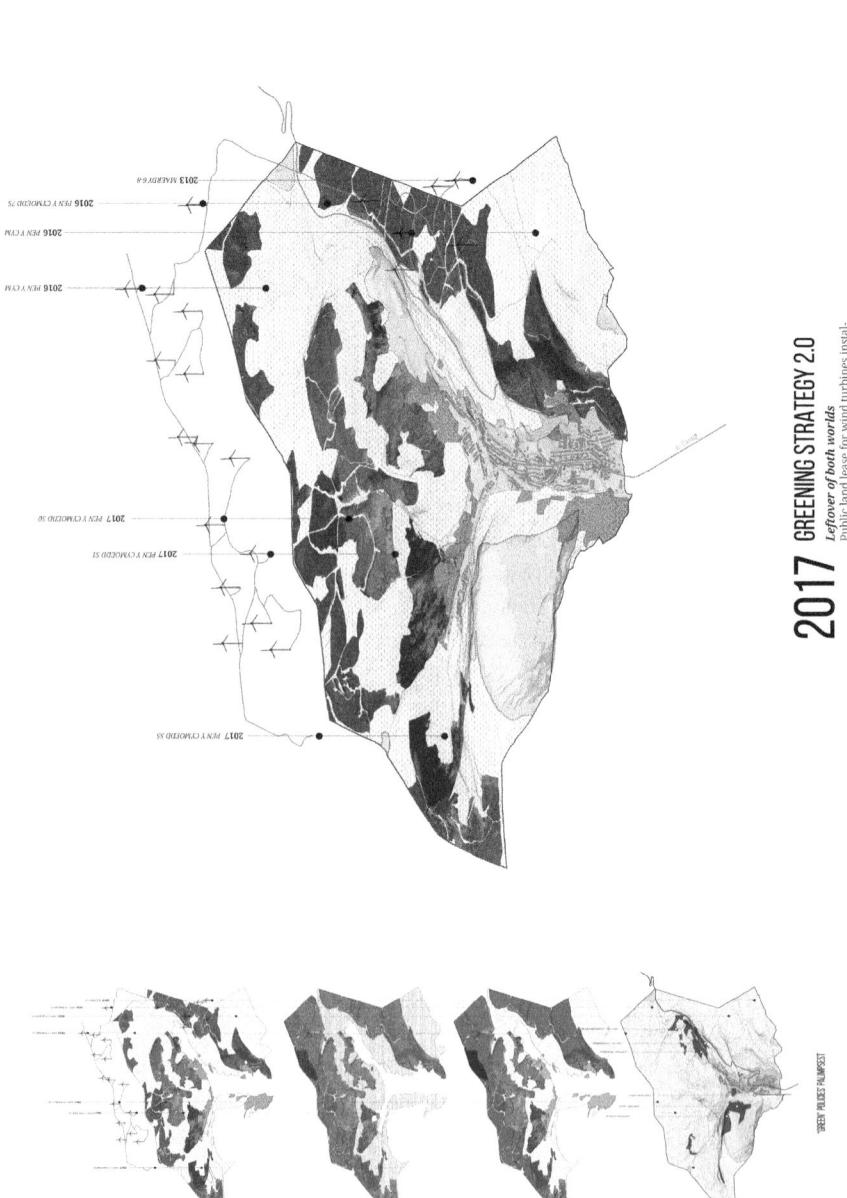

FIGURE 21.3   Axonometric and maps showing the history of policies and their impact in the landscape around Treherbert within Rhondda Valley. Rafael Martinez, Yasmina Yehia and Elena Luciano.

the town. The acquisition of land expanded to secure stocks of timber production for war purposes. The existing forest resembles more a factory than woodlands. Conifer trees were planted three metres apart, preventing light from going through and effectively producing a monoculture land with little biodiversity and impoverished, acidic soils, best described as dead dirt. This forest was consolidated with the expansion, in 1985, under banners of biodiversity and sustainability, of new forested areas safeguarded with a legal zone established in 1994 (SINC, Sites of Importance for Nature Conservation), further enclosing the land and banning access to and use by local communities. By 2017, renewed green mentalities led to a new green strategy where land was cleared to allow the construction of windmills, managed by a transnational company to extract, once more, energy from the valley. In short, the valley has been historically shaped by these green and forestry policies that have prevented the deprived local community from accessing or benefiting from it.

Despite these conditions, a local small organisation called Welcome to Our Woods has been fighting for years for the opportunity to manage four hectares of forest to open alternatives for a different community/forest/planetary relation. Welcome to Our Woods managed to beat a labyrinthine set of bureaucratic procedures to obtain the right to manage the land and convert the wood factory (that is, the existing forest) into a live and biodiverse community-managed forest. The community is active by thinning trees, using wood materials for biofuels, organising community activities within the forest, and building micro dams from existing streams, among other small interventions that generate green local jobs.

Rafael, Elena and Yasmina learnt from this experience and thought to expand it to the whole landscape of Treherbert to the Rhondda Valleys in Wales. They created a niche where they could gear their design agency towards envisioning a series of policies that can support and incentivise local landscape community reappropriations, while supporting Welcome to Our Woods' collective vision and knowledge sharing. This model could even serve as an example for transforming forestry policies that could potentially reconfigure UK landscapes into a "Common Landscape Policy" through the lens of a local community, at the heart of historically unjust energy transitions (Figure 21.4).

In this sense, their project can be understood as planetary, as it aims to transform this valley from being a wood factory forest and wind energy producer for transnational enterprises, into a forest managed by the local community with the capacity to create a community life dependent on its surrounding landscapes. Thus, envisioning the planetary entails learning from Welcome to Our Woods' existing practices, the production of cartographic media to communicate the range of existing local skills, small landscape interventions and the visualisation and design of a set of common landscape policies that can make the project transferable to other parts of the UK, and with a potential planetary impact.

Using as starting points the existing dominant landscapes, an overstocked coniferous land and barren grassland, the project sets out to reach a managed

FROM WELCOME TO OUR WOODS'S 4 HA   TREHERBERT PUBLIC LAND   TO UK'S PUBLIC WOODLANDS

FIGURE 21.4 Just Transition proposal to escalate the project from a local, to regional, to countrywide application. Rafael Martinez, Yasmina Yehia and Elena Luciano.

status where a rich, diverse and healthy forest can be established over time, and appropriated by Treherbert's community through a set of designed interventions (Figure 21.5).

These interventions start by applying community thinning, which in the short term is less profitable than ongoing mercantilist, and so-called more productive, clear-felling. Clear-felling produces macro-patches that are visually defined and cut at once with heavy machinery, at the expense of biodiversity and human access. In comparison, access to forest is crucial in community thinning for harvesting, monitoring and recreation, but it is also a main factor for soil and biodiversity protection. To minimise impacts, students propose reorganising the path system for selective thinning, using an algorithm that minimises the number of tracks, bringing both logs and people together. They aim to combine this proposal with less heavy machinery, and traditional methods suggested and developed by the community.

So, how can the community transition to these green jobs and economies? How can they build the skills and knowledge? Associated with policy redesign, students also coproduced a handbook with Welcome to Our Woods. The handbook is divided into two parts. The first is an inventory of the tools and infrastructure required for the community transformation. So, the long-term success of the plan is guaranteed, the handbook's second part considers the transfer of

FOREST EVOLUTION TIMELINE

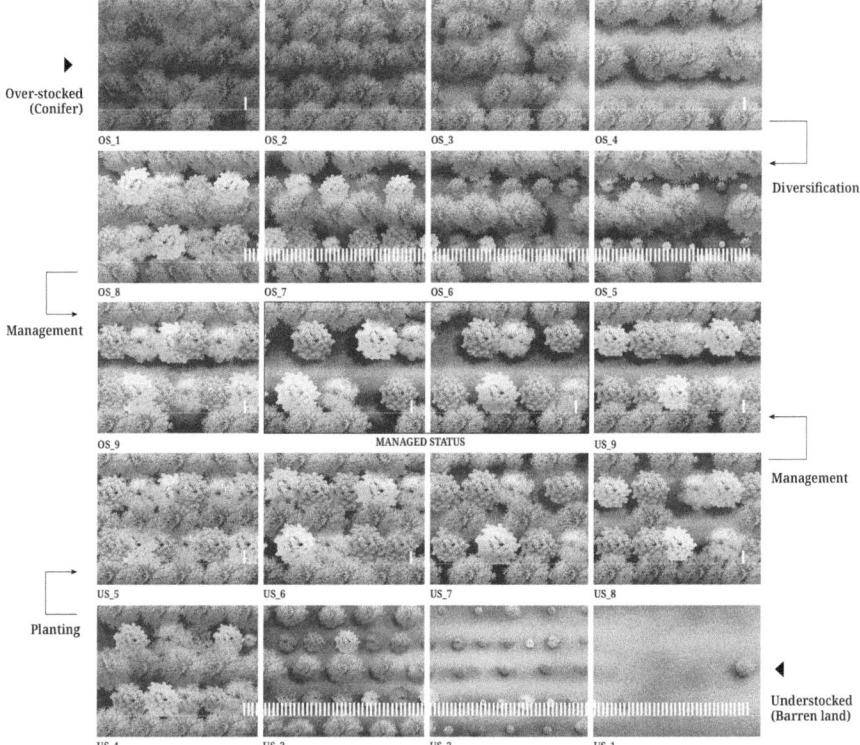

FIGURE 21.5 Top: Forest Evolution, from over stocked monoculture status (top left) and understocked barren land (bottom right) towards a manage/biodiverse status (centre). Bottom: Screenshots of a woodland dynamic's simulation. It portrays a process of natural growth, human intervention and design articulation in the management of woodlands. Rafael Martinez, Yasmina Yehia and Elena Luciano.

knowledge to future generations about the interventions needed to develop and improve the conditions imposed on the landscapes.

To envision a forest that is the result of both community interventions and natural growth rates, students developed a digital model to simulate woodland dynamics. The model is based on growth, reproduction and tree mortality. It helps document in time actions and decisions taken and adapts support incentives for this model. The digital model includes human interventions to exemplify how removing trees and building the path system will interact with the growth of the forest (Figure 21.6).

The model is run several times to form what is called a Shifting Plan which was applied in detail in a 200-hectare site beside Treherbert. It is structured in

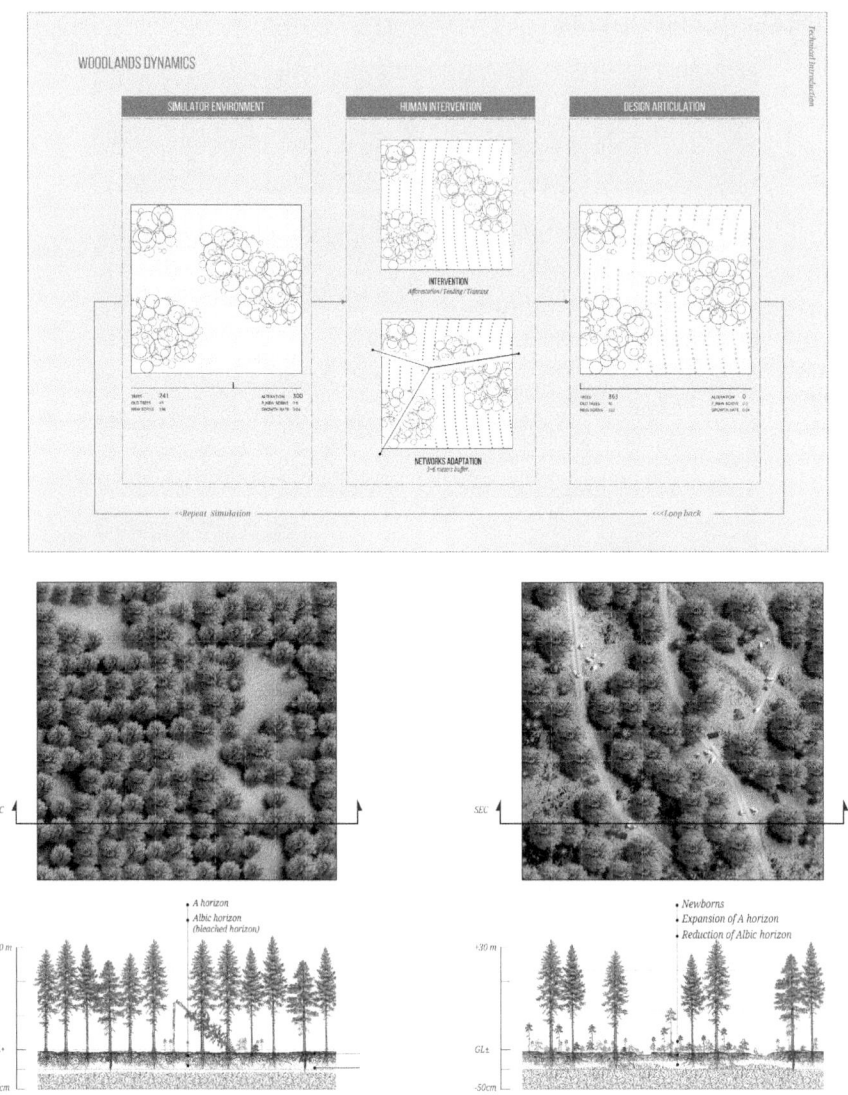

FIGURE 21.6 Visualisations that describe the processes of natural growth in the woodland and its integration with human activities and programme in the woodlands of Treherbert. Rafael Martinez, Yasmina Yehia and Elena Luciano.

three phases lasting 40 years. The Shifting Plan includes a full woodland coverage, starting from areas closer to the community in the lower parts of the valley, then moving towards the historically neglected areas at risk and the upper parts.

To achieve this, the project proposed the transformation of existing policies into a Common Landscape Policy. This policy supports economically, through

FIGURE 21.7 Visualisation of the woodland's restoration in Treherbert after 40 years of community management. The image envisions a biodiverse forest with integrated community activities reciprocal and reinforcing each other. Rafael Martinez, Yasmina Yehia and Elena Luciano.

FIGURE 21.8 Top: Section manifesto that envisions the transition from barren land into biodiverse and community-active woodlands. Bottom: Second part of the section manifesto that envisions the transition from a monoculture wood factory towards biodiverse and community-active woodlands. Rafael Martinez, Yasmina Yehia and Elena Luciano.

a variety of subsidies and grants, and technically via the community handbook, the implementation of the overall strategy and local interventions such as: maintenance of established woodlands, the afforestation of barren land and of riparian areas, a new system of paths, community infrastructure (community hub and sawmill), sediment traps, woodland terracing and tourism infrastructure, all developed over a span of 40 years (Figure 21.7).

Finally, students designed a long section as a cartographic manifesto in which they document the co-dependency their project aims to create between the local community and the ecologies of the woodland, as an exemplary common landscape that puts forward new human–planetary relations.

The section documents how existing barren land and overstocked pines require more intense and long-term intervention and care from the community. It also depicts a landscape that offers an alternative to connect the marginalised communities of the Rhondda valleys, and the diminished and neglected non-humans. In this sense, this section is a way to envision the planetary, not by depicting a totalising object that can be managed and programmed, but by bringing to the fore those humans, non-humans, materials, technologies and politics that are usually hidden behind traditional planetary images such as the blue marble (Blue Marble Next Generation, 1972) (Figure 21.8).

References

AA Groundlab and Common Wealth. (2020). Green New Deal: Glasgow. Retrieved from https://www.common-wealth.co.uk/interactive-digital-projects/gnd-glasgow.

Brenner, N. (2016). The hinterland urbanised? *Architectural Design, 86*(4), 118–127.

Fleming, B. (2019). Design and the green new deal. *Places Journal.* https://doi.org/10.22269/190416.

Gabrys, J. (n.d.). Becoming planetary. *e-flux Architecture.* Retrieved from https://www.e-flux.com/architecture/accumulation/217051/becoming-planetary/.

Heisse, C., Luciano, E., & Yehia, Y. (2019). The British Government is fuelling climate disaster. Retrieved from https://tribunemag.co.uk/2019/12/the-british-government-is-fuelling-climate-disaster.

Hutton, J. (2013). Reciprocal landscapes: Material portraits in New York City and elsewhere. *Journal of Landscape Architecture, 8*(1), 40–47.

Klein, N. (2020). *On fire: The (burning) case for a green new deal.* Simon & Schuster.

Lawrence, M. (2019). Road map to a Green New Deal: From extraction to stewardship. Retrieved from https://www.common-wealth.co.uk/reports/road-map-to-a-green-new-deal-from-extraction-to-stewardship.

Martinez Caldera, R., Yehia, Y., & Luciano Suastegui, E. (2019). *Just transition.* Architectural Association.

NASA. (1972). Blue marble next generation. Retrieved March 10, 2021, from https://earthobservatory.nasa.gov/features/BlueMarble/BlueMarble_history.php.

Oloriz Sanjuan, C. (2019). Introduction. In Clara Oloriz Sanjuan (Ed.) *Landscape as territory* (pp. 5–16). ACTAR.

Powell, D., Balata, F., Van Ler Ven, F., & Welsh, M. (2019). *Trust in transition.* New Economics Foundation. Retrieved from https://neweconomics.org/2019/11/trust-in-transition.

Public Practice. (2017). Retrieved from https://www.publicpractice.org.uk/about.

Purdy, J. (2017). Understanding environmental law as public provision. Retrieved from https://lpeproject.org/blog/understanding-environmental-law-as-public-provision/.

Spencer, D. (2019). Going to ground: Agency, design and the problem of Bruno Latour. In C. Oloriz Sanjuan (Ed.), *Landscape as territory (150–157)*. ACTAR.

UK Government. (2008). *Climate Change Act 2008*. Statute Law Database.

Wainwright, O. (2017). Spectacular buildings for everyone: How to stop ugly housing and fix urban planning for good. *The Guardian*. Retrieved from https://www.theguardian.com/artanddesign/2017/nov/09/architecture-housing-design-urban-planning-local-government.

22

A SENSED LANDSCAPE

Craig Douglas

Introduction

In response to the rapidly shifting global climate, and the consequent climate crisis in which we find ourselves, fundamental environmental factors should, and can, be given greater consideration in relation to the social, political and economic drivers that shape the landscape. This enquiry engages a disciplinary responsibility to explore rapidly emerging sensing technologies to understand how they might extend design agency of the discipline in relationship to the global climate crisis.

The Dynamic Landscape

The urban landscape is a palimpsest of conditional processes and properties, one that is open-ended, flexible and adaptable, displaying a self-organising uncertainty and dynamism. It is defined and shaped by a collection of material processes that reflect environmental conditions that change over time. The resultant formal composition of these phenomena inherently describes the forces that have shaped them and will continue to do so as form translates the material registration of force as "a network of enveloped material processes".[1] Sensor technology is explored here for its capacity to identify and measure these processes to a greater accuracy, and in doing so point toward new opportunities for design intervention.

The enquiry commences with the elementary notion that sensors have the potential to extend our degree of observation and measure beyond that which is visible to the human senses – this involves making the invisible visible. Gabrys makes the case that since the time of the Sputnik satellites (1957), the potential for "monitoring to aid in the management of the environment, suggesting that

DOI: 10.4324/9781003145905-27

this could not only reveal undiscovered dynamics within nature but also extend to identifying 'resources for extraction, and monitoring land use and living patterns' has been clear".[2] The use of sensor technologies within the context of the climate crisis makes "visualising" and "re-making" relationships between people and the environment possible by yielding "processes for making new environments not necessarily as extensions of humans, but rather as new con-figurations"[3] that can grow across "technologies, practices and non-human entities".[4]

Inherent to utilising sensor technologies is to explore how the landscape can be made "sense-able", how might it be operationalised, what kinds of "smart landscape" can be shaped, and to what effect?

House Zero/Plot Zero

This research was made possible thanks to the support of the Harvard Center for Green Buildings and Cities, which "promotes holistic change within the built environment, namely through the creation and continued improvement of sustainable, high performance buildings and cities".[5] To this end, the Center's "House Zero" model house, facility and laboratory – a pre-1940s typical "triple-decker" Cambridge house retrofitted into an ultra-efficient, healthy, positive energy structure – is both the inspiration and the catalyst for this work. House Zero contains

> nearly five miles of wiring that capture 17 million data points per day. Some are critical to the operation of the building: for example, controlling the system of windows and shades in response to inputs about temperature, rain, wind direction, and indoor CO2 levels and airflows.[6]
>
> *Harvard Center for Green Buildings and Cities*

The aspiration of the "Plot Zero" research – with respect to the typical Cambridge plot that accompanies these homes – utilises continuous in-operation sensors externally to gather data specific to dynamic environmental parameters (e.g., solar and infrared radiation, air temperature, humidity, wind speed, carbon dioxide levels and significant air pollutants) to build a dynamic visual model capable of translating and visualising the data in space and time. The intention of this work is to make it possible to analyse and visualise the complexity of the landscape system and identify its propensities in the form of its key operational characteristics. It aspires to understand the impact of the composition of outdoor spaces on the energy efficiency of building operation. It aims to understand the atmospheric agency of the landscape to inform the design of spaces that contribute to the sustainability of the city and improve the health and wellbeing of its citizens.

To meet the challenges of energy efficiency and a healthy environment for a city's inhabitants, there is a need for innovative ways to design and manage the thermal performance of the indoor and outdoor spaces of the urban fabric, and the interconnected relationship between those spaces. This work recognises the

> imperative [to develop] new observational strategies that are linked directly to innovative modelling approaches that directly address the most potent feedbacks in the climate structure. Because it is the feedbacks in the climate structure that set the time scale for irreversible change.[7]

The physical composition of the urban fabric acts to absorb, produce and trap heat, resulting in sustained higher temperatures, 1–3 degrees Celsius warmer than neighbouring rural areas. Heat generated in the city, including waste heat, is trapped along with airborne pollutants generated by vehicles, transport infrastructure, commercial enterprises and industry. This condition adversely affects water and air quality, and the health and wellbeing of citizens. Energy demands simultaneously rise due to the prolonged and increased use of mechanical ventilation and air conditioning in response to the hotter temperatures that strain energy resources and further contribute to the production of global emissions.

The landscape, and specifically the landscape "plot" of House Zero, encapsulates a range of disparate narratives, including that of the geophysical and the geospatial, the material and the immaterial, natural processes and social regimes. These studies suggest new modes of analysis and how they might enable the visualisation, modelling and simulation (Figure 22.1) of complex atmospheric

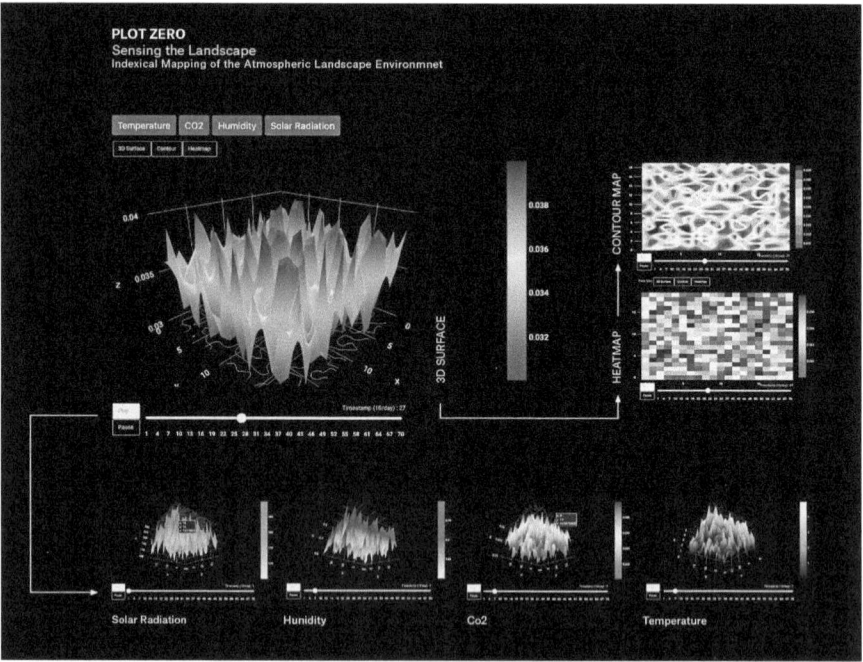

FIGURE 22.1   Plot Zero, real–time atmospheric data visualisation. Craig Douglas and Zhaodi Wang.

conditions that explore the material relationships at the intersection of natural systems and the built environment.

The intention to "start small" – at the scale of the typical urban plot (less than 1,000 m^2) – was intended to challenge the trope of the sensor network's capacity to make the invisible visible, in order "to measure and elaborate on a quantitative description of hidden urban characteristics".[8] What became apparent was the capacity to define differential space as "a field of rapidities and slownesses, and via these infinitesimal relations"[9] that describe space as a complex dynamic manifold. This supports the notion of the landscape as a dynamic space of flux, and simultaneously brought to bear a distinction between "extensive" and "intensive" conceptualisations of space.

Borrowing from the field of mathematics and thermodynamics – and more specifically non-equilibrium thermodynamics – the distinction between the extensive and the intensive space, as considered by Deleuze, characterises "intensive spaces, as the site of processes which yield as products the great diversity of extensive spaces".[10] This is a motivation for understanding and mapping space with extensive boundaries defined by limits that describe an artificial "evenness", together with productive zones of intensive difference defined through critical points of abrupt transition: "wherever one finds an extensive frontier (e.g., the skin which defines the extensive boundary of our bodies) there is always a process driven by intensive differences which produced such a boundary".[11] The approach facilitates a more fine-tuned assemblage of extensities to be described with the structure of the intensive systems that maintain the extensive boundaries. This structural system of flows, forces, energy and matter holds the possibility for generating difference that has the capacity to shift the extensive boundaries operationalised through design intervention.

With this in mind the goal is to extend the agency of the discipline by employing sensor technologies that augment the visual field and construct a deeper understanding of the dynamic material composition of the landscape. This process will reveal its constraints and engender creativity through operable models of representation that enable design intervention and predict and test future scenarios.

Visualisation, and consequential translation, of the sensed landscape data (generated at its source in the binary form of zeros and ones) shifts away from conventional graphs toward three-dimensional dynamic models. These technologies intend to "help us approach entities not as detached objects for our subjective sensing and contemplation, but rather as processes in and through which experience, environments, and subjects individuate, relate, and gain consistency".[12] These models create a representational bridge to current digital design modelling techniques. They are themselves tools for thinking. They are models "for" ideas that set the stage for new possibilities to emerge and be remade. In short, they are models designed through processes of relating, by which entities are parsed and persist in environments.

Sensor Technology

This course of study questions, describes and challenges how we, humanity, might imagine ourselves living in the future with the biological processes of the natural world that support life on the planet. Among the many themes that arise, there is the "decentring of the human in its relations to other species, machines, and the material world".[13] Papadopoulos explores this idea as a "beginning" (one of two) that engages with the unfolding 21st-century vision of "a more-than-human world", in which humans may no longer be in control and instead exist along a "human–non-human continuum".[14] In this instance, the global activity of humans, understood as their interaction with technology and the intertwinement of scientific and technological development "to create new nature-culture hybrids"[15] reflects the "continuous folding of science, technology and the everyday into each other".[16]

As the network of sensors grows, a "new contextual and cultural layer within the landscape"[17] is emerging at the confluence of the digital and the physical. Initial urban projects in cities around the world have, so far, utilised the gathering of sensor data to identify and solve problems. In these cases, the sensor networks engender evidence-based design solutions that leverage authority from the well-measured and the measurable. The concern here is that this approaches a form of positivism, or positivistic design, that has a tendency to produce quasi-naturalised landscape design responses that might more accurately be described as biological experiments.

The increased capacity to rapidly extract measures and identify "world parameters that enable an effective synoptic level of control has yielded demonstrable performance assessments within the field of civil engineering".[18] Robinson describes this as a powerful design methodology that creates

> the hazards of employing a inevitably narrowed vision of the world required for administrative management, one that crops out far more of a phenomenon that it contains, compounded by how we build to feed these measurements, structuring a world that maximises select parameters and neglecting a world outside the brackets.[19]

The danger of such a "model of effective design science"[20] is that it erodes the qualitative characteristics and tendencies of a landscape cherished by landscape architects, which are difficult to measure with this technology. Yet, simultaneously, and perhaps counterintuitively, this approach may engender a greater authority for the landscape architectural profession.

In the fields of agriculture and forestry, sensor-based technologies are rapidly reducing production costs in response to the pressures of greater demand and the need to promote a more rational use of natural resources. These technological solutions are created "with the aim of increasing production or accurate inventories for sustainability while the environmental impact is minimized by reducing the application of agro-chemicals and increasing the use of environmentally friendly agronomical practices".[21]

Smart farms employ an array of sensors, including a fundamental set that measures soil moisture, temperature, rainfall, light, wind speed and direction, leaf wetness, barometric pressure, evapotranspiration and air quality. These are augmented by rapidly developing optical sensors – technologies used for: detecting weeds; soil analysis and identifying soil fertility; water conservation; yield; precise and minimal deployment of pesticides and fertilisers; the detection and classification of crops, weeds and fruits; weed control; positioning, navigation and safety; and pest control, extending further into taste and odour detection in the form of an electronic tongue and nose.

Not only are the sensor-based technologies evolving rapidly in these industries, they are also reshaping their own operations and processes of production. These industries have achieved success in many areas; they have reduced costs, increased production, they have reduced environmental impact, and they are easily and clearly defined. It is less clear how they might be used and measured within the discipline of landscape architecture, and this is therefore a point of departure for exploration in the teaching and learning framework described in the following projects.

Whether it be in the agriculture industry or the field of landscape architecture, the capacity to continuously monitor specific conditions is inherently related to tracking and measuring change over time – form is understood as only a "stable moment in the system's evolution".[22] The complex material, spatial and temporal relationships that describe the phenomena of form-making are engaged in a perpetual exchange of information engendered within their own material parameters in negotiation with external conditions. Using sensor technologies enables a description of phenomena, which may include a wide range from the mundane to the extraordinary, and from the benevolent to the catastrophic, by translating the material registration of force as information. The intention here is to reveal the "motor of matter, the modulus that controls what it does".[23] How we then visualise this information, translate and re-present it, augments our observational field beyond that which is visible to the human eye; instead, new forms of engagement with the landscape are shaped. This may hold potentials to inform new modes of practice relevant to the changes that the climate crisis brings, not simply in the next stable state of being, but one that may be in an even more radically dynamic state of change.

Projects

The following projects were conducted by students in the Landscape Architecture Program at the Graduate School of Design, Harvard University. The projects demonstrate a number of key themes that exemplify the diversity by which sensor technologies may inform and reshape the practice and discipline of landscape architecture.

Uncommon Ground

Initially interested in the sensor technology's capacity to make visible the invisible forces at work in the landscape, the students moved quickly to realise the potential to not only engage with the landscape in new ways, but also the possibility of working with less conventional types of landscapes uncommon to the landscape architect's wheelhouse. Colin Chadderton's project, titled "Reveal–Transform–Respond", endeavoured to engage with a post-agricultural extracted peatland – usually the domain of engineers and industry; it was a project that lent itself to using sensor technology because of the peatland's remote location and fragile condition.

Continuous Engagement

Regarding the climate crisis, and the issue of carbon sequestration, peatlands represent the world's largest natural terrestrial carbon store. In a relatively small area, they hold more carbon than all the world's forests combined; however, due to lack of meaningful regulation, they are highly vulnerable. In a deceptively flat terrain, microscale changes, such as water levels, have a macroscale effect on peatland ecology because of the capacity of keystone species to survive and thrive. The continuous monitoring and responsive capabilities integrated into the design contributed to a more rapid recovery of the peatland ecology post extraction. In contrast to a conventional contract's endpoint at the "completion" of a project, the enquiry evidenced a continuous form of engagement (Figure 22.2) that made it possible to set strategic goals and work dynamically with them over time, as required. This both shifts the roles and responsibilities of the landscape architect and allows for response and intervention over a greater timeframe.

Comparatives

The proposed sensing network of the extracted peatland was further extended to a natural "untouched" (unharvested) peatland site in the region. This was established as a "control" site (Figure 22.3) to compare and contrast growth rates of keystone species within each peatland system, and to understand the effects of external systems, such as global mean temperatures, on internal systems.

Non-human

The remote peatland site was designed less as a space for humans, than as an ecology for non-humans from which everyone benefits. Physical human interaction is not discouraged yet is not required for the site's development, which conventionally has been a major financial hurdle for existing industry-lead responses that essentially sees them leaving the post-extracted site to its own (devastated)

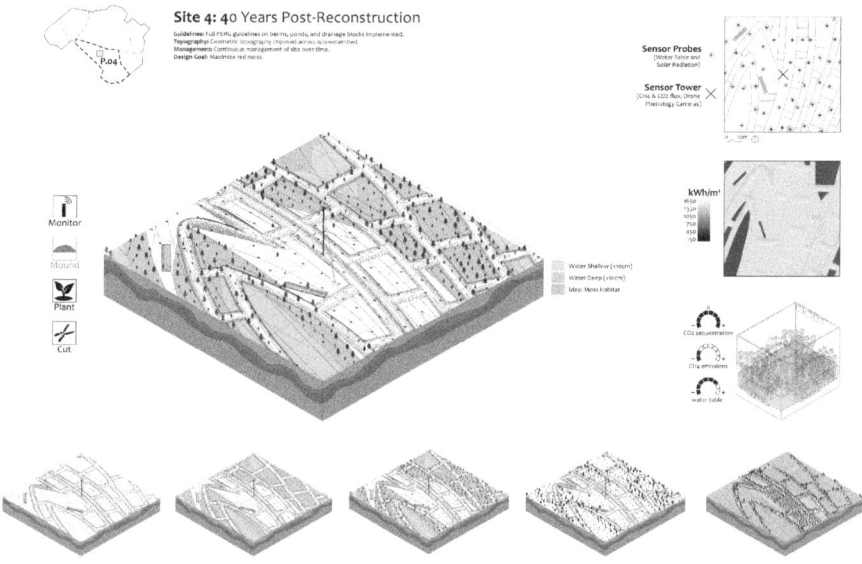

FIGURE 22.2 Continuous response of change over time, "Reveal–Transform–Respond". Colin Chadderton.

devices. Anyone who does manage to venture to the remote location can "see" the less visible complexity and phenomena of the landscape on their personal digital device such as their phone (Figure 22.4) through an application that links to the sensed landscape. The virtual landscape is deployed further afield on a public scale, through online websites and experiences, which educate and bring public attention to these vulnerable sites, provoke government responsibility and hold industry accountable. The work suggests realms of co-existive territories beyond a human-centric endeavour, and prompts efforts to assess the economic value of ecological functions.

Memory

Zhaodi Wang's project emerged from an interest in utilising sensor technology and associated actuators to shape new senses of time, geography and memory in and through the landscape. The sites selected for this investigation were the towns of Pripyat, abandoned during the Chernobyl nuclear disaster, and Slavutych, the purpose-built township to which the people of Pripyat were resettled.

The project titled "Mutual Dependence" proposed a set of living memorials to address the deep connection of the community to their former homes. It did so by using sensor networks to connect the new town of Slavutych to the former town of Chernobyl. Conditions within the former town were recorded,

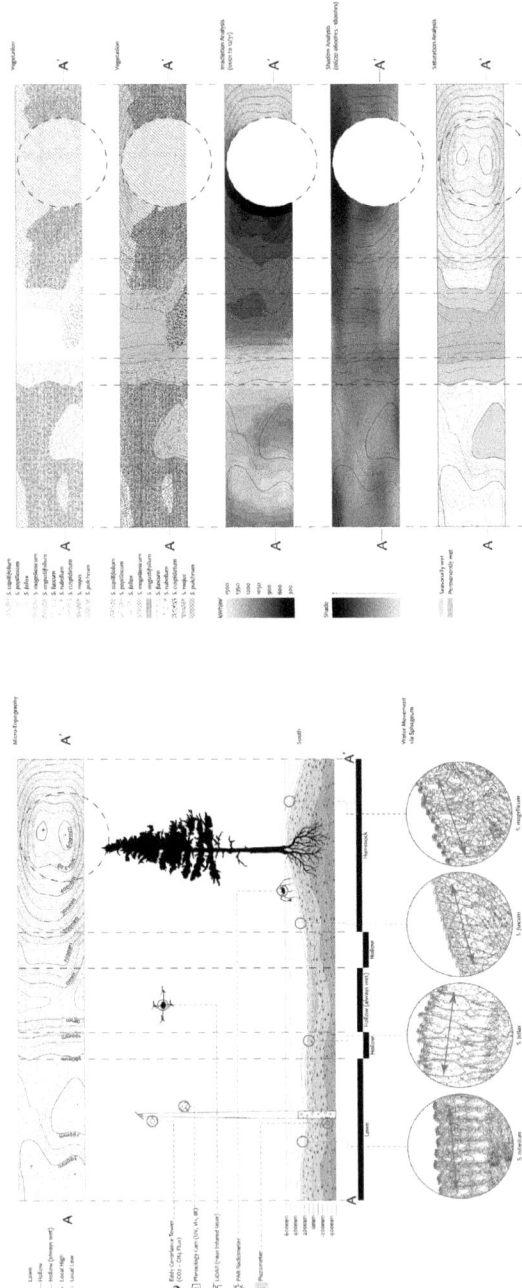

FIGURE 22.3 Comparative site, sensing, analysis and visualisation, "Reveal–Transform–Respond". Colin Chadderton.

FIGURE 22.4   Making the invisible visible, "Reveal–Transform–Respond", Colin Chadderton.

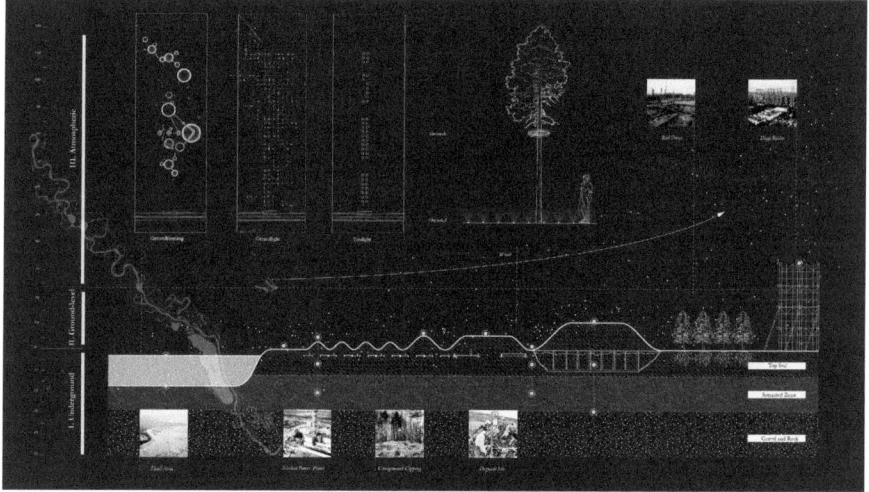

FIGURE 22.5   Sensed landscape of Chernobyl, "Mutual Dependence". Zhaodi Wang.

decoded, translated and transferred, thus linking the towns and relinking the community across the geographically imposed gap.

A range of lighting forms and effects was carefully composed and integrated with landform and vegetation strategies. (Figures 22.5) Together, they create in the new town a unique, visually rich sequence of spaces out of the translated sensor data from Chernobyl.

Next Steps

Sensors are instruments that may enable us to expand our myopic gaze. They are reality amplifiers. They do not stand as a promise to produce the final theory of everything. The unknown, and perhaps unknowable, will continue to exist as necessary gaps in our knowledge that engender and catalyse translation and conjecture and, therefore, creativity and invention.

The "Sensed Landscape" moves toward a "world of ambient intelligence, happening around us on the periphery of our awareness, where our environment is not a passive backdrop but an active agent in organising daily lives".[24] Agency resides in the challenge of not thinking of the socio-political and the ecological as separate and divided worlds. Instead, agency "seeks to explore how humans and non-humans collectively and in emergent ways construct a region of objectivity by being implicated in networks of connectivity".[25] This framework encourages assessment of the value of the ecological work of nature, shifting our perception of it from an amenity to plunder, to an active agent accounted for as natural capital which ultimately equips the landscape architect with tools to observe and design alongside change in ways that inform new forms of agency and practice.

Notes

1 Massumi, B. (1992). *A user's guide to capitalism and schizophrenia: Deviations from Deleuze and Guatrrai.* MIT Press, p. 10.
2 Gabrys, J. (2016). *Program Earth, Environmental Sensing Technology and the Making of a Computational Planet.* University of Minnesota Press, p. 3.
3 Ibid, p. 4.
4 Ibid, p. 8.
5 Harvard Center for Green Buildings and Cities, Harvard University, led by Founding Director Ali Malkawi, https://harvardcgbc.org.
6 Ibid.
7 Anderson Research Group, *Chemistry and physics of climate and earth system change, energy and climate,* Harvard University, https://www.arp.harvard.edu/context-energy -climate.
8 Sayegh, A., & Andreani, S. (2016). Experiencing the built environment: Strategies to measure objective and subjective qualities of places. *Open Geospatial Data, Software and Standards, 1,* 11, https://doi.org/10.1186/s40965-016-0013-0.
9 DeLanda, M. (2005). Space: Extensive and intensive, actual and virtual. In Buchanan, I., & Lambert, G. (Eds.), *Deleuze and space* (p. 83). University of Toronto Press.
10 Ibid, p.,81.
11 Ibid.
12 Gabrys, J. (2016). *Program earth, environmental sensing technology and the making of a computational planet.* University of Minnesota Press, p. 9.
13 Papadopoulos, D. (2018). *Experimental practice, technoscience, alterontologies and more-than-social movements.* Duke University Press, p. 1.
14 Ibid, p. 3.
15 Latour, B. (1987). *Science in action: How to follow scientists and engineers through society.* Harvard University Press.
16 Papadopoulos, D. (2018). *Experimental practice, technoscience, alterontologies and more-than-social movements.* Duke University Press, p. 1.
17 Phelps, B. (2018). Beyond heuristic design. In Cantrell, B., & Mekies, A. (Eds.), *Codify, parametric and computational design in landscape architecture* (p. 197). Routledge.
18 Pajares, G., Peruzzi, A., & Gonzalez-de-Santos, P. (2013). Sensors in agriculture and forestry. *Sensors, 13,* 12132–12139, https://doi.org/10.3390/s130912132.

19 Robinson, A. (2015). Owens Lake rapid landscape prototyping machine reverse engineering design agency for landscape infrastructures. In Gerber, D., & Ibanez, M. (Eds.), *Paradigms in computing: Making, machines, and models for design agency in architecture* (p. 349). Actar.
20 Ibid, p. 351.
21 Pajares, G., Peruzzi, A., & Gonzalez-de-Santos, P. (2013). Sensors in agriculture and forestry. *Sensors, 13,* 12132–12139, https://doi.org/10.3390/s130912132.
22 Allen, S. (1992). Landscapes of change. In Assemblage (19) (p. 59). *The MIT Press.*
23 Kwinter, S. (2006). The Judo of cold combustion. In Reiser, J., & Umemoto, N. (Eds.), *Atlas of novel tectonics* (p. 12). Princeton Architectural Press.
24 Crang, M., & Graham, S. (2007). Sentient cities ambient intelligence and the politics of urban space. *Information Communication and Society,* (p. 790).
25 Papadopoulos, D. (2018). *Experimental practice, technoscience, alterontologies and more-than-social movements.* Duke University Press, p. 148.

23
CONVERSATION WITH BRADLEY CANTRELL

Could you define ecology in the context of your current research and teaching?

I oscillate within a very broad definition of ecology, positioning it in relation to Timothy Morton, where it's focused on ecological thought: a focus on relationships. I find this definition productive for my work, but I also find myself sliding back into ecological science and the kind of normative ecological thinking at other moments. I would say it's typically in this broader range, and this broader idea of ecology and relationships. What's important for me about that is trying to find ways of working that are not reductive in nature. It's about finding ways to access complexity and being able to understand it and being able to use that as a generative design tool. This concept of relationships is important, but it's less about trying to build the model of it and more about using that as the framework to understand what I'm working within.

Could you elaborate on the nature of your teaching and research practice?

I'm an unashamed academic. I've pretty much fallen into that realm at this point. That said, the practice side of things is more about consulting. It's about how my body of research, which I've been working on for the last decade and a half, finds its way into moments of practice. I have an idea that in the next 15 years or so there is space for my research in a design practice. I'm seeing a demand from people doing work that is similar to my research. We've been teaching students for a long time, and we convince them somehow! Now they're at the right places in firms, so they want us to figure out how to work with them. And the research is mature enough, so I believe there is a space for it.

My research started off about representation – 20 years ago – about how we are able to expand our lexicon or forms of representation, how we begin to move beyond a modernist regime of documentation and look more at a relational quality of representation and broader temporal aspects. But in the last ten years my research has focused more on erasing representation and exploring a direct

DOI: 10.4324/9781003145905-28

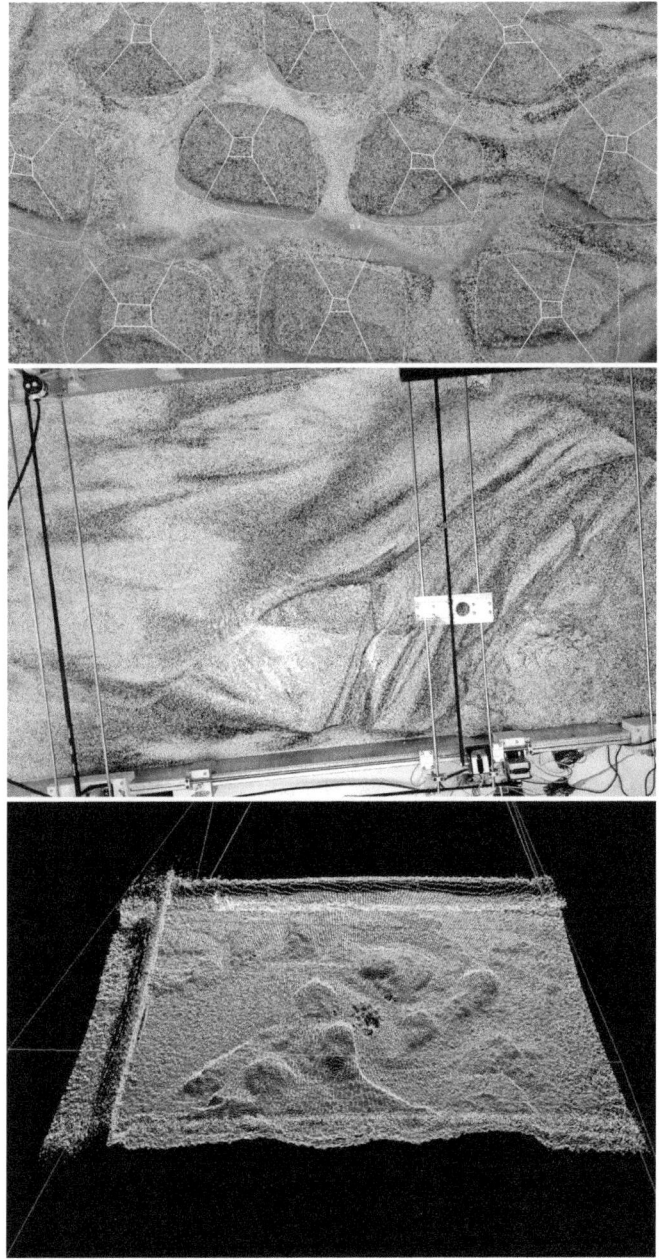

FIGURE 23.1 Top: Sediment patterning abstraction and underlay, centroid tracking. Middle: Geomorphology table view with sediment patterning and sensor arrays. Bottom: Real-time point cloud of surface morphology generated via Microsoft Kinect. Bradley Cantrell.

connection between procedures, and the construction and maintenance of the landscape. How does the original representational project start to evaporate and become the landscape that we work within?

My work really took off when I was in Louisiana looking at the Mississippi Delta. It's just such a rich place for all the things I was trying to understand at the Harvard Graduate School of Design (GSD) during my time there as an MLA student, and then in my subsequent work. It just started to foment there in terms of this really rich, liquid landscape. I'm trying to understand how it transitions to other landscapes, so all of my work these days has been about that basis for these dynamic spaces; how that actually has a foothold in other landscapes that essentially we've reconstructed to be static, but the change is occurring around it. How do we chisel it apart and use these kinds of ideas to build more resilience, particularly in a coastal landscape? Virginia has been a really interesting place to begin to think about that, just because the coast here has been so well-constructed over time, so well held in place.

For me, the most fruitful part of that research has been less about the design speculation and more about the speculation on tools. That's where I found myself to have the most impact: what tools do we need to develop? What tools am I seeing? What are the weaknesses in them? How do I feel like I could augment and extend them or just reinvent them completely? And that's been through my work with the physical models in connection with computational models. I believe the rhetoric of dynamism and flux, and all these things we've talked about for the last couple of decades, but what are the tools we need to actually do something like that?

My teaching in recent years, on the foundation curriculum here at The University of Virginia (UVA), has been about taking this concept of how we begin to actively interact with a landscape. It's been about thinking of it through the lens of gardening but how does that actually happen at scale? What does it mean for us to actually modify the territory in this way?

Could you please elaborate on what you see as the relationship between the tools and the material? Why are the tools important in this relationship?

The term tool is really broad. It requires some definition. If we think about tools from the perspective of representation, what's important is that we are typically taking another discipline's tools – such as civil engineering architecture, even modelling in ecology – and we're then adapting them to the way landscape architects want to think about the world. My take is that we need to drive that process a bit more. We don't need to completely reinvent those representational tools, but we do need to go a little bit deeper. Rather than taking another discipline's tools and just saying we can make a picture of what we want, we need to think about how those tools drive that process. If we believe in the agency of representation, then I think we need to think about how it's generative, more than (or as much as) what the product is. That's where those tools become important to me (Figure 23.1).

On the other side, those tools become important because they're not just images or models we produce. They also have the capability to directly impact or modify the environment. Those tools are being driven now by a mediated process, and for me it's about compressing that process and making it tighter between what we're imagining and what we're constructing. There are different ways to think about that, but the one I've been most interested in is the tools that allow us to continually make a modification, understand its impact, and then re-evaluate and make the next move. That's the space I've been interested in working in, so material-wise it's been very much about getting closer and closer to the landscape material, to climate, to hydrology, to soil, to the plant material. How can we get closer to the medium and have a less mediated relationship between our thoughts, our representation and our methods of construction?

Could you discuss how your research and pedagogical focus has changed over time, and what has influenced those changes over time?

My current work has been very much about advancing this concept of tools. Part of that has been finding a better relationship between disciplines. Here at UVA, it's been working closely with Environmental Science and Geomorphology, and the models they've been using of landscape systems. What's so interesting to me about geomorphology is that these models are looking at landscape evolution over millions of years. Those models are essentially painting a picture of how we understand the geology of the Earth, or even other planets. How do we take the models they're starting to construct and, as landscape architects, think about our generative models? How do we begin to find a middle ground there? Some funded research we've received here lately has been about that synthesis. What is that connection between design and geomorphology, and particularly around this question of models?

The lab we're forming at UVA – myself, Andi Hansen, Brian Davis, Zihao Zhang, Xun Lui and Marantha Dawkins – is really about design tools collaborating in Environmental Science and Geomorphology where they've started developing a landscape evolution modelling lab. Then we have this middle ground where we're building a series of models that look at what that hybrid is between those two; so it is making that connection almost directly between the disciplines, and using the model as the space for that.

This research has been primarily around this concept of interface, a topic one of my PhD students, Xun Lui, is really interested in. What are the interfaces we start to build so it becomes less about the model itself, and more about how we harvest the data? How are we then to visualise that, interface with it, alter it? How are we building those interfaces? Brian Davis and Alexander Robinson wrote an interesting essay for *Codify*, the book Adam Mekies and I put together. The essay is all about this concept of interface, and really thinking about that as a mechanism for how we open up these really complex modelling systems to ourselves as designers as well as a broader public (Figure 23.2).

The other is a project we've been working on here at UVA. It's connected with a range of people in landscape architecture, in the School of Architecture more broadly. We're calling it The Land Lab. It's an old military airfield the school owns and which we have access to. We're beginning to develop ways of

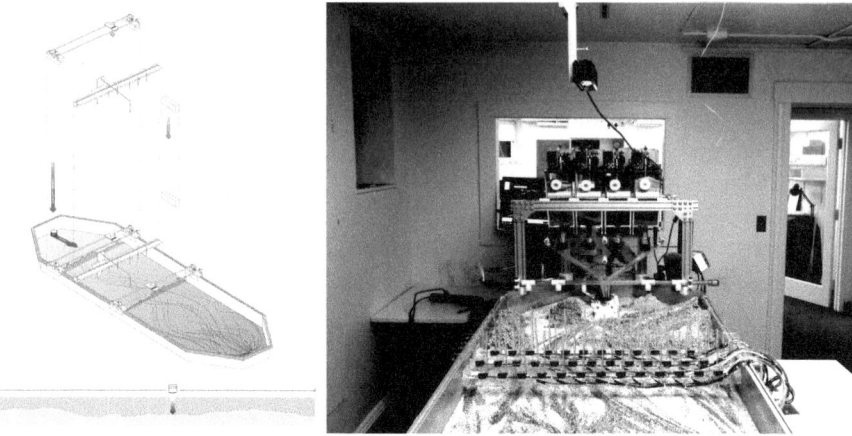

FIGURE 23.2   Left: Diagram of geomorphology table sensing devices. Bradley Cantrell, Justine Holzman. Right: Geomorphology modelling table and towards sentience robotic infrastructure. Leif Estrada and Bradley Cantrell.

doing long-term prototyping, long-term modelling in this large site, as well as one-to-one landscape construction and fabrication. That is a long-term project for us here – one that plays to our strengths being a rural school and rural research facility. Basically, we're compressing representation and landscape into a single space and beginning to work with that in a real way.

Could you discuss what you see as the relationship between sensing, adaptation and climate disturbance?

I'm most interested in the concept of responsive landscapes. How do we, as designers, begin to think about how we understand the environment, how we modify it and how we have a continuing relationship with it? It is what is most interesting for me in this idea of sensing and feedback, and how that plays out then in ideas of resilience or adaptation.

We're attempting to understand as much as possible about the world around us. We often call that observation or site observation, but we're trying to understand as much about the world as possible. We then try to build representations or models of that world. We make a proposition through those models, and then we build a set of documents to construct and maintain that new landscape. With sensing and feedback, rather than pretending we're ever going to know enough, it might be possible to just begin and then learn. By doing this, we then start to develop a new set of epistemologies that I would say start to produce a concept of an adaptive epistemology: we think we know the truth, and as we begin to alter the world, we slowly evolve that truth.

This is a most encouraging idea. It's situated within how these technologies might actually change our view of how we construct, of how we start to build in the world. It folds in the idea that we will never know enough, and that what we really need is a much more robust knowledge network. We need to be adaptable in the way we're maintaining these infrastructures, or whatever we are going to call them at this point. I think it has a whole new language which starts to build up from it.

I got a little stuck on this idea of physical modelling and simulation. What I'm coming to grips with is the idea that none of those things are going to answer the problems we have in front of us. These are huge, intractable problems. They really ask for a different way of interfacing the world. I think this idea of shifting how we begin to design and construct in the built environment, that's where I see the most promise.

Could you discuss the significance of landscape as a medium of material or ecological qualities, and the importance of having coded techniques of measure, analysis and form-making?

I think this then plays back to your original question about ecology and how I see ecology. It's the idea that soil with its chemical properties, its formation, it is its own ecology in itself, which is then attached to another set of relationships. It's about the way hydrology plays a role in that and, in the end, builds all of these endemic qualities in landscape systems that we see effects from.

The medium of landscape is, for me, the most important aspect of what we do. There is no kind of spatial quality that's landscape specific. There's no experience that's landscape specific. It's the medium. To me, that's all there really is. It's for us to re-mix, re-curate, re-choreograph that medium and begin to create these new spaces, these new experiences. The first-year curriculum here at UVA is all about understanding that. What is the medium of landscape? How do you manipulate it? What are the effects you can start to pull out of that?

What you find is that it's infinite. History might say there are only certain things landscape architects do, yet there's a whole range of things we really can do with this medium. And that is for me really, really, really exciting. I always go back to those things. I'm someone who loves water and dirt. I put robots with those two things and that's my happy place! So the medium itself – that's where it all is (Figure 22.3).

What do you see as the future adaptations or extensions of your current research as a means for responding to the climate, and let's say associated crises at the moment?

I don't know if I have a direct answer. I'm trying to be humble about the way I'm working in this area because I feel like my work just chips away at a tiny place. This is why I focus on what I call tools. That is important to me because I can begin to take the larger issue and think about how we make small differences. My hope is to take the growing computational power we're developing as human beings, and tie that together more clearly to the physical medium of the planet we live on, or the universe, not to a virtualisation of the world. How do we tie this weird virtual computational space we're making, which Bratton would refer to as a stack – how do we tie it to the actual materials? Because right

FIGURE 23.3   Top. Robotic sediment gates. Prentiss Darden and Bradley Cantrell. Bottom: Geomorphology modelling table instrument visualisation. Jeremy Hartley and Bradley Cantrell.

now they're not so closely tied together, and I think it's problematic because it does not address true complexity.

It's important that those two things are more tightly intertwined. That's another underlying project for me: those tools I'm developing, the ways of seeing I think that we're working on, are really about getting our hands dirty in some ways. Without that connection, we're slowly floating ourselves farther, farther and farther away from the crisis itself. I think that's actually a problem. It allows us to feel a bit superior, as though we're going to solve it, and it also allows us to have more inequality in the work. I think that's another big issue — that as we move away from the crisis and start to virtualise it, we can also start to ignore the inequality we continue to put into the world.

24

ARCHITECTURE OF ECOLOGICAL ATTUNEMENT

Environment, Form and Feedback

Dana Cupkova

Founded on the premise that architecture is part of a larger planetary ecology, *Environment, Form and Feedback* is a design studio focused on architectures of extreme urban environments (Figure 24.1). The post-industrial context becomes a testing ground for new urban forms and infrastructures within a landscape of perpetual flooding, projections of rising waters and extreme weather events, further intensifying the site's proximity to the river. Grounded in concepts of ecological thinking, the studio operates between landscape and architecture, encouraging students to develop environmental sensibilities and a deeper understanding of environmental ethics relative to resource sharing and systems of spatial justice. According to Thompson,[1] the issues of environmental ethics, ethical design practice and ethical theory have been undermined historically by being grounded in the theory of evolution. Introducing frameworks of socio-biology within design (Figure 24.2), this studio probes at entanglements of social and biological through its spatial propositions. Situated at the cusp of discrepancies between land ownership and climate-based landscape behaviour, architecture is positioned as a form of *ecological restoration*, conceived within a territory of ecological attunement that intends to function beyond the environmentalist, solution-oriented paradigm.

Ecology works across boundaries imposed by social and political systems. Historically, large-scale ecological patterns have been disregarded by practices of architecture and urban planning. Modernist design thinking, as inherited from the era of industrialisation and master planning, has been largely co-opted by ideologies of capital that organise social systems according to political and economic engineering, rather than equitable access to resources or cycles of resilience. To conceive of an architecture of ecological restoration, multilayered descriptions of landscape behaviours become central to positioning design tactics. The maps constructed as part of this process bring forward multidimensional

DOI: 10.4324/9781003145905-29

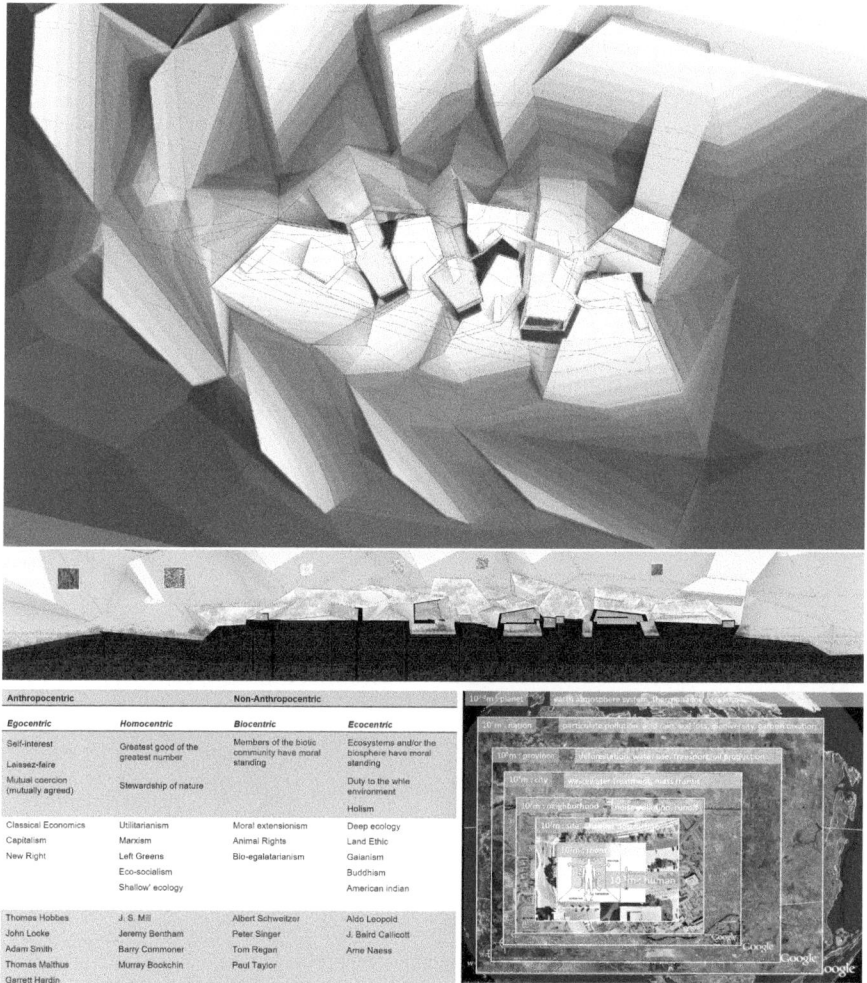

FIGURE 24.1 Top: EEF – Water Communities Studio, 2018, Six-mile Island transformation. Project by Gil Jang. Co-instructor: Nina Chase. Bottom: A Typology of Theories and Positions in Environmental Ethics created after Thomson (1995), and diagram of scalar interconnectivity relative to data information queries to environmental patterns. Diagrams by Dana Cupkova.

representations that question the traditional definition of the site as a reductive formal diagram. They intend to communicate the dynamic nature of littoral patterns, flood dynamics, shifts in geology and the effects of existing grey infrastructures relative to pollution patterns and soil-water toxicity (Figure 24.3). The large-scale maps intend to form a deeper knowledge that incites drama, captures experience and solicits design reaction across scales (Figure 24.4).

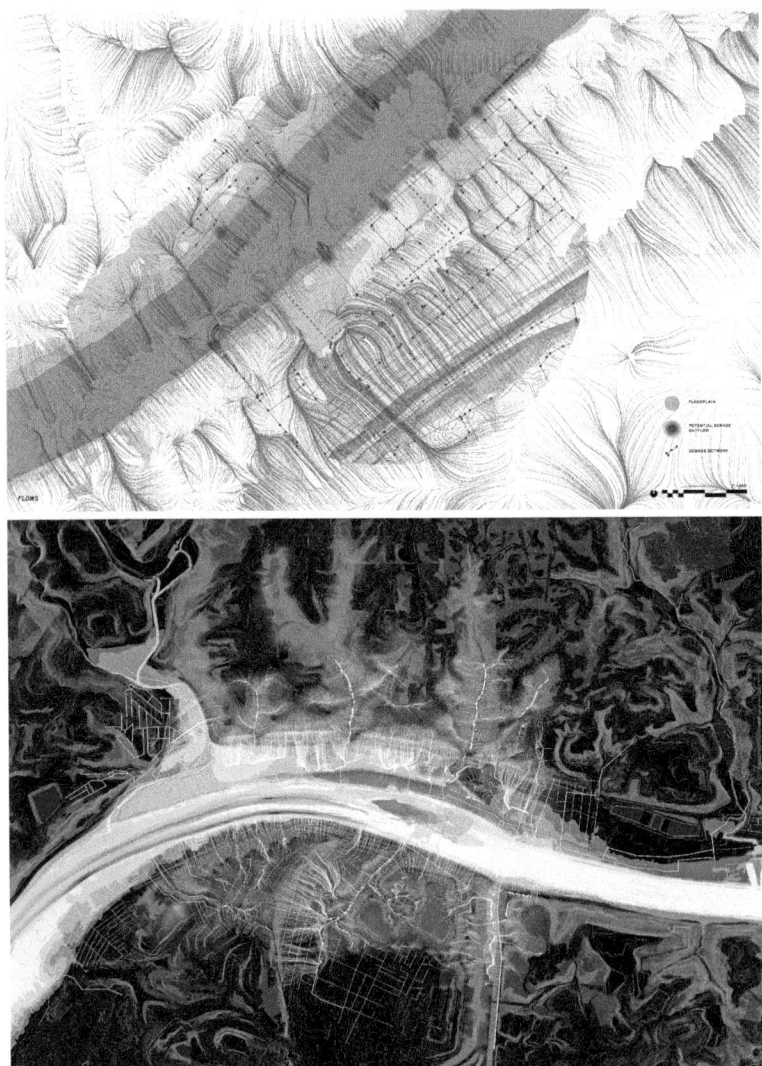

FIGURE 24.2 Top: EEF – Modular Eco-Morphologies Studio, 2016, Strip District Waterfront transformation, site and climate map, showing potential patterns of toxic distributions of sewer overflows in conjunction with floodplains, urban watershed and drainage basins. Project by Adam Kor. Co-instructors: Gretchen Craig and Eddy Man Kim. Bottom: EEF – Water Communities Studio, 2018, Six-Mile Island transformation, site and climate map showing ecological patterns within a large-scale territory combining historical shifts of terrain morphology, flood zone information, topography, natural watershed, water surface flow and underground water sewer system. Project by Ryu Kondrup. Co-instructor: Marantha Dawkins.

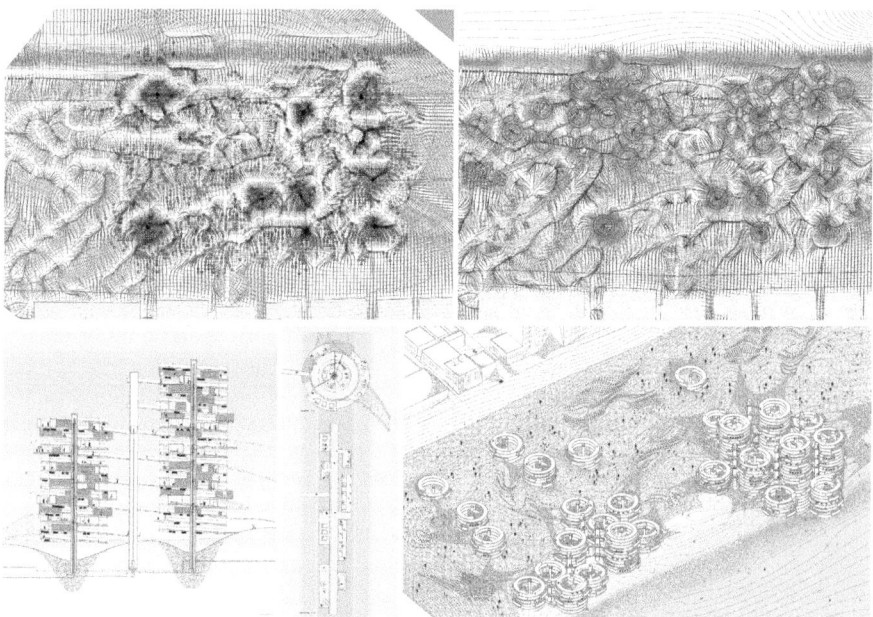

FIGURE 24.3 Top: EEF – Modular Eco-Morphologies Studio, 2016, Strip District Waterfront transformation. The Pollution Outflows drawing series identifies the points of conflict between underground combined sewage network systems and natural watersheds. Not tied to ownership of land, or territory of urban rights, this investigation promotes shared forms of habitation based on negotiating collective social logic implied by environmental urgency of waste and water flow patterns. A new design proposal emerges out of concentrating pollution intensities into architecture of grey water remediation. Project by Adam Kor and Timothy Khalifa, supported by Urban Wetware drawing workshop with Marco Polletto. Co-instructors: Gretchen Craig and Eddy Man Kim. Bottom: EFF – Modular Eco-Morphologies Studio, 2016, Strip District Waterfront transformation. The Water Coils operate like interconnected vertical streets, allowing for plug and play programs, connecting circulation to higher grounds on the unflooded land connected to a network of bioswales for water filtration. The filtered water is extracted underground through the infrastructural coils that vertically transform into a collective housing scheme, while using gravity as a basic force for water distribution and filtration. Project by Adam Kor and Timothy Khalifa. Co-instructors: Gretchen Craig and Eddy Man Kim.

The overarching pedagogical ambition is to examine architectures that inquire into energy and matter relative to the history of industrial landscape formations, while simultaneously re-examining the role of site as a descriptive force. Performative models in architecture have exhausted their breadth and fidelity by reducing context to a fixed information set, generally queried through measured

FIGURE 24.4   EFF – Water Communities Studio, 2018, Six-Mile Island transformation. Embracing the ecological volatility of site, amphibious houses rise and fall with the flooding cycles of Pittsburgh's Allegheny River. The geometry of the houses enables beaches of silt to accumulate and dissipate according to the temperament of weather. Project by Ryu Kondrup. Co-instructor: Marantha Dawkins.

simulation sets at a singular moment in time and space. Promoting a shift away from purely data-driven rationales, the desire here is to engage in a design process framed by environmental ethics and sensory subjectivities as part of our collective aesthetic and ecological experience. Environmental aesthetics and aesthetics of nature are branches of philosophy that study the appreciation of the world at large, as it is constituted, by relating to the environments themselves, not simply by particular objects. Engaging the process in which the representations of not-obviously-visible contingencies of natural behaviours within landscapes are visualised helps us to investigate their histories. Focused on patterns of transformation connected to resource extraction and its impact on ecological wellbeing, these drawings try to tease out and imprint new patterns of care upon future development. Environmental empathy is rooted in concepts of otherness and difference through better understanding of that which may be invisible. Design grounded in environmental empathy leads to more diverse paradigms in redistributing resources, new forms of co-shared domesticity, intrasubjectivity, as well as social equity within our collective urban space. At the same time, design remains closely entangled within its ecological functions. In this studio, the process of visualising and understanding a larger set of multidimensional relationships aims to enable a projective design imagination tightly linked to creating biosynthetic and natural, multispecies environments.

Resisting the current trend towards ever-larger human–ecological footprints, this studio employs design techniques founded in a rejection of reductivism, borrowing methods of drawing, data harvesting, simulation and mapping from cartography and other landscape descriptions. Moving towards an evidence-based, data-rich design framework, the architecture is positioned between landscape and material systems that embrace ecological restoration. By discovering and engaging social and environmental patterns within the large-scale landscapes, the hope is for students to develop a design intuition to identify and integrate microclimatic variations within the general climate. The aim is to entangle public and biotic life to foster the creation of new forms of spatial democracy, provoking novel architectural and urban aesthetics. Central to this workflow is to use environmental simulation to encode layered drawings as a multimodal tool for discovering spatial entanglements – for example, bringing forward patterns of toxicity and human settlement (Figure 24.5) – and moving towards the design of architectural interventions that respond to those patterns and translate across scales (Figure 24.6). The studio projects aim to give a new shape to the contemporary city, in the form of these entangled biosynthetic and natural, multispecies environments.

Contemporary conditions of climate change, increasing sea level rise, land degradation and lack of resources pose questions about habitation along the waterfront. *Environment, Form and Feedback*, situated within the proximity of

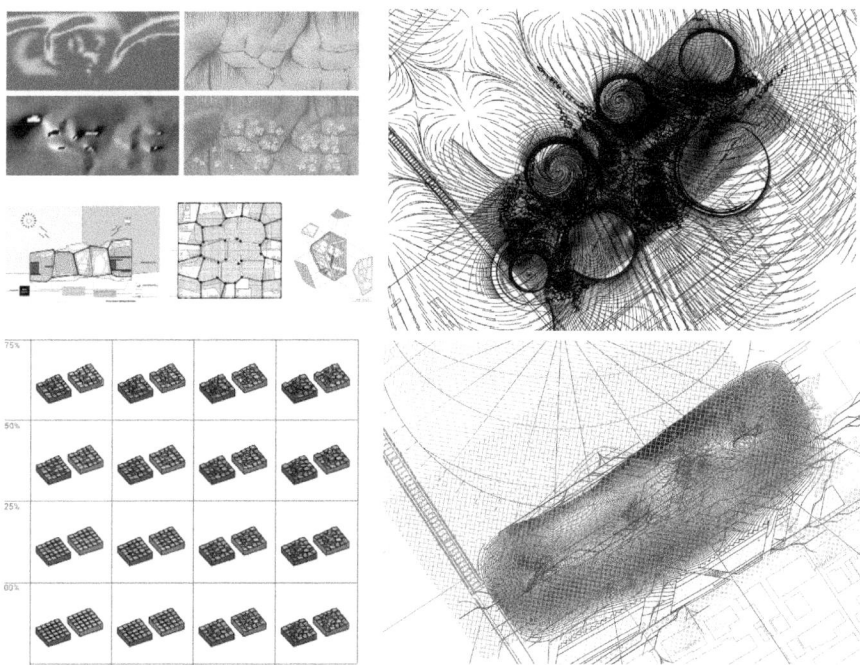

FIGURE 24.5 Between Simulation and Abstraction. The role of simulation in design process is not focused on optimisation, or validation of specific parameters. Rather, it focuses on building intuition towards creative understanding of natural behaviours within multivariable environments, and the ability to translate data frameworks into geometric instances that help shape spatial qualities of new architectural microclimates. Drawings by: Brandon Darreff, Nicole Lee-Park, Elizabeth Levy, Kelli Mijares and Kyle Wing.

FIGURE 24.6   EEF – Museum of Weather, 2019, Six-Mile Island transformation. Curating Flood project takes cue from erosion patterns that shape architecture of floodable landscapes. Project by Cheng Zhou and Bingjie Sheng. Co-instructors: Eddy Man Kim and Emek Erdolu.

waterfront, is rooted in the notion that the environment is a fluctuating force and architecture is an adaptive system that can help to process and redistribute environmental resources. A design of collective typologies responds to current and projected water flow patterns created by aging infrastructure and extreme topography and mediates conflict with the land ownership subdivision. Using environmental data and topographical flow analysis, the goal is to propose the network of new collective living settlements while negotiating the constraints of a unit-based architectural system. Based upon the distribution and capacities of water systems, closed water loops, bioremediation and water conservation, this prompt is designed to develop a deeper awareness of architecture as a part of a larger ecosystem. Promoting the notion of collective living with shared resources, the notion of sustainability is questioned within the framework capital. In this shift towards social-ecology, new forms of land subdivision, based in natural behaviours and building systems, facilitate construction of biotic multispecies spaces. The interdisciplinary goal for the project is to consider global socio-ecological issues such as improvement of water quality through measures that could benefit the construction of natural habitats, support of public recreation, engagement with local food production, and quality of life enhancement at waterfront and upland communities.

Through a semester-long design project, students are introduced to both analogue and computational techniques that integrate form-making with environmental simulation in search of individualised architectural language. In this process, architecture emerges as a contingent object, negotiated between its internal typological constraints and ecological forces of site. Such design thinking is rooted in a systemic approach yet acknowledges the disciplinary autonomy of the creative design process. Students negotiate spatial relationships among the site, water systems, infrastructure, local ecologies and a proposed tectonic logic that accommodates an environmental and programmatic agenda which changes year to year.

Two main concepts linked to design strategies are introduced: visualising the invisible (through using simulation) and eco-machines (a concept of architecture directly linked to living and processing systems, through carefully considering resources).

The invisible is investigated through simulated analysis of dynamic processes and drawing abstracted information to define the character of the site. Students are introduced to a series of computational techniques and environmental simulation methods through which they can speculate about the representational abstraction of large-scale information sets. This is an iterative process where analytical feedback situates datasets within a computational framework in search of formal trends, or solution spaces. The use of drawing and simulation continuously straddles across empirical and metric-based design input.

The notion of the eco-machine is tightly coupled with the context of the invisible. The eco-machine serves as a conveyance system, a processing model to enable distributions of resources and matter between organic and inorganic constructions. The corresponding architectural system is not conceived of as an optimised object, but as a flow model for producing and embodying operations that can self-sustain and transform over time. This concept is rooted in Howard T Odum's[2] work and the understanding that all natural systems aspire towards effectiveness, in search of new forms of quality, while expending minimum energy. The eco-machine always serves a dual pedagogical purpose: it is an object whose creation facilitates dexterity with the computational techniques that inform its evolution, and it is a rhetorical device that can be used to question the artificiality of natural systems.

Terminology of the Studio

The terminology of the studio is structured into three parts across the four phases of the semester project development. Terminology builds awareness supported by lectures and debates, and project-based phases focus on design agency and architectural competencies.

Three Parts of the Studio Project

Part One: Environment

Environment is situated within a framework of interdisciplinary research in ecology, the social sciences and medicine that supports the point of view that dense, urban, biotically diverse environments are also better human environments. Major post-industrial cities across the United States are revitalising the water's edge to bring nature back into cities and to integrate green infrastructure with public access. Working at the boundary of two distinct environments, between ground and water, offers close investigation of strategies for entangling non-human biotic systems, civil infrastructures of transportation, and waste management, with cultural, political and functional patterns of human life at, along and beyond the water's edge.

Part Two: Form

Form in its definition here moves away from the design of discrete city forms. It favours instead the development of architectural intervention through negotiating a series of contingencies that move across different scales and enable definition of a shape for a specific urban community. *Form* is invested in the definition of architecture as closely entangled with infrastructural and landscape operations that inform potentials for actively restructuring the public environment with provision for shared public water access and environmental filtration.

Part Three: Feedback

Feedback is rooted in technique learning. As recent advances in computation, visualisation, simulation, material science and fabrication technology have begun to alter our theoretical understanding of general design principles, they require deeper reformulation of architectural design processes. Our interest lies in shape-making, computational composition, methods for investigating the inter-relationships of parts to whole, emergent organisations and pattern systems at multiple scales and applications. The evolution of digital media has enabled both new techniques in modelling and fabrication, alongside new ways of understanding material organisation as a function of physical properties and potential for hybrid assemblage.

Four Phases of the Studio Project

The studio project's phase is divided into four phases, each with a theme grounding the production of specific drawings and models.

Phase 1: Site and Climate

Students are asked to analyse the site's climate, environmental characteristics, topography, demographics, urban physical relationships, socio-economic and political organisation, ecological potential and hydrological patterns in the context of riparian and wastewater systems. The goal is to gain a comprehensive understanding of larger ecological forces, and to develop analytical drawings and new operative site diagrams to facilitate the negotiation of formal parameters in future proposals. Digital modelling and parametric techniques will be introduced throughout the first half of the semester to help with the development of drawings and diagrams.

Phase 2: Eco-Machines

In this phase, students research, document and diagram the functionality, geometric rules and operative logic of a particular eco-machine: a bio-integrated architectural assembly, processing system or passive strategy that is climatically responsive to the site. The goal is to understand the system's metrics and perform-ative parameters, and to leverage its potential integration into an architectural design proposition. Within the framework of the eco-machine, architecture is used as an applied ecology that underpins its behaviour and form-making. The desired outcome is to gain knowledge that helps to inform future creative design development.

Phase 3: Hybrid Typologies

The goal is to develop a mixed-use typology (a hybrid chunk of a whole) that can be shaped into a future green infrastructure proposal, while creatively resourcing the work developed in Phases 1 and 2. The intention is to form a proposition for a specific typology conceived as a hybrid ecological infrastructure, an environmental filter and a climate and site-specific eco-machine with the ability to support both biological and human systems. The geometry of the hybrid chunk facilitates "eco-machinic" functions, spatial compositions and material organisations that accommodate human habitation programmatically, and address issues of spatial justice underpinned by resource–human–ecology relations.

Phase 4: Proposal

This is a final phase of the semester-long project, with the expectation of delivering a complete architectural/infrastructural proposal, embedded in a hybrid logic that integrates the landscape urbanisation and architectural typology. The goal is for students to understand the larger urban, ecological and structural forces as criteria for developing specific aggregate urban form.

The Goals of *Environment, Form and Feedback*

As the land becomes overpopulated and susceptible to the climatic fluctuations of extreme weather, this pedagogy embraces the larger ecology of the site, followed by a careful development of architectural objects, and their contingencies in shaping public spaces and social interactions at the urban scale. The interdisciplinary goal for the project is to consider global socio-ecological issues such as: improving water quality through measures that could benefit the construction of natural habitats, urban access and equitable distribution of resources, design of material assemblies that reduce or eliminate reliance on carbon infrastructures and exploitative use of natural resources and support of public recreation, especially in ways that help to raise discourse on human–ecological relationships.

Environment, Form and Feedback is invested in architecture that sources its intelligence from landscape behaviour. It encourages students to negotiate spatial relationships through sourcing contested histories of development, while attuning to natural patterns of the site. Proposals are rooted in understanding architecture as a form of projective restoration, an architecture that has a more productive relationship with the environment while being tightly linked to its landscape sensibilities.

Acknowledgements

Architecture of Ecological Attunement: Environment, Form and Feedback is a design pedagogy developed and implemented by Dana Cupkova at the Carnegie Mellon School of Architecture, starting in 2015. The studio sequence is structured as a coordinated content and workflow framework. It involves collaborative teaching sections connecting environmental ethics theories with computational design strategies, supported by an embedded simulation workshop unit. Collective input and the collaborative nature of the studio has been supported by the invaluable contributions of Eddy Man Kim and Matthew Huber through workshop development and section instruction. The additional studio section instructors over the past five years included Gretchen Craig, Matt Plecity, Nina Chase, Marantha Dawkins, Nicolas Azel, Emek Erdolu and Heather Bizon. The 2016 sequence included a digital drawing workshop led by Marco Polletto, and a 2018–19 sequence modelling workshop led by Rhett Russo.

Notes

1 Thompson, P. (Ed.). (1995). *Issues in evolutionary ethics*. State University of New York Press.
2 Howard T. Odum introduced the concept of Emergy as a measure of quality differences between different forms of energy, diagram mappings of which are based on Alfred J. Lodka's predator-prey diagrams. See also Odum, H. T. (1995). *Environmental accounting: Emergy and environmental decision making*. John Wiley & Sons, Inc.

25

FROM GRAIN TO THE TERRITORY

Ana Abram and Maj Plamenitas

This chapter describes landscape architecture as a critical design discipline that deals with processes and changes on a range of spatial, temporal and operational scales. Some of these are demonstrated through relationships developed between projects conducted in the Amphibious Lab and research practice Linkscale, and the works of selected design studios and research clusters that we lead at the Bartlett School of Architecture, University College London (UCL). We outline some of the approaches and methodologies used in our designs and the design research studios as a means of demonstrating the relationship between our research and innovation-focused practice for teaching.

> To suppose that the evolution of the wonderfully adapted biological mechanisms has depended only on a selection out of a haphazard set of variations, each produced by blind chance, is like suggesting that if we went on throwing bricks together into heaps, we should eventually be able to choose ourselves the most desirable house.
>
> *CH Waddington*

> The role of the architect here, I think, is not so much to design a building or a city but to catalyse them, to act that they may evolve. That is the secret of a great architect.
>
> *Gordon Pask, Evolutionary Architecture*

The Cross Scale Design (CSD) teaching and research cluster at the Bartlett School of Architecture stems from innovative research conducted within the Linkscale design practice. Our approach expands the effective agency of design decision-making by incorporating intricate multiscale system dynamics, environmental

DOI: 10.4324/9781003145905-30

conditions, material cycles, forces, economic fluctuations and complex social and ecological dynamics as vital components of the design process.

It is a process of engaging with multi-objective questions, complex conditions and non-standard contexts, utilising non-linear scale-specific relations that allow interactions that can affect, accelerate, alter or inform the link between perpetual processes and discrete interventions. This creates the capacity to curate and influence broad spatial, temporal or operational ranges in landscape and territorial scale design projects

The intention is to introduce a radical shift towards increased resilience. This is achieved by substituting design, production and automation of finite-state outputs with the creation of landscape structures, systems and territories that are capable of adaptation and development during a given lifecycle. This substitution captures the potential for evolution between multiple sedimentary cycles as a model for design orchestration (Figure 25.1).

Design Orchestration is a process that enables design control by establishing a relation between embedded and encoded information, behaviours and contextual fluctuations, to challenge and redefine the established linear and finite

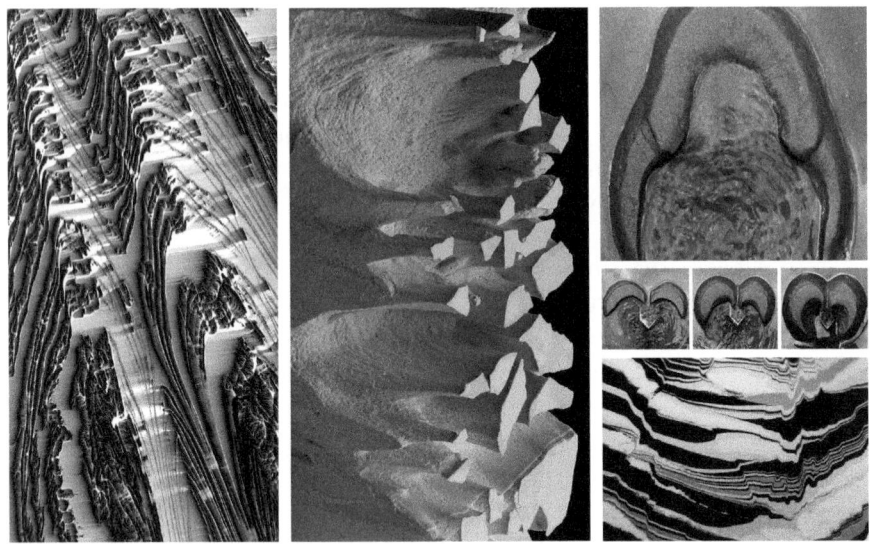

FIGURE 25.1   From left to right: 1. Multi-Scale Flow Map (Maj Plemenitas – Linkscale, 2010–ongoing), 2. Amphibious Interface – Specificity through Exposure and Programmable Erosion (Maj Plemenitas and Ana Abram – Amphibious Lab, 2012–18), 3. (top right) Growing Islands – Evolution of the Surface Morphology as a result of interaction with water currents (Maj Plemenitas). Bottom right: Growing Islands – Deep Ground Model (Maj Plemenitas, Linkscale, Amphibious Lab, 2010–ongoing).

separation between design, construction and lifecycle. This approach also engages with questions of large-scale automation without robots, as demonstrated in the project "Growing Islands and Resilient Shorelines". This example of ongoing research introduces the expanded operational scale range and enables a novel territorial production protocol "from grain to the territory". The project utilises environmental forces and materials present on site. It interfaces with them through strategically positioned nucleation nodes. The nodes locally redirect currents and trigger multi-material sedimentation and structural aggregation, resulting in substantial geomorphological changes but without a construction site's negative impact on surrounding ecosystems.

The research branches of the Amphibious Lab and Linkscale Research have a feedback loop with the "teaching lab". There, together with students, we further investigate specific ideas about tensions between materiality and computation, multiscale structures and systems, and automation and system dynamics. Our ambition is to develop dynamic multiscale structures and territories. The collaboration and tension between these agencies enable critical evaluation and development of innovative forms of design and research that synthesise, and transition from, concept to application.

Landscape and Design with Natural Forces

Landscapes are inherently dynamic, "live" and subject to change.

Landscape architecture is fundamentally distinct from other design disciplines due to its primary agency of design with dynamic agents (we design with them and for them). Some of these agents are, first, alive (such as plants, animals), or second, exhibit dynamic behaviours (like sand and water) and respond to, and influence, broader contextual dynamics. Landscape architecture can effectively respond to challenges associated with accelerated and intensified climatic, ecological, economic and social changes. This is coupled with an investigation into additive, subtractive, substitutional and metamorphic qualities of territories, structures and materials. This can include both anthropogenic and non-anthropogenic forces and agents which drive and orchestrate ecological potentials.

Research and teaching do not focus solely on an isolated single problem-solving approach. Instead, research and teaching focus on the comprehensive, multitrophic nature of coexisting matter and information-based systems that enables their latent generative potential to manifest. By integrating the design process into the external and internal relationships that inform the territorial machine, this outlook on landscape introduces the fundamental shift from isolated, non-contextual design and production methods and finite-state outputs. It substitutes them by focusing on creating a lifecycle of the project. This is achieved by employing a subtle orchestration of agents, and by deploying strategic interventions that are utilised and operated at specific operational scales. Interventions on a targeted operational scale trigger change that affects more extensive spatial, temporal and operational scale ranges. The research and academic projects demonstrate this redefinition of strategic interfaces between the multiscale structures, and the hydrological and geomorphological systems with inherent adaptive, developmental and evolutionary capacities.

Research, Practice and Teaching: Key Projects Which Influence and Inform Our Teaching

Design studios and research clusters at the Bartlett School of Architecture UCL span three distinct programmes: MA/MLA Landscape Architecture, Bartlett Prospective Urban Design (MArch UD) and Bartlett Prospective Architectural Design (MArch). These programmes operate on the expanded scale range. However, all focus on the evolution of the role and agency of design capable of dealing with questions related to accelerated and intensified environmental and climatic dynamics, as well as other social, economic and political changes which can increase the resilience of our built environment.

In 2012, an exploration of distributed territorial production through sediment manipulation was introduced to students. Material and digital simulation tools (such as geo/hydro-morphology tables) allowed students to test, simulate and evaluate their projects within the studio. These tools provide digital and physical feedback during the design process, which students use to assess their evolutionary design proposals (Figure 25.2).

Key Research and Teaching Approaches and Methodologies

(Figure 25.3)

Methods in design and teaching explore cross-scale design to propose and implement novel design protocols and strategies capable of activating an expanded operational and design range. These novel approaches are intended to have a capacity to effectively engage with complex, dynamic, emerging conditions and contexts beyond the effective scope of traditional design practice. The projects in the design studios often focus on hydrological, geological and geomorphological aspects of territories. They position them as multiscale systems affecting agents and resources with shifting identities. They act throughout a lifecycle as vital generative forces that interact with other landscape, urban, architectural or other anthropocentric and non-anthropocentric systems.

They record gradual behaviours, acute events, and even non-linear changes in the strata visible in the organisation of landscape structures and elements and biological agents that inhabit the landscape. Design decisions leave traces in the earth's surface, its ground, water and atmosphere. Together with its cultural, material and ecological specifics, continuous cycles of anthropogenic and non-anthropogenic territorial agencies inform and influence the additive, subtractive,

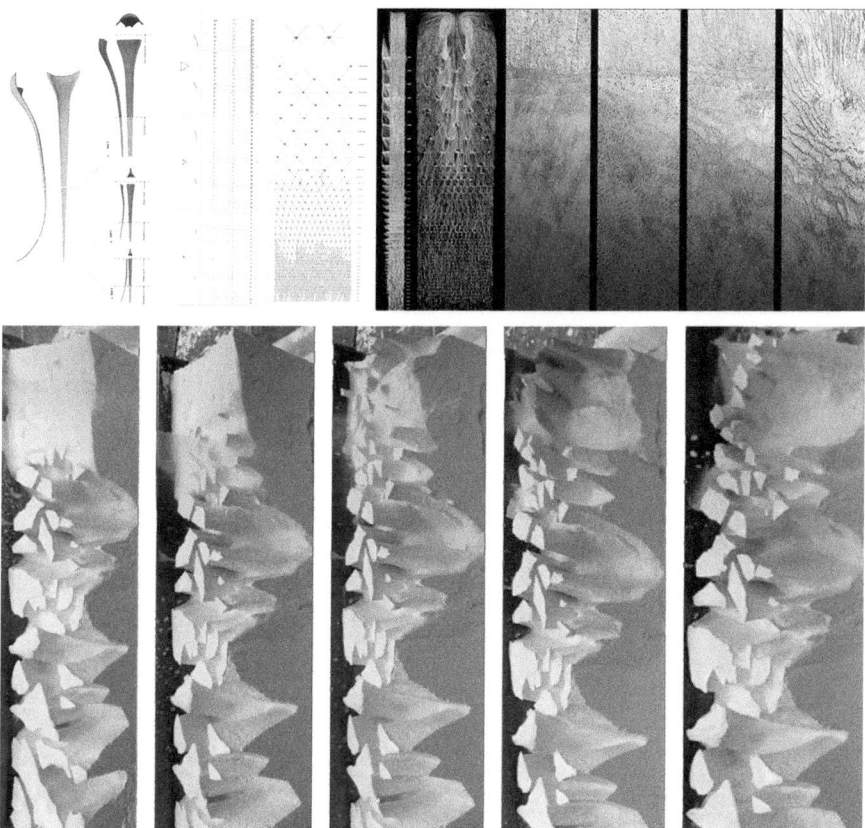

FIGURE 25.2   Top: In-Situ Network – Growing Islands and Resilient Shorelines: Type A001 Island Nucleation Nodes, Nucleation node Network and Morphogenesis of the new landforms (Maj Plemenitas and Ana Abram – Linkscale and Amphibious Lab 2012). Bottom: Amphibious Interface. Amphibious interface is an innovative design and production strategy that investigated the developmental capacity of building skin systems during their lifecycle. The research project explores the principle of specificity through exposure. This is particularly important as it enables use of generic building components that during their lifecycle develop site-specific qualities base on the exposure they receive. The method includes environmental forces as a vital productive agent. (Maj Plemenitas and Ana Abram – Linkscale, Amphibious Lab, 2012–ongoing).

metamorphic and substitutions processes that shape and inform the landscape's specific character.

Within the studio, students envision, prototype, design and produce scalable and transferable design outputs. These range from objects and structures to strategies and systems that operate across the scale range with the capacity

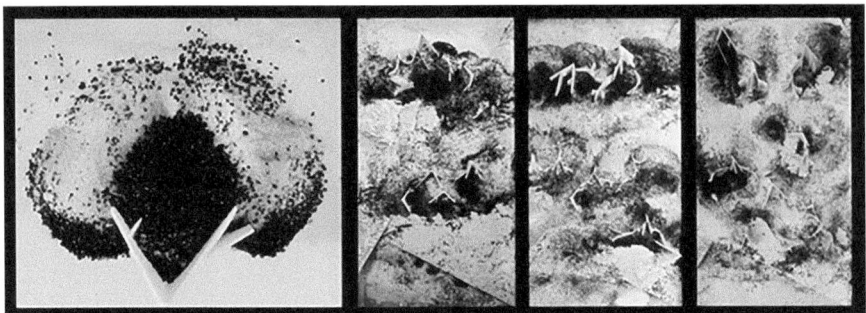

FIGURE 25.3   Physical simulation of material flow in the lab, with the use of hydro-geo-morphological tables. Monitoring the process from the top is as important as understanding landscape evolution in the cross-section. Additive Morphology - Project for Manaus, Brazil, BPro MArch UD Urban Design – 2012, Yang Yang and Shruti Shandhir.

to adapt, develop and evolve. An essential part of the methodology that the research implements is development, testing and fine-tuning of specific computational tools and techniques that enable an approach with three integrated characteristics.

A. Simulation based on generative and analytical data and matter.
   A1. Multiscale approach.
   A2. Digital simulation tools are crucial instruments that allow us to experiment rigorously with processes, forces and interactions. We use computation for understanding and controlling processes and behaviours, and not primarily for representation or form generation. Physical simulations in the hydro/geomorphological tank, with embedded sensors, allow us to experiment and evaluate morphogenetic processes over time.
B. Automation and autonomous territorial production.
   B1. Deep ground models.
   B2. The hydrological and geomorphological apparatus allows us to capture and read the formations well beyond the surface condition. This allows for an understanding of relations between fine grain and large-scale dynamics – a fundamental principle of cross scale design.
C. Analytical evaluation through embedded and remote sensing and representation.
   C1. In situ embedded and remote sensing are valuable assets for understanding how, if and when controlled lab experiments are transferable and scalable.

A1. Multiscale Approach

We investigate the expanded spatial, temporal and operational scale ranges. More specifically, in the context of water that is a constant in our work, its multivalent character can be perceived as a force, material or an agent. It has the ability to change, flow, mix, dissolve, solidify and vaporise, while simultaneously possessing biological, economic, political or symbolic value. Water has the ability to trespass and to establish boundaries and has potential to create new boundaries.

These complex, coexistent, yet sometimes contradicting, qualities call for a fundamental shift from singular problem-solving, and isolated, non-contextual design and governance methods. They call for a turn towards multiscale relations between objectives, territories, ecologies, constructive and destructive processes, external and internal constraints. We use multiscale synthesis as one of the vital strategies for understanding and unlocking the latent capacity that enables engagement with complex questions of our time and the near future that any individual discipline cannot tackle. Cross-scale relations refer to processes at one spatial, temporal or operational scale interacting with processes at another scale. The interactions affect, alter or promote relationships between processes and patterns across scales. These interactions can result in small-scale processes that can influence a broad spatial extent, or large-scale conditions that can interact with small-scale processes to curate complex system dynamics. Within and between these systems and structures, information is translated, embedded, encoded, stored, processed and retrieved.

A2. Digital and Physical Simulation

Digital and physical simulations play a vital role in our research and pedagogic process. Digital simulations are usually used in our lab to simulate behaviours and territories which would not be directly scalable in the lab environment. We use the hydro/geomorphological table for material-related morphogenetic processes when testing material movement, porosity, sedimentation, erosion processes and state-changing materials.

This information exchange through different tools is particularly evident in projects about territorial printing developed by Amphibious Lab and Linkscale. The behaviour of the nucleation node arrays, and material erosion and deposition, are tested in the lab environment to determine their position, orientation and overall configuration. Digital simulation enables production and testing of radically diverse design options before a 1:1 test site is constructed, as in the case of the In-Situ Network's Growing Islands and Resilient Shorelines project (Amphibious Lab, Linkscale, 2010). This was a pioneering project which established foundations for a new approach to territorial scale automation and is used as one example in our academic work (Figure 25.4).

"When the Barrier Falls" is a project that proposes solutions for areas of London under threat of sea level rise, which causes tidal floods. Currently, the Greater London relies on a mechanical solution – "the Thames Barrier" – which

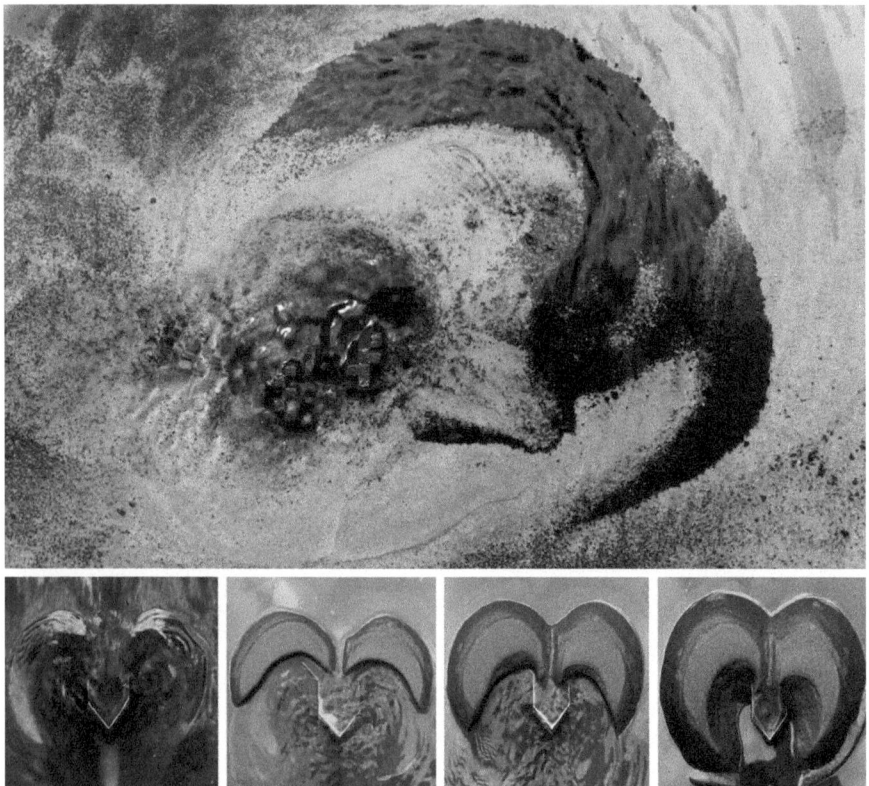

FIGURE 25.4 Growing Islands and Resilient Shorelines – In-Situ Network. Physical simulation of the nucleation node Type C7a53 and bi-directional flow – sediment deposition and deep aggregation. (Maj Plemenitas – Linkscale, Amphibious Lab, 2010-ongoing).

has a limited lifespan and high maintenance cost. The project proposes an alternative approach: deep pocket spaces that provide space for water in the event of flooding. Instead of creating conventional expansive floodplains, which are not an option in the confined urban area, the project proposes new articulated ground which during the non-flooding period is used as a public space, sports ground or play area. In the event of flooding, it can take water and act as a prominent water feature. During the design process, physical prototypes were used to simulate water behaviour and differentiate models which are most appropriate to direct, store and discharge water.

"Tidal Wharf" is a project representing a system of aggregated multi-objective components that form the core of the structure. The system acts as both a flooding prevention mechanism, and the port. The components consist of a variety of constituent units programmed to address a range of objectives. Each constituent is embedded with actuators. When any actuator is in contact with water, it triggers specific behaviour. The behaviour is linked with the porosity designed into the constituents and the material embedded within each of them. The system is designed to redirect water, and to inform sediment deposition by using the geometry assigned to every component. The structure can, therefore, change its programmatic qualities depending on weather conditions: in regular conditions as a port, and during tidal floods as a flooding prevention system. Digital and physical simulations were used to test how the constituents and the structure perform during different water level events, as well as demonstrate the performance of the floating structure.

B1. Deep Ground Models

We use deep ground models to investigate the morphology and structural composition of landscape as a dynamic multidimensional system. The approach creates a bridge between two domains of significant influence – atmospheric and geological. Modelling these deep ground models enables us to understand the landscape's membranes, layers, materiality, wetness and dryness, and hidden forces underneath and above that shape it.

"Fire Barriers in Madeira Island" is a project proposing strategic incisions into the slopes of the island of Madeira which prevent the spread of destructive natural fires. The size, position and shape of cuts take into consideration wind direction, geological ground composition, slope inclination, and type of vegetation. Concave landforms accumulate orographic rainfall which occurs daily on this volcanic island with steep slopes. The water pools establish gaps in the landform and prevent the spreading of fires. Deep ground models represent areas of the land with an impermeable ground composition where creating these natural water basins is feasible (Figure 25.5).

"Shifting Ground" is a project that identifies the shifting, subsiding and sinking landscape of the Nottingham region as an urban process, due to the inextricable role of activities in accelerating these processes. By understanding these geological processes, the project proposes a responsive and dynamic approach for the ecological and infrastructural development of the landscape in the Nottingham area, UK. The approach is based on strategies which, alternately, stabilise and accelerate to create new modes of urbanism across the shifting ground. After digital simulation of the deep ground, dynamic physical models were built to understand and simulate the processes of sinkhole creation.

B2. Hydro/Geomorphological Apparatus

In our design studio, research cluster and laboratory work, students use several hydro/geomorphological apparatuses equipped with actuators, environmental control systems and sensors. These enable students to explore the role of water

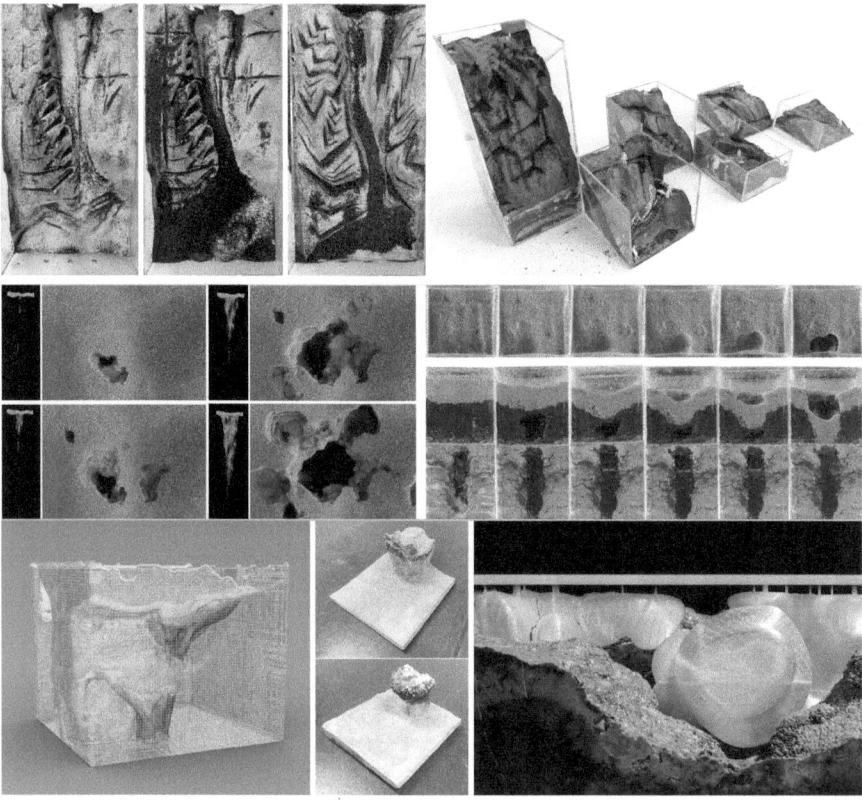

FIGURE 25.5 Top: Deep ground models of test sites for "Fire Barriers in Madeira Island". (MLA Year 1, Studio One, Xiao Mo, 2020). Bottom: Design process transitions between different media and software, from digital to physical simulation, 3D scanning of physical simulation, creating physical cast of the physical experiment final state, to projected design proposal (a point in time) considering all accumulated spatial information. (BPro UD, Project "Shifting ground". Student team: Berta Garriga and Min Yeong Cha, 2018).

as a force, material, effecting agent, resource and a multiscale system, with multivalent identify that holds a vital role, especially when interfacing other landscape, urban, architectural, political, cultural, exo/endo/eco systemic and non/anthropocentric systems.

The work demonstrates the potential for autonomous territorial scale production of resilient islands and shorelines with strategic use of in situ resources. It uses strategically positioned nucleation nodes that interact with water currents and wind to control their velocity, and to trigger sedimentation and aggregation.

This amphibious interface developed in the research practice, and then translated to teaching, enables students to explore the principle of specificity through exposure. This is particularly important as it enables the use of generic build-ing components that develop site-specific qualities based on the exposure they receive during their lifecycles.

Dynamic physical simulations, with simultaneous real-time data capture and analysis, are conducted in laboratory-controlled conditions. The simulations include programmable temperature, water and wind velocity, tidal movement and directional currents. The apparatus is an automated amphibious test environment that allows us to input data from the site, including tidal changes, water and wind velocity and directions, material compositions, and temperature. Since 2010, the automated, programmable apparatus has been a crucial instrument in our experi-mental research. It enables critical investigation, furnishes metrics, enables data capture of the physical simulation processes, and supports evaluation of their poten-tial for real-life application, alongside scalability and transferability (Figure 25.6).

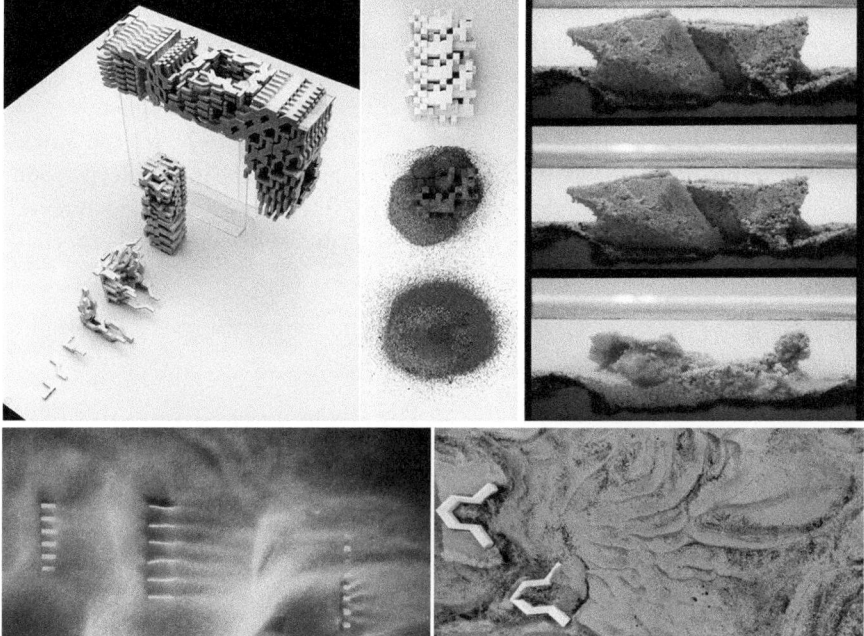

FIGURE 25.6 From left to right: 1. Assembly and disassembly with changing material states (Maj Plemenitas, RC15 with researchers Ian Tu, Oliver Ledesma 2017). 2. Material lifecycle: From grain to part to object, and back to grain. (Maj Plemenitas, 2016–17). 3. Unit made out of program-mable material, which with contact with water changes its state and function (BPro March UD, project "Tidal Wharf", Research Cluster 15 – Maj Plemenitas with researchers Mario Santaniello and Peerapol Jirawattanaek, 2018).

C1. In Situ Application and Remote Sensing

Students investigate how the common distinction between the construction/ production phase and the lifecycle, as separate phases, can be challenged through computation, multi-objective simulation, and 1:1 production with in situ forces and materials.

Furthermore, students explore how indirect design orchestration and production with environmental forces reflect the notion of change through continuous adaptation, development and evolution throughout the lifecycle.

They investigate how design computation, big data, AI and autonomous territorial production facilitate innovative approaches and relevant relationships between research, design practice and industry (Figure 25.7).

One of our research focuses is developing and applying transferable and scalable Minimal Input–Maximal Output design and production strategies. The purpose is to address emerging questions and opportunities associated with accelerated climate change and its symptoms (such as seawater rise, water scarcity, pollution and associated destruction of ecosystems).

In situ work is a vital component of our research. Along with the students, we use embedded and remote sensing and drones that capture visual, IR and thermal spectrums. We regularly use mobile floating laboratories – RIB boats of various sizes equipped with acoustic and thermal sensors, and multifrequency sonars that can penetrate river, lake or sea bottom. With these sensors, we can quickly deploy, capture and generate our bathymetry maps and 3D models, deep ground point cloud models of the underwater structures and deep ground models revealing morphogenetic processes which expose structural and behavioural properties.

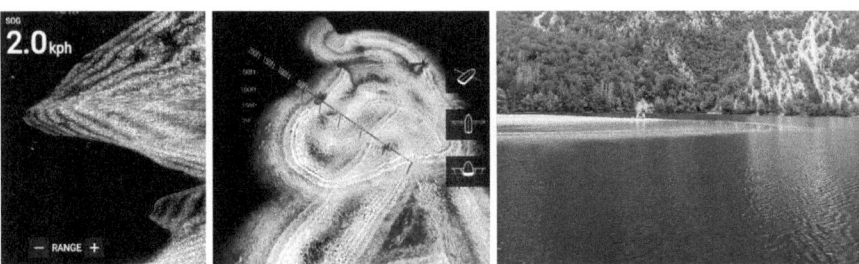

FIGURE 25.7   Images from the rapidly deployable floating laboratory. Remote and in situ underwater sensing from vessels equipped with multifrequency sonars, transducers and other sensors. These allow us to produce constantly updateable bathymetry maps, deep ground point cloud models of the bottom structure configuration, and high resolution 2D and 3D sectional and top view models of the bottom and structures on the bottom, including nucleation node arrays and new morphological features produced through controlled sedimentation, among others. (Amphibious Lab and Linkscale, 2010–ongoing).

The fundamental components of investigations are transitional loops and information exchange between analytical and generative physical and digital models and simulations. These use machine learning and in situ, embedded sensing for complex landscape patterns and behaviour recognition on the level of landscape morphology and the context of deep ground and water.

The design studio explores the relationship between water and built/grown environments on a scale ranging from material to components, structures, buildings, cities and landscapes. The role of water, and its relation to the designed environment, is fundamentally important for the future of humanity. This is the base of our work through which we address challenges and questions with increasing global impact and importance. Globally, a majority of large urban areas are in the effective range of various hydrological systems; however, they have not been designed to cope with the accelerated and intensified dynamics of these systems. As a result, a multitude of lives, assets and cultural products are at risk.

The methods and principles used in these projects explore and question the role of adaptive, developmental and evolutionary capacity of grown and built environments as a way of increasing resilience. They also engage with questions about automation, multi-objective design and using human and non-human intelligence in the context of the designed environment. The research and design projects question how to plan for multiple species, and other biological and non-biological agents that collectively play their part and contribute to territorial dynamics.

As the landscape and its processes constantly change, our research and teaching perpetually adapt, develop and evolve.

26

LONGITUDINAL LANDSCAPES

Justine Holzman

The broad project of modernity, assisted by techno-optimistic engineering visions, plumbed rivers around the world into fast-flowing single-benefit concretised channels and opened up floodplains for development. As cities have transformed around these simplified hydrologic systems, the dual pressures of urbanisation and climate change are exposing the limitations and failures of engineered designs. Persistent and pervasive flooding in river cities has catalysed large-scale river modification efforts to address growing flood capacity needs. This is particularly true on the coast of the United States where sea level rise pressures intensify fluvial flooding. The limitations of these engineered systems to support their technical, ecological and human dependents have captured the public's collective imagination through the work of artists, activists and environmentalists. Landscape architectural intervention, and collaboration to unmake and re-envision fluvial landscapes, challenge technical understandings of design performance, offering multi-benefit alternatives that prioritise varied forms of social and ecological infrastructure.

Along these lines, two iterations of the "Longitudinal Landscapes" "option studio" examined processes of revision and unmaking of channelised rivers as spaces for social and ecological vitality. The studio was taught at the Daniels Faculty of Architecture, Landscape and Design (Daniels) by Justine Holzman in the autumn terms of 2018 and 2019. In the autumn semester of the third year of study in architecture, landscape architecture and urban design, the curriculum at Daniels brings together graduate students to focus on a contemporary design prompt, which usually includes a weeklong site visit. In each iteration, across multiple sites, *sensing* was positioned as an important aspect of designing for adaptive management and for public engagement. Broadly conceived, sensing ranges from sense perception and the sensorial to physical sensors – analogue and digital devices used to measure environmental phenomena (Cantrell

DOI: 10.4324/9781003145905-31

and Holzman, 2016). The first iteration, "Longitudinal Landscapes: mud, monitoring and mobilisation", built on the momentum of the international design competition, *Resilient by Design: Bay Area Challenge*, and the work of the Public Sediment team to develop monitoring and design strategies for tributaries around San Francisco Bay.[1] The studio was supported by Autodesk's residency programme at its Technology Centre in Toronto, which provided software and hardware access and tutorials alongside one-on-one prototyping and fabrication assistance.[2] The second iteration of "Longitudinal Landscapes" designed proposals for two different engineered waterways. The first, was a rapid visioning exercise for Waterfront Toronto's ongoing project to re-naturalise the mouth of the Don River, integrating ecological monitoring, public outreach and stewardship. The second, was a design prompt for a section of the Los Angeles River in downtown LA to inform the proposed US Army Corps of Engineers (USACE) pilot project to introduce ecological functionality and habitat into the concrete flood control channel.

Each studio project differed in site and scope. However, the shared objectives of the studio were to:

1. Understand contemporary dialogue, design theory and design practice surrounding the role of landscape systems and ecological infrastructure within urban fluvial landscapes.
2. Design using concepts for living infrastructure, change over time and adaptive management.
3. Understand and design for public engagement.
4. Develop a conceptual understanding of internet of things physical monitoring infrastructure for fluvial landscapes.
5. Utilise fabrication software and/or machinery to realise and prototype design concepts.

Fluvial Monitoring and Design Interventions for Bay Area Tributaries

The research and design work of the Public Sediment team, led by SCAPE/ Landscape Architecture, revealed the slow and difficult-to-perceive crisis of sediment scarcity and wetland drowning in the Bay Area. The team advocated for "designing with mud" to develop new methods and implementation pathways for sediment management and sea level rise. An overarching ambition of the Public Sediment proposal was to reconnect the Bay Area's tributaries (and their sediments) to the baylands: the marshes and mudflats that host important ecologies, protect communities along the Bay and retain carbon. This approach reoriented coastal risk and resilience design narratives that have long focused on the coastal edge as a "first line of defence" for understanding hydrologic systems *longitudinally* – connecting rivers to their coasts. For the Bay, this meant connecting upland, lowland and coastal communities through their historic and contemporary hydrology. During a weeklong trip to the Bay Area, students studied the Public Sediment team's proposal, "Unlock Alameda Creek", which proposed unmaking a channelised and disconnected flood control to reconnect sediment, people and fish. Understanding that working with living systems requires careful monitoring and adaptive management, the studio produced site design strategies and designed monitoring infrastructure for the Bay's tributaries, inspired by the Public Sediment team's insights and the challenges facing the Bay Area.

For the first half of the studio, students designed monitoring prototypes for upland, fluvial, tidal and coastal reaches using the support of the Autodesk Innovation Center (Autodesk software and fabrication tools). Making a working monitoring device was a difficult task, despite students being given tutorials in Arduino (microprocessor), Eagle (circuit board design software) and Fusion (integrated industrial design and fabrication software). To ensure students had ample opportunity to think through important questions of site deployment and design, they were provided with a working monitoring device designed by Cy Keener and Justine Holzman for which they could design and fabricate a casing. The device measured temperature, pressure to approximate water depth and light to approximate turbidity using off-the-shelf sensors on a custom-designed circuit board shield for an Adalogger (microprocessor). Students worked in groups to develop sensors for each reach of a tributary, learning important ecological indicators and solving for deployment and installation (Figure 26.1).

FIGURE 26.1 Photograph of the longitudinal landscapes tributary monitoring device designed by Cy Keener and Justine Holzman. Photograph by Justine Holzman.

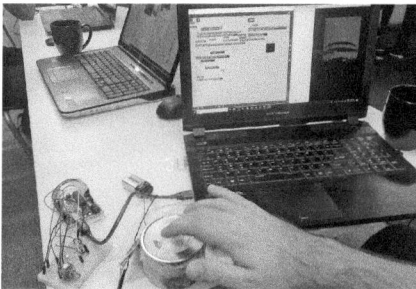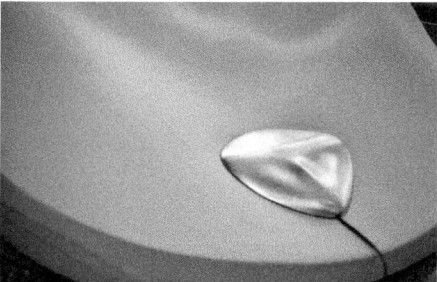

FIGURE 26.2   Left: Photograph of Devin Tepleski prototyping components for their group's (Aaron Hernandez and John Nguyen) "hydrophone" sensor (2018). Photograph by Justine Holzman. Right: Photograph of "Hydrophone" by Aaron Hernandez, John Nguyen and Devin Tepleski at the final review and exhibition. Photograph by Devin Tepleski (2018).

Rather than design a unique casing for the provided fluvial sensor, the upland group (Aaron Hernandez, John Nguyen, Devin Tepleski) designed a custom sensor to monitor sediment bed load by recording audio vibrations (an especially important measure in upper reaches). The group called the device "The Hydrophone" after research in fluvial monitoring technology, illustrating the use of hydrophones as a less expensive method for monitoring bed load and sediment stuck behind dams. The device was designed to use hydrodynamic force to hug the bottom of the river – part stingray and part horseshoe crab – with a milled aluminium casing that registers the vibrations of sediment particles as they pass over its surface. To show proof of concept, a piezo microphone installed in the interior records vibrations with an optional Arduino controlled depth sensor. In addition, their design offered the potential for public interaction through amplifying and visualising underwater sound. Students designed an abstracted topography model with particles that exhibition viewers could drop faux sediment over, triggering real-time visualisations programmed in Max/MSP/Jitter (Figure 26.2).[3]

For the second half of the studio, students designed site interventions that would be supported by, or relate to, their sensor design. Two students in the coastal sensor group worked together on complementary site interventions, titled "Underwater Meadows", to design a soft coastal infrastructure to restore eelgrass meadows and promote public access and legibility. Hadi El-Shayeb designed a modular 1 × 1-metre adaptable boardwalk with interchangeable rugose bottoms to support marine habitat, and colourful interchangeable tops to create public and social spaces, while marking the particular ecological restoration goal below. Peggy Wong designed an "Underwater Meadows Toolkit" (inspired by Public Lab's Balloon Mapping Kit). The intent was to bring awareness and curiosity to the biodiversity and complexity of eelgrass meadows, while encouraging citizen

FIGURE 26.3 Top: Section of "Underwater Meadows" by Hadi El-Shayeb and Peggy Wong (2018). Bottom: Photograph of the toolkit and instructional manual by Peggy Wong and modular boardwalk model by Hadi El-Shayeb. Photograph by Devin Tepleski (2018).

science involvement in eelgrass restoration projects throughout the Bay Area. The Toolkit provided a manual and the materials to construct a Buoy-Deployed Seeding System (BuDS) as a bottom-up method to work alongside restoration ecologists and marine scientists in developing thriving eelgrass meadows. The studio culminated in a public exhibition of student work at the MaRS Centre Toronto, and an advanced discussion of how monitoring infrastructure can assist in the design, adaptive management and understanding of urbanised coastal watersheds, and at the same time provide opportunities to connect publics and democratise data (Figure 26.3).

"Bridging the Ecologist/Technologist Divide" at the New Mouth of the Don River

For decades, undoing the ill-conceived, engineered L-shaped terminus of the Don River has occupied the minds of Toronto landscape architects, urbanists and architects. The Don River is one of two major rivers that flanks the downtown metropole of Toronto, and it drains a substantial amount of the city's storm water into Lake Ontario. Constructed with a large adjoining land reclamation project, the mouth of the Don River was to become a new industrial hub of the Great Lakes region. However, with global shifts in industrial production, these economic hopes never materialised. Its canalisation, and especially its sharp turn, has since caused significant flood risk and disconnected positive ecological processes between river and lake. Waterfront Toronto has led the massive and ambitious effort to recreate a floodplain and naturalise the mouth of the Don River, with Michael Van Valkenburgh Associates as the landscape architecture firm taking the unique role of managing the project's engineering. The Port Lands Flood Protection and Enabling Infrastructure (PLFP) Project is a provincially defined comprehensive plan for the south-eastern portions of Toronto at risk during a Regulatory Storm Event. This project targets 290 hectares for revitalisation, in which two new outlets for the Don River create engineered green space that includes the Parks, Public Realm and River (PPRR) Project. With a variety of programmes, this effort creates large ecologically robust areas for public access. Waterfront Toronto has recognised the potential value of custom monitoring and sensing for these landscapes to assist in understanding their performance, while also creating *social infrastructure*.

Well under construction, and in a unique phase of onsite experimentation at the time of the studio, Waterfront Toronto was interested in pursuing design concepts for an expanded set of ecological monitoring methods that would reach the public. The use of human perception alongside other forms of sensing might assist in a number of ways: education and awareness, activating play and exploration of all ages, addressing the nature deficit, stewardship and volunteerism, governance and public support, landscape infrastructure performance and climate change resilience. Additionally, an advantage of approaching the design of monitoring infrastructure as landscape architects (as opposed to framing it through just an environmental science or engineering lens) provided ample opportunity to consider how aesthetics and experience might alter or enhance understanding. With this task in mind, the studio worked with Waterfront Toronto to generate concepts for a "technology overlay". Technology (considered broadly) was framed as a catalyst for various forms of related management and engagement to "bridge the ecologist and technologist divide", a phrase Waterfront Toronto used to capture its intent.[4]

The students were asked to consider how monitoring, sensing and human perception might support a robust social infrastructure for this large and newly constructed landscape. Using a rapid design process that included multiple in-depth lectures, a tour of the construction site and a daylong workshop with Waterfront Toronto, each student developed three concepts that would contribute to a "technology overlay" for the proposed landscape.[5] One goal for the design exercise was to produce concepts relevant at different durations of the project's lifespan: *initial construction, understanding change* and *enduring into the future. Initial construction* addressed performance, set baselines and introduced this landscape to the public. *Understanding change* anticipated what metrics, forms of measure or frameworks, would be necessary to understand change over time, and which might also garner public interest. Finally, *enduring into the future* considered how change, growth, evolution and adaptation (ecological or cultural) could be understood over a long period while encouraging lasting stewardship. The projects were varied and included concepts such as: piped telescopes for observing with different senses from afar (Wenpei Fang), a secchi disc on a measure-marked rope for observing turbidity from a bridge (Avery Clarke) and a float lab designed to monitor water quality and host species (Hillary DeWildt) (Figure 26.4).

More than Mono-Functional Pilot Projects for the LA River

The second project of the studio focused on the yet-to-be-realised revitalisation project for the Los Angeles River. Much of the design work on the Los Angeles River has focused on a few marquee, large-scale projects along its length. Augmenting the concrete-trapezoidal profile of the river opens up important opportunities for unmaking and thinking *longitudinally* in a manner that situates the river channel as important connective infrastructure from the mountains to the coast. The students' working design introduces ecological function and habitat through a series of parallel linear concrete terraces without access for people. This approach is far from the vision of the many activist organisations, artists, and designers that have worked towards a revitalised river: an LA River with public access, cultural production and placemaking alongside ecological enhancements and flood protection.[6]

Students were asked to propose an alternative vision for the pilot project, considering it as a constructible site that engages the public and decision-makers in its experimental qualities.

As a "pilot project", the studio had the chance to explore constructions of prototype, model, test and experiment as they appear in engineering as well as landscape architecture – an opportunity to consider both qualitative and quantitative understandings of landscape performance and change. To encourage this overlap, students were asked to consider aesthetics, measurement and analysis as important instruments of both ecological management and hydrosociality – taking inspiration from the many past and ongoing cultural interpretations and engagements with this iconic infrastructural landscape. The studio took a weeklong trip, led by Alexander Robinson (Associate Professor of Landscape Architecture at the University of Southern California), to experience the geography of the Los Angeles River, conduct fieldwork and engage with local designers and stakeholders. As a collaborative effort, Alexander Robinson and Justine Holzman led an experiential tour of the river, fieldwork and one-to-one design charrette with University of Southern California and University of Toronto students. In Robinson's words,

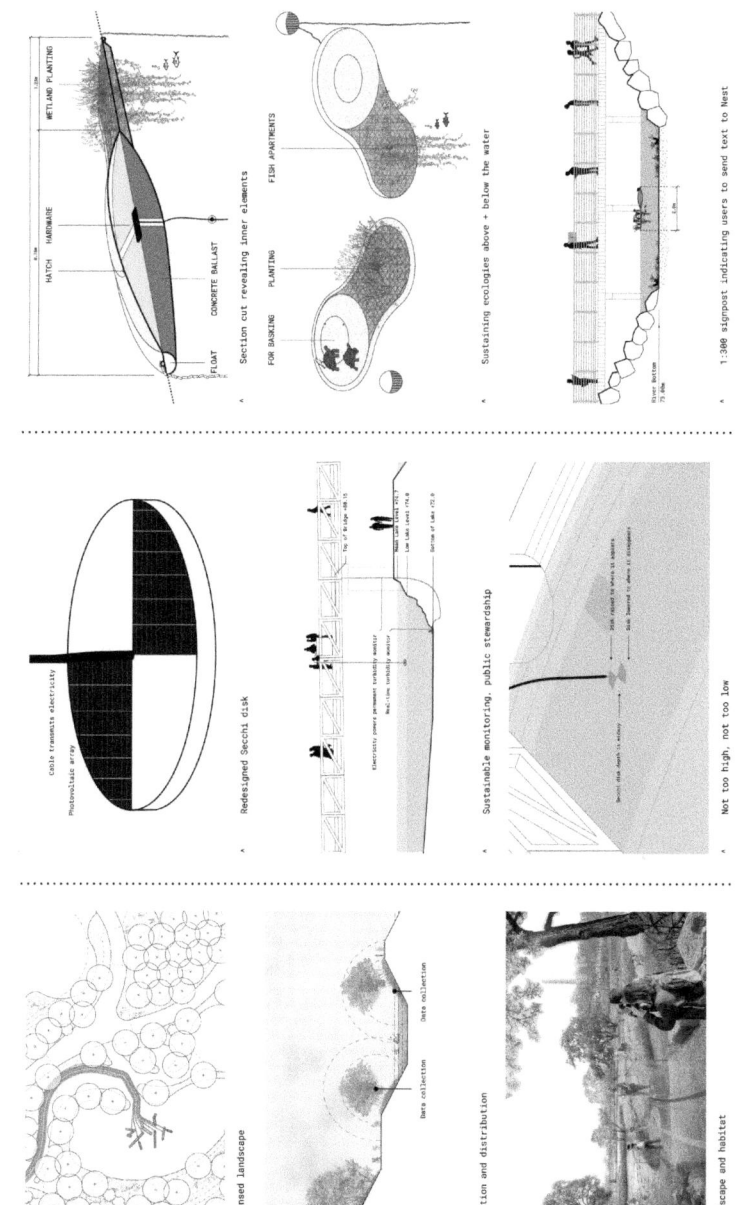

FIGURE 26.4 A triptych of three design concepts associated with three durations of time. Drawings on the left by Wenpei Fang, middle by Avery Clarke and right by Hillary DeWildt (2019).

our relationship with the River will begin with studies of perception, experience, and consciousness, exploring how the particular genius and morphology of the Los Angeles River has produced a powerful artistic and activist resonance, with particular emphasis on the role of the arts, alternative practices, movement, and aesthetics.[7]

Alexander Robinson

The majority of the trip was spent within or alongside the river. Upon initially encountering the Los Angeles River, the studio began with a meditative walk to engage students' senses in noticing *without* using devices or forms of recording. During subsequent visits to the river, the studio collaboratively embarked on two one-to-one design exercises. The first, a cross segment of the USACE-proposed design for vegetated concrete terraces on one side of the river was marked out with chalk to enable understanding of the scale and imagine the experience of the design. The second exercise was generative. Using long segments of rope, students iteratively marked and divided one side of the channel to imagine possible water diversions, plantings and paths. Given the massive infrastructural scale of the channel, these exercises established the human experiential scale for students and provided a starting point for their designs. In addition to meeting with USACE and the landscape architecture firms involved in revitalisation efforts, the studio toured the artist-activist practice, Metabolic Studio, and the sites of their ongoing efforts to irrigate adjacent parklands with treated river water (Figure 26.5).

While satisfying the USACE pilot project's goal of introducing habitat and ecological function into the river channel, a primary design goal of the studio

FIGURE 26.5 The 2019 studio having lunch in the LA River during the 1:1 design. Photograph by Justine Holzman.

was to illustrate the possibility and potential of access and experience of the channel. Students were asked to take a stance on ecology and aesthetics, placing their proposals on a spectrum between feral (no maintenance) and full ecological restoration. With their stance in mind, students were tasked with considering the implications for constructability, management, labour and maintenance. The emphasis on site experience, and the ongoing work on the river by artists, activists and designers during the field trip, was present in the students' final projects, with proposals that firmly situated and braided practices of maintenance, organising and experience.

Conclusion

Contemporary concerns in the field of landscape architecture about design within fluvial and coastal landscapes necessitate collaboration with science and engineering. The design, maintenance and management of landscapes that have specific (and often shifting) ecological and engineering goals require adaptive management and ecological monitoring. What role can sensing play in multi-benefit and multifunctional landscape infrastructure if imagined in collaboration with landscape architects? How can broader conceptualisations and understandings of sensing support the aspirations of landscape architectural interventions beyond ecological goals?

Participating in the design of monitoring strategies, devices and practices is an intervention into how landscapes are known, made and managed. Proposals prepared by landscape architecture-led collaborative teams in fluvial and coastal contexts (which include formal monitoring strategies and plans) are becoming more common. Many collaborations are pilot projects, considered experimental, and require forms of measure and knowing to inform larger projects and adaptive management plans. This work contributes to novel approaches and expertise in technical and experiential aspects of design strategies, while encouraging public participation through designated spaces and frameworks for citizen science. Within these endeavours, the design of monitoring strategies, devices and types of sensing have the potential to challenge, add or extend typical ways of knowing environments. As landscape architects seek to understand how to unmake and revise engineered waterways and coasts, it is worth asking who this data/knowledge is for, who can produce it and who can make decisions about it.

Engineered works and environments are designed in relation to specific histories and cultures of knowledge production and measurement. Entering into this landscape as collaborators and designers means understanding how engineering and scientific cultures of knowing and their path dependencies shape the built environment, its management and our knowledge of it.[8] Landscape architecture design studios might draw from historians of science and science and technology studies to better understand the social and cultural contexts of science and engineering.[9] Approaches in the studio might range from purely speculative drawing to field-tested working prototypes and landscape installations.

The studios presented here are cursory attempts to intervene and expand the epistemic practices of landscape architecture through the design of sensing and monitoring strategies. They endeavoured to conceive of sensing and monitoring broadly in landscape architectural design to extend existing aspirations in landscape architecture. Pedagogical approaches to *sensing* explored here cut across different disciplinary practices of fieldwork, knowledge production and landscape maintenance. They extend them to the public realm – to human connection, curiosity and care.

Acknowledgements

Thank you to the students who participated in the first iteration in 2018: Hadi El-Shayeb, Aaron Hernandez, Alexandra Kalman, Krystal Kramer, Vinaya Mani, Reesha Morar, John Nguyen, Anton Skorishenko, Neil Phillips, Devin Tepleski, Peggy Wong and Gloria Zhang. And, to the students who participated in the second iteration in 2019: Alexandra Walker, Alexandre Dos Santos, Avery Clarke, Bonnie Chuong, Chiling (Alan) Zhou, Hillary DeWildt, Wenpei Fang, Yuanyuan (Dorothy) Ye and Zixin (Sing) Chen.

Thank you to the Daniels Faculty at the University Toronto, especially Liat Margolis (MLA Director at the time) and Richard Sommer (Dean at the time). A large network of people to whom I am incredibly grateful supported both iterations of the Longitudinal Landscapes studio. I am thankful for the support of: the Public Sediment team and SCAPE Landscape Architecture, especially Cy Keener and Gena Wirth who generously workshopped with students, and Margaret Ikaeda, Adam Marcus and Evan Jones of the Buoyant Ecologies Team who met with the students at California College of the Arts; the Autodesk Innovation Centre in Toronto, and the efforts of Matthew Spremulli, Chris Chekan and Ellen Hlozan; the Bay Observatory; Waterfront Toronto, especially Christopher Glaisek and Pina Mallozi; the collaboration and support of Alexander Robinson in Los Angeles and Toronto; Charles Dwyer and the Los Angeles USACE, Metabolic Studio, Jessica Henson at OLIN, and Margot Jacobs at MLA Studio for our inspiring time in Los Angeles.

Notes

1 The Public Sediment team was one of ten teams selected to participate in the yearlong design process of Resilient by Design. The Public Sediment team was led by SCAPE Landscape Architecture. It included the engineering firm ARCADIS, faculty from the UC Davis Departments of Human Ecology and Design, local landscape architects TS Studio, the California College of the Arts-based Architectural Ecologies Lab, artist Cy Keener and the Dredge Research Collaborative (DRC) of which the author is a member.
2 The Autodesk Technology Centre in Toronto is one of Autodesk's four technology centres. Other locations are in San Francisco, Boston and Birmingham (UK). Each centre has resident programmes for individuals and teams. For the 2018 studio's duration, students were residents at the Toronto Autodesk Technology Centre, with

access to designated workspace, equipment, software and technical support. The director, Matthew Spremulli, and shop technical supervisor, Chris Chekan, were wonderfully supportive and provided workshops and individual project assistance throughout the studio's duration.

3 Students took inspiration from the work of Mathieu Marineau et al. (2015), part of the US Geological Survey.

4 The intent to "bridge the ecologist/technologist divide" came out of a Waterfront Toronto internal workshop. After Chris Glaisek, Chief Planning and Design Officer of Waterfront Toronto, joined the final review and exhibition of the first iteration of longitudinal landscapes supported by Autodesk, he offered this concept as a potential studio project for the second iteration.

5 Christopher Glaisek, Pina Mallozi and Meggan Janes, with Waterfront Toronto, generously gave the studio a tour of the construction site. Students saw the process of soil and technical engineering, landscape material and design testing, and aquatic habitat design and ecological monitoring of submerged, repurposed trees.

6 There is a substantial amount of research and design on the revitalisation of the Los Angeles River. Planning processes led by the city and the county of Los Angeles over the past several decades, the work of landscape architecture firms, particularly Mia Lehrer and the Los Angeles-based MLA Studio. The current masterplan led by OLIN presents the most detailed and comprehensive document to date with extensive planning and detailed landscape architectural visioning. The studio was fortunate to visit MLA Studio and discuss their Los Angeles River design work and to have a lecture and discussion on the current masterplan with Jessica Henson from OLIN.

7 At the time, Alexander Robinson was teaching a seminar and studio on the Los Angeles River. These are Robinson's words describing the shared approach of the two studios for becoming acquainted with the river. The approaches to infrastructure, ecology and aesthetics explored in the studio were influenced by Robinson's historical, theoretical and design work in *Spoils of Dust* (2018).

8 Kristina Hill's practice and scholarship has contributed greatly to understanding landscape architecture's relationship to science and engineering, especially to ecology and its closely associated fields; "Nexus: Science, Memory, Strategy" (2017) and "Shifting Sites" (2005) are keen examples of this thinking.

9 For example, *Environmental Sensing Technology and the Making of a Computational Planet* (2016) by Jennifer Gabrys was particularly useful for situating environmental sensing in the studio.

References

Cantrell, B., & Holzman, J. (2016). *Responsive landscapes: Responsive technologies in landscape architecture*. Routledge.

Gabrys, J. (2016). *Environmental sensing and the making of a computational planet*. University of Minnesota Press.

Hill, K. (2005). Shifting sites. In C. J. Burns & A. Kahn (Eds.), *Site matters: Design concepts, histories, and strategies* (pp. 131–156). Routledge.

Hill, K. (2017). Nexus: Science, memory, strategy. In C. Girot & D. Imhof (Eds.), *Thinking the contemporary landscape* (pp. 185–195). Princeton Architectural Press.

Marineau, M., Minear, J., & Wright, S. (2015, April 19–23). *Using hydrophones as a surrogate monitoring technique to detect temporal and spatial variability in bedload transport*. 3rd Joint Federal Interagency Conference, Reno, NV.

Robinson, A. (2018). *The spoils of dust: Reinventing the lake that made Los Angeles*. Applied Research + Design Publishing.

PART 5

Expanded Ecologies

Narratives

The 160 km journey from the Victorian state capital, Melbourne, to the regional city centre of Latrobe City, Gippsland, describes a progressive transformation of the urban fabric of Australia. Here, one can see material transformations and distribution networks that move from the city centre to the suburbs, and then extend from the rural to the hinterlands. This typical transect illustrates the metabolic processes of the urban landscape and its extension into its regions. The landscape shifts from a dense urban accumulation of skyscrapers, retail outlets, supermarkets, food outlets, and a sprinkling of sandstone cultural institutions, to suburban homes, large-scale industrial warehouses, and factories. It then transmutes into a scattering of sprawling outer suburban towns, interwoven with agricultural pastoral landscapes that supply the city. Large gaping excavated craters in the earth, surrounded by mounds of overburden and processing residue, interrupt the rural sublime. These craters form centres of coal extraction, accompanied by towering carbon polluters; coal-fired power stations. Near to these disruptions on the ground are small townships that were established to support these extractive landscapes with a labour force. The townships themselves are physically configured as extensions of the power industry, and exist to fuel the metropolis and its surrounding regions. However, extraction has its price as these small townships now suffer from energy poverty, degraded health conditions, and negative environmental aftereffects.

The Australian energy resource landscape, once demarcated as a government asset, supported the growth and transformation of the nation. In this form, the asset was not only concerned with generating power; it was also an agent for servicing and building community. As a result of the transformation and privatisation of the industry, it is now at the beck and call of market economies and fluctuating electricity prices. Consequently, the mines and power stations have become stranded assets, an anthropocentric legacy of industrialisation, progress and a

DOI: 10.4324/9781003145905-32

continual desire for growth. These are the outcomes of political and territorial implications driven by multinational tensions and political territorial claims.

Intense moments of extracted ground reveal rich gradations of soils and colours that hark back to the formation of the ground itself in the Miocene period. Venturing down into the pit, into the depths of the ground, we find the coal strata that was laid down around the time that the Australian continent detached from Antarctica. The coal deposits provide a view into a world of lush vegetation that eventually formed large swathes of subterranean coal seams covering vast territories of the southern state.

Moving forward in time to more recent histories, the ground is marked with a different set of extreme transformations, a landscape of massacre of the Gunaikurnai people, indigenous First Nations people killed in land grabs by British pastoralists during the Australian frontier wars. These lines of conflict translate and describe the tension of economic agendas that are still present through land-use patterns and resource rights.

Moving along the 160 km transect between the "service" town and the capital, we now journey along the right of way, a continuous clearing defined by utility easements that is intermittently dotted by large pylons stepping down from 500 kV, 380 kV, 220 kV, to 16 kV. The transmission towers shift in scale and voltage through the pastoral landscape and eventually extend all the way back to the city, supporting heavy industry, institutions, homes and commercial tower blocks. These vast tangible and intangible assemblages of productive metabolic flows and material processes/circulations (Swyngedouw, 2005) reveal a ground thick with geological, ecological, social and political histories.

Definitions

The predicament of climate change is clearly related to the energy issues. When acknowledging this relationship, we can relate this not only to problems associated with extraction, production, distribution, consumption, and the waste of energy; we can also relate it to larger societal values of production and use of power that drives the economy and shapes the world in which we live.

The aim is to shift the industrial paradigm to a holistic worldview based on an expanded ecological ambition of relations and interconnections. In light of this, each moment of design – whether of cities, homes, or landscapes – is responsible for an overarching desire to conserve the earth's resources. Each design aims to act as co-creators with other species and systems.

The term "Expanded Ecology" frames a set of complex adaptive systems that embrace the relationships between living and non-living beings; it encompasses social, political and economic systems. This represents an interwoven condition that shifts away from the dichotomies of humans versus non-humans, and between industrial versus environmental systems.

Expanded ecologies posit the potential of entanglements between biological, geological, social, political, and associated infrastructural materials. They draw on the ability of these materials to produce new landscape configurations that can absorb, adapt, and facilitate the transition of our current urban environments to meet the challenges of the global climate crisis. This includes techniques of inoculation, seeding, and grafting of geological substrates to the design and implementation of catalyst sub-structures for growth and expansion. These additive processes explore how biological processes vital for living organisms and geological dynamics have the potential to mediate between matter and environment through a range of design approaches and techniques.

An "Expanded Ecology" relies on a deep understanding of how material processes, material flows, and biological systems act as agents and agencies that form and configure the landscape to collaborate and curate potential growth and decay patterns. In particular, Expanded Ecology examines the notion of weakness– drawing from a biological understanding of "weak bonds" (Jun-Ichi Takahashi 2015). This understanding is crucial in forming and stabilising biological molecules. Moreover, it gives us insight into how we might conceptualise the formation of a vulnerable and mutable landscape that is encoded with the potential for ongoing change.

The range of projects included in this section investigates entanglements/ "imbroglios" (Haraway, 2016). These occur in places and situations where human and non-human processes become irrevocably intertwined, generating an ecological perspective on the communal effects of indeterminable landscapes. The entangled landscapes of these projects provide an opportunity to reference and incorporate into a design continuum the design research processes (technologies, processes, and representations) covered in previous sections. This can be achieved by using the instruments and ways of "looking" that are modified to generate, influence, and make-with the production of worlds.

Pedagogies

Expanded Ecologies argues for a rewiring of the relationships between the social, ecological, and political systems that contribute to shaping the urban fabric. With this in mind, it is important to identify the climate crisis as more than a consideration of the measure of carbon in the atmosphere. This is because it is also an outcome of the deep histories and complex entanglements of socio-political and techno-economic tendencies associated with periods and paradigms of growth and development in the city. Adopting such an approach enables the shift towards an ethical world-making approach imbued with the flows and processes of people, material, natural systems, and energy, including their implied agents and agencies. This shift represents an attempt to engage with the social and economic inequities associated with, and amplified by, the climate crisis.

This shift also characterises a pedagogical ambition that frames the urban as an assemblage. It considers that the city is composed of multiple heterogeneous parts and networks. These are intertwined with less tangible metabolic and material processes that describe the "natures" of the city through its indeterminable characteristics. The ambition frames the landscape beyond a singular surface condition, moving towards an understanding of a "thickened-ground" condition that inherently carries with it the histories of its past. Unless they are radically transformed, these histories will continue to shape the city's futures.

Indexical techniques are used as a subjective device of the landscape that defines a specific meaning within a context. This furnishes the ability to embrace scientific, political, and material informational systems to inform multiple states of making the urban.

The "assemblage of entanglements" (Swyngedouw, 2005) is a dynamic set of complex systems that form interconnections scaled through diverse formations of time and symbiosis. This enables a world-making that folds together multiple positions, needs, and effects through human and non-human agents.

The teaching and learning ambition proposed here positions the student and the institution as one of many agents in the learning and world-making process. It offers them the ability to reveal and question explicit biases and power structures in the studio environment.

References

Haraway, D. J. (2016). *Staying with the trouble: Making kin in the Chthulucene.* Duke University Press.

Swyngedouw, E. (2005). Metabolic urbanization: The making of cities. In Nik Heynen, Maria Kaika, & Erik Swyngedouw (Eds.), In the nature of cities: Urban political ecology and the politics of urban metabolism. Routledge.

Takahashi, Jun-Ichi. (2015). Weak bonds. In Muriel Gargaud, William M. Irvine, Ricardo Amils, Henderson James (Jim) Cleaves II, Daniele L. Pinti, José Cernicharo Quintanilla, Daniel Rouan, Tilman Spohn, Stéphane Tirard, & Michel Viso (Eds.) Encyclopedia of astrobiology. Springer-Verlag.

Bibliography

Shiva, V. (2008). *Soil, not oil: Climate change, peak oil and food insecurity.* Spinifex Press.

Tsing, A. L. (2015). *The mushroom at the end of the world: On the possibility of life in capitalist ruins.* Princeton University Press.

27

ASYMMETRIES AND URBANISATION

Elisa Cristiana Cattaneo

Bruno Latour, in his proposition for a neo-Humboldtian pedagogy, presents an epistemological framework for which we need a "radical reorientation: what used to be called extension, outreach or pedagogy is no longer the last but the first frontline and alongside which all actions of the future university will be evaluated" and "It is thus very important that intellectuals, artists, statesmen, activists, begin to sketch the landscape that we will have to inhabit" (Latour, 2016: 9, 10).

The following methodological process is framed in light of Latour's proposition of reorientation. That entails embracing the need to reverse the logic, the aims and the points of observation of the design process, and stressing concepts such as project timing, the relationship between data and design, and the concentration on geosciences as a theoretical place for disciplinary experimentation. The aim is to define a new temporary imagery able to respond to the *asymmetries* of contemporaneity – like climate change – as positive and distinctive assumptions of the present day and constituting incidents in the new interpretation and planning of the city in its broad definition.

We, therefore, move away from a dramatising reading of the question of climate change. We shift the discipline of landscape to one of being an exploratory field of research concerning ongoing global dynamics, and capable of testing new methods and new codes in the world of contemporary design.

Representing the concepts of hyperobjects (Morton, 2013) and the new climatic regime (Latour, 2018), and understood to offset and fall upon forms of urbanisation, these asymmetries are also expressive of new ecological, social and spatial imbalances. These imbalances change the stable conditions of urbanised territories. In particular, they generate new complexities that break the well-tested consequentiality of scale, the dislocation between causes and effects, the

DOI: 10.4324/9781003145905-33

spatial locations of the projects and the assumption of a long-term definition. They open up a design scenario whose codes, forms and processes are to be explored toward new forms of urbanisation, in which the global (the ecology question) and the local (the specific/endemic condition) need to find new measures of complex and generative interactions.

Specifically highlighting the critical question of ecology, the asymmetries need a new and appropriate answer from design processes. Right now, these processes are not fully able to convert either the languages and tools demanded by global responsibility or changes in the ecological paradigm. The process of dissolving the canonical compositional categories seems, in fact, to require a new epistème for the world of design, one which traces a new misaligned genealogy to those culturally ascertained, capable of responding to the "flat ontology" (Harman, 2002) typical of contemporary times.

Since the asymmetries require abolishing the metalanguage and changing our phenomenological structure, requiring a new *speculative realism*, realism and imaginative speculation become the axes of the research.

It is in light of this speculative realism that we look for new forms of *aesthetic functionalism*, framing nature as performativity that escapes any deterministic approach that countenances new geo-philosophical figures. In light of these double and opposing terms (functionalism and aesthetics, or performativity and imagination), our taxonomy refers to the positions of Edgar Morin on the epistemology of complexity and transdisciplinarity. In particular, we refer to his definition of nature as *technonature*. It seems even more effective if the landscape project is related to the recent flagrant delicacy of contemporary "ecological trauma" (Baserau and Morin, 2002).

If dragged into the world of design, Baserau and Morin's conception could define a union between cultural and historical opposites (as nature and city), which is particularly useful in the urban context. Technonature is an experimental way of interpreting/designing urban shapes according to a renewal of the idea of nature and rooted in the philosophy of science. It is capable of overcoming both dichotomous relationships and any attempt at vernacularism, immanent hibernation or conciliation, such as the aesthetics of disappearance, camouflage, metaphor and, above all, of every deviation from sustainability or environmentalism as ideological or restorative systems.

By amplifying this assumption, the project acquires a perspective in which nature expresses the continuous renewal of its capacity for genesis and creation. Above all, it demonstrates adaptation, "absorbing artificial matter as its component". Finally, by unfreezing conceptually, it "dissolves from the immanent hibernation that has always haunted it" (Ponty, 1995).

Therefore, it is possible for design science to specify itself according to ecological strategies that are instructed operationally in consideration of nature as having a symbiotic relationship with culture and, above all, as performativity: to identify its theoretical-methodological potential (as a new way of thinking about the design tools at different scales) and its operational (as a viable and non-metaphorical strategy) and linguistic (as a transdisciplinary hybridisation) setup as a possible response to the advancement of new ecological (and formal) questions.

The vision we want to introduce, therefore, is that of an urban project based on a double passage of the term nature: both as a system in crisis (environmental issue) and as a device capable of acting in its crisis through new shapes. In purely design terms, nature can be considered as an active device in producing a critical space of contemporaneity. It is able to drag the project to new spatial thoughts, patterns and figures capable of prefiguring symbiotic spaces. It is able to produce differentiated effects of synergy and not of scale. It can develop differentiated space-time sequences. It is, therefore, a project that passes through natural sciences on its way to find new aesthetic figurations.

We look at urban and territorial urbanisation in light of the previous hypotheses – the neo-Humboldtian framework and technonature as an assumption. We are testing whether the project can become a cognitive and experimental discipline, a new transdisciplinary process and a response that formulates new scenarios and imagery more capable of synchronising with current urban phenomena than the "strong" schemes of high and recent modernity.

Thus, the ultimate purpose is to elaborate and experiment with an open, dynamic and visual taxonomy, and in so doing, to generate an advanced practice and a new aesthetic for the project at different scales.

Research Methodology and Approach

Developed from within Weakness Theory Research (Cattaneo, 2015), the teaching activity I have pursued at the Polytechnics in each of Milan, Turin and Genoa has been developed primarily by students in more than 150 master's theses since 2013, and in the courses I teach. The teaching now collects thousands of critical places in the world, particularly developing countries, recovery landscapes, refugee camps, as well as shrinking cities and consolidated cities.

My teaching activity and its outcomes are intended as a transversal architectural course. In this approach, students are called to specify solutions related to contemporary/global issues as a methodological, theoretical and figurative strategy for creating a short circuit in our typical idea of design.

In a way, "weakness" is a pretext for building cognitive meaning for a new world scenario.

I denote the term weak as a crucial point; it is most productive in terms of research on contemporary urban and regional design. Its etymology (infirmitas), if applied in design practices, can accurately represent contemporary spatial situations as a combination of synthetic and systemic categories. In more general terms, it appears to be composed of fractured positions that need to be merged through a designative and definitive approach to urban research, as was sought in the majority of architectural theories in the last century. The paradigm of research paraphrases the *weak thought theory* (*pensiero debole*, in Italian) of the philosophers Pier Aldo Rovatti and Gianni Vattimo (1984), in which they refute the latest all-embracing legitimisations within a postmetaphysical and postmodern framework.

The purpose of this pedagogical method is to experiment, within the construction of contemporary cities, with how a weak approach can defy conventional urban studies and urban planning methodologies.

In accord with this approach, we experiment through an abductive process, working on feedback loops instead of cause-effect systems. In particular, the process is eradicated in two ways:

- Theoretical: through continuation of the critique of modernism and synthetic-deterministic approaches, and through transdisciplinarity as a scientific development of continuous regeneration.
- Methodological: through hybridising ecological logic with design logic (plan, programme, urban design and architecture), dispensing with the dual/dialectic concept in favour of a relational/dialogic concept, as suggested by the theory of complexity (that is, to dissolve binary positions such as culture/nature, nature/city, figure/background, small/large and so on).

In this sense, students are called to define a cognitive process for their research projects so that their thinking is relational and synergistic, not dichotomous. In particular, they are called to interrelate the imaginative/conceptual process of the theory with the specific and contingent reality of the data.

Quantitative approaches, such as geographic information system (GIS) software, satellite images and databases, are tools capable of generating a spatial display of the project conditions. In particular, they allow correlation of mappings and data, which is, in fact, essential for the complexities of the agents in an architectural/urban/territorial scale, going beyond simplifications and permitting the display of heterogeneous factors in many directions (Figure 27.1).

This theoretical methodology is finalised in a visual/cognitive approach rooted in five steps:

Step 1: *Geography of influence/theoretical-design frame.*
   A scientific/theoretical platform of reference (projects, theories, methods) related to the specific place/topic, intended to inform the student's theoretical/conceptual position.
Step 2: *Re-discovering.*
   Formulate new design codes focused on data to be tested. Design new comparative matrix/pattern/maps in light of heterogeneous data, with a focus on pedology, geomorphology and topology.

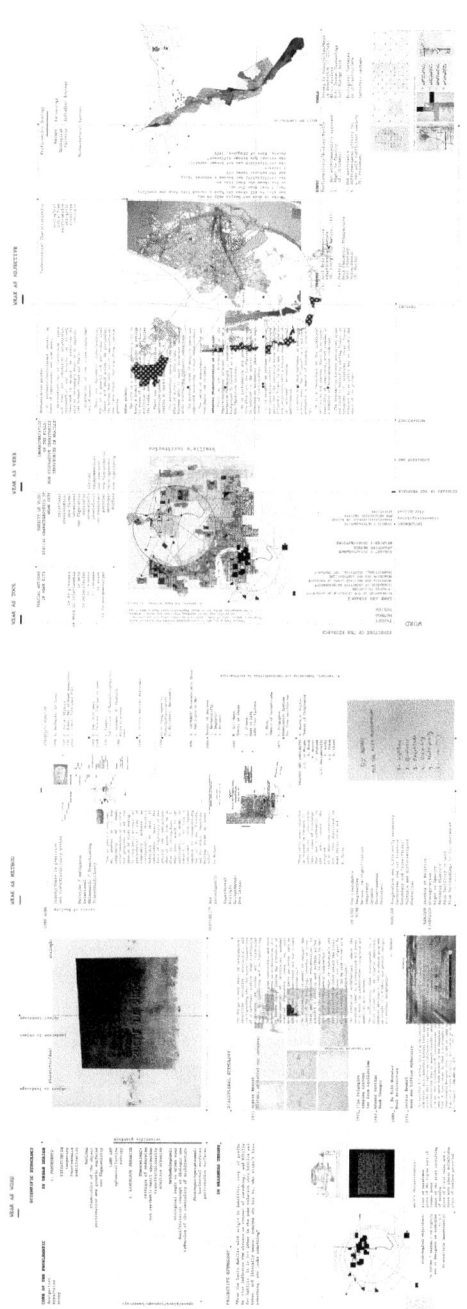

FIGURE 27.1 Cognitive map, Weakcity (2015). Elisa C Cattaneo.

Step 3: *De-territorialisation*.
  A critical point of imagination in which design becomes the point of discontinuity for the status quo, externalised in dynamic masterplans.
Step 4: *Overlapping time/space*.
  To introduce the four dimensions of ecology in urban strategy, finding
new immersive and atemporal spaces.
Step 5: *Astrolabia or landscape machine*.
  The potential to experiment with projects in a situation that is not static
but dynamic and continuously evolving. Students are called upon to develop
their projects within a conceptual cognitive map, a landscape machine,
able to work as a new idea of the conceptual/physical model and which
introduces an active-retroactive processual project design.

Within the general framework of asymmetries as an area of experimentation,
the previous methodology aims to test if, through dynamic/open systems, it is
possible to experiment with new processes in response to the global problems
(Figure 27.2).

Accordingly, ecological logics become essential to establishing symbiotic,
performative and resilient languages and forms which are capable of both
renewing our idea of composition, and of taking into account global and complex
contemporary spatial interactions.

These new inputs, that lead the project to geographies in which the effect
of the Lorenz butterfly – for which, in a deterministic non-linear system, there
are dependencies between small initial modifications and global final effect – is
reversed, requiring new spatial imagery for the discipline of the project (Figure
27.3).

Just as György Kepes attempted to formulate images responding to new
scientific and territorial conditions in the new landscape in art and science (Kepes,
1956), an attempt at experimental renewal of the design discipline should be
made, starting primarily from a scientific experiment that can be transferred into
formal results. Thus, an elementary process for eliciting results from scientific
experiments (data processing) could be transformed through aesthetic processing
to render possible a new spatial model for contemporary geography.

This process generates new patterns and an unusual methodology of making
them active: they, therefore, become preparatory and propedeutic not only to
understanding the global phenomena but also for a method of reading urban
contexts and for a new design methodology in a general sense.

Therefore, they can be considered original, as new starting conditions for
experimenting as yet unexpressed methods or the current ecological conditions
in progress.

In this light, the patterns become a critical shape in the search for new
ecological aesthetics. Critically, it is an alternative to the figure and drawing, as
typical of 20th-century logics.

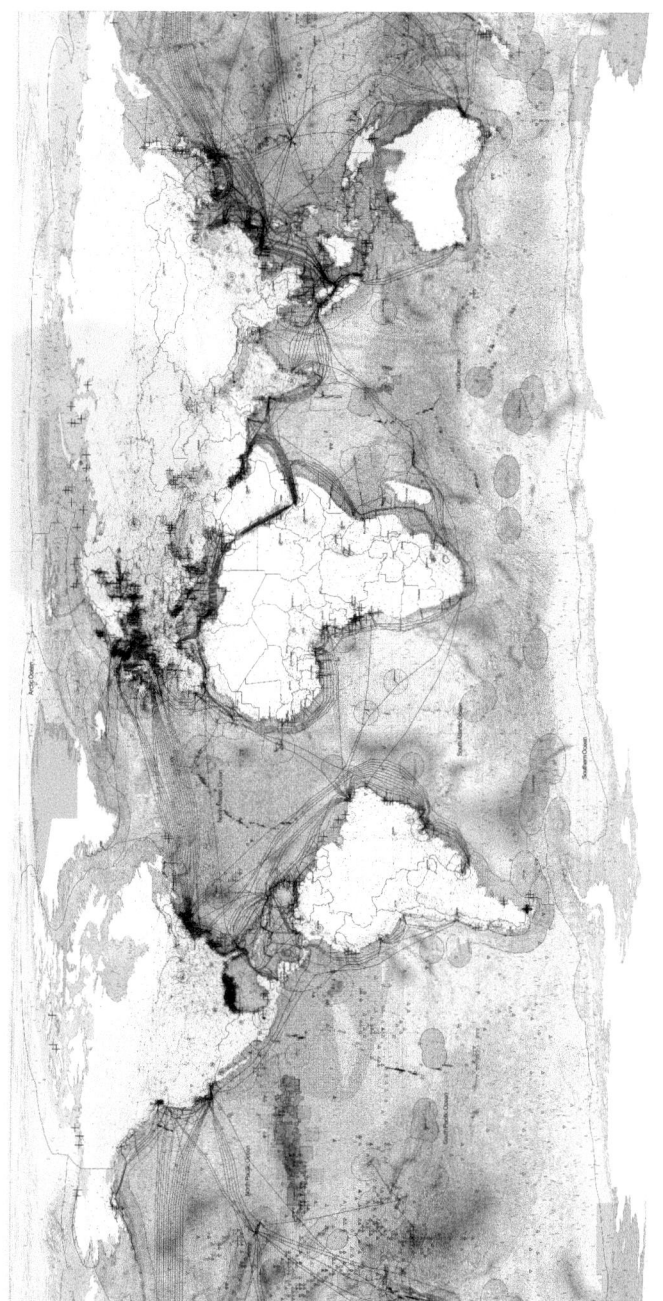

FIGURE 27.2 What about the other 71%? Ocean as a performative landscape, Ocean Masterplan 2020. Master's Thesis, Polytechnic University of Milan (2020). T Malchiodi Albedi, and I Furbetta.

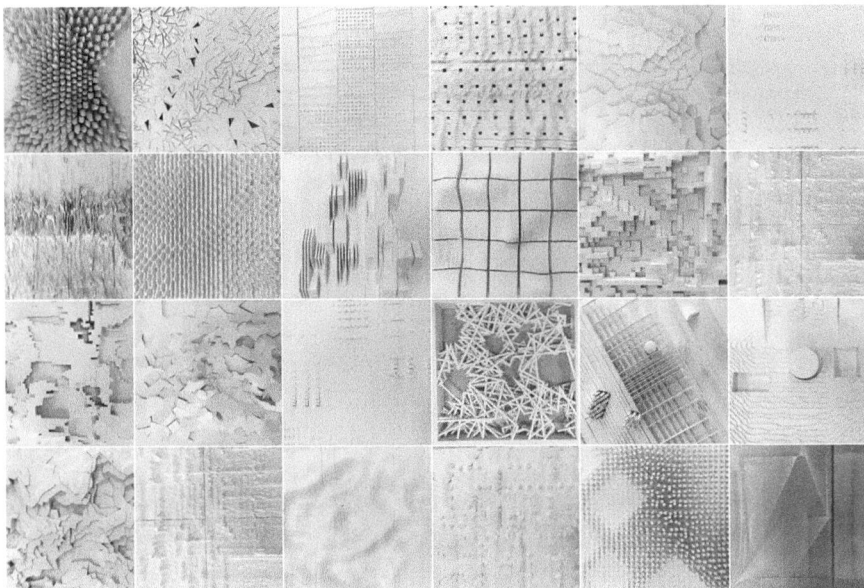

FIGURE 27.3 Ecological patterns. Landscape course: Nature through the mirror. Course leader: Elisa C Cattaneo, Polytechnic University of Milan (2016).

In fact, unlike the concept of the figure, the pattern acts as an open, dynamic, never fixed regulation between heterogeneous shapes. Above all, it does not seek the stability of a form. Instead, the instability – as characteristics of the dynamic shapes – becomes a necessary condition for the logics of the method and not a condition to be solved.

In fact, within the theoretical dimension of topologies or topological spaces (Kwinter, 1992), the pattern acquires the content of dynamics, disturbances and symbiotic interactions between internal and external spaces through the hybridisations of differentiated ecological data. Moreover, it reveals time as the determining factor in the formation of contemporary ecological shapes (Figure 27.4).

In this sense, the pattern is the new synopsis in which we find a synthesis between open fields versus stable figures. It reveals the dominance of time over the concept of space. As outlined by Kwinter (1992), the contemporary ecological pattern overturns the project's level of consistency by expressing temporal dynamics before formal/spatial ones. It does so to the point of promoting a continuous diachrony, a long- and short-term temporal overlapping, thus generating visions never unique and fixed, but multiple, dynamic and serial.

As expressive of complex topologies, the pattern reveals the morphogenetic process of a shape. As morphogenetic topology, it lights dynamic spaces and diachronic times as generative of original forms of the present time.

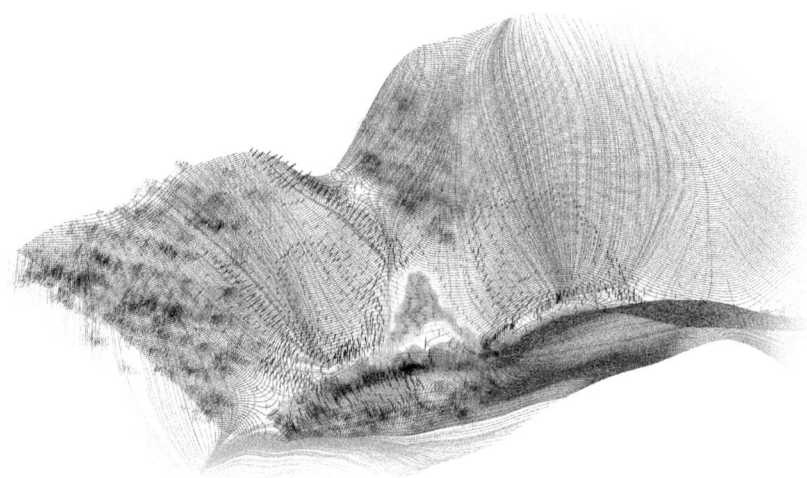

FIGURE 27.4   In between disturbances. Urban Design Studio. Course leaders: Elisa C Cattaneo and S Giostra. Polytechnic University of Milan (2018). E Dinarama.

In this sense, the pattern is the short-circuiting shape of contemporaneity. It expresses differentiated and propagative forms. It reveals the transition from the concept of 20th-century *logos* (identitarian and repeatable) to the *topos* of contemporaneity (differentiated and fluid) through the singularity of its geo-localised points (Crotti, 1988).

Expressive of the ecological logic, the patterns also represent the paradox of the ecological spaces: to build invisible space. In this sense, the pattern generates maps of invisible but perceptible forms, bringing ecology and the landscape toward a concept of immersion, not planarity.

However, these forms need new codification to be elaborated and processed through the project (Figure 27.5).

In this light, the whole project is oriented toward new taxonomic and visual formulations, which reweave new genealogies instructed – for the first time – by a line of research not yet fully disclosed and investigated. By reactivating a theoretical movement not properly esteemed by our conventional project idea, this project brings to light studies and research by authors such as D'Arcy Thompson, Norbert Wiener, György Kepes, Andreas Feininger and Naum Gabo, among others.

In a nutshell, from a theoretical point of view, we attempt a new formulation of space in contemporaneity. In this context, ecology becomes the generative discipline for an epistemological leap, bringing the natural sciences into the field of formal spaces (Figure 27.6).

From a methodological perspective, we broke the concept of synthesis and led it toward an active-retroactive dynamic process. From this point of view, we

FIGURE 27.5 Unavoidable Landscape. Landscape Course: Nature through the mirror. Course leader: Elisa C Cattaneo, Polytechnic University of Milan (2016). T Croattini, G Tagliente and E Triantafyllidou.

substitute the concept of the figure with that of the pattern. In search of a new ecological image of inhabitable space, we summarise multiple spaces and differentiated times in new, ecologically informed, immersive/experimental forms that aesthetically express the performativity of nature in symbiosis with the city.

Perhaps it is in the transition between ecology and landscape that we interfere in a creative process and are capable of generating new spaces and new aesthetics yet to be investigated. The identification of formal data structures, their transformation into immersive forms that are not scalar-consequential, and their re-territorialisation in displaced territorial areas, would in fact allow us to identify new hybridisations that deliver an integral blend of art and science.

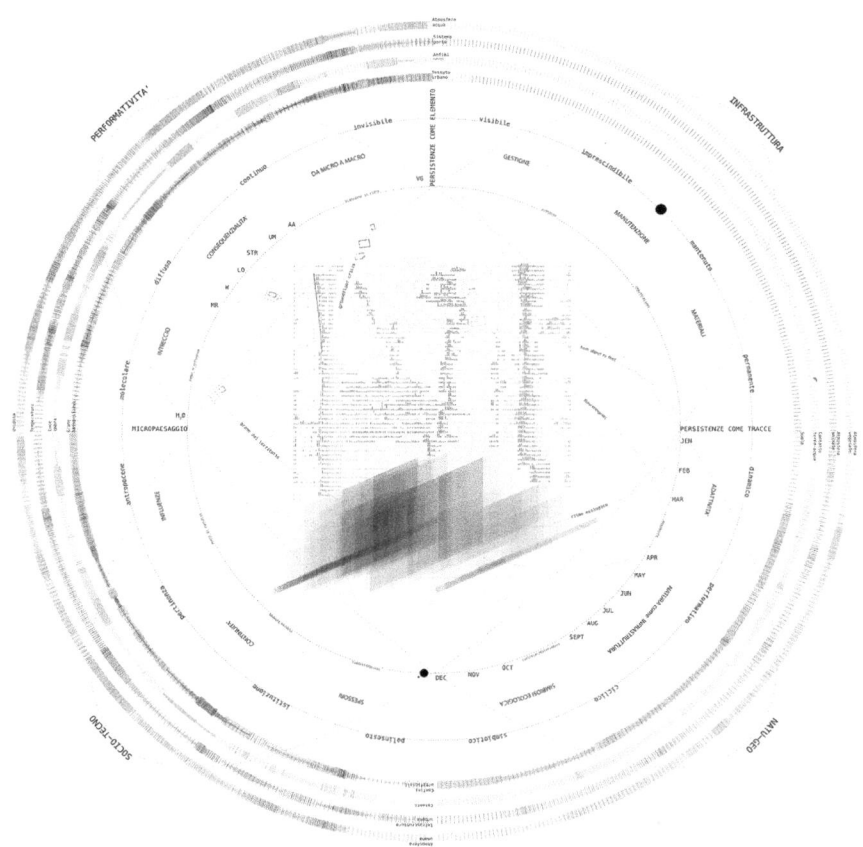

FIGURE 27.6 Final Astrolabia. Urban Design Studio: Course leaders: A Armando, Elisa C Cattaneo, M D'Ambros and P Simeone. Polytechnic University of Turin (2020). F Pessotto, L Rebolino and E Prataviera.

References

Baserau, N., & Morin, E. (2002). *Manifesto of transdisciplinarity*. State University of New York Press.

Cattaneo, E. C. (2015). *Weakcity*. Listlab.

Crotti, S. (1988). Dal logos al topos. In Patrizia Zanella, Alberico B Belgiojoso, and Bianca Bottero (Ed.), *Morfologia dello spazio urbano. Vol.29.Milano* (p. 162). Franco Angeli.

Harman, G. (2002). *Tool-being: Heidegger and the metaphysics of objects*. Open Court.

Kepes, G. C. (1956). *The new landscape in art and science*. P. Theobald.

Kwinter, S. (1992). Landscapes of change: Boccioni's Stati d'animo as a general theory of models. *Assemblage, 19*, 50–65.

Latour, B. (2018). *Down to earth: Politics in the new climatic regime*. Polity Press.

Latour, B. (n.d.). *Is Geo-Logy the umbrella for all the sciences? With a few hints for a New University* | Conference, Cornell University, 25th October 2016. Retrieved from http://www.bruno-latour.fr/node/702

Merleau-Ponty, M. (1995). *La nature. Notes de cours du Collège de France*. Seuil.

Morton, T. (2013). *Hyperobjects. Philosophy and ecology after the end of the world*. University of Minnesota Press.

Rovatti, A., & Vattimo, G. (1984). *Il pensiero debole*. Feltrinelli.

Bibliography

Kwinter, S. (2008). Wildness (Prolegomena to a New Urbanism). In C. Davidson (Ed.), *Far from equilibrium: Essays on technology and design culture* (pp. 186–194). Actar.

28

THE TERRITORY AS A SUBJECT[1]

Paola Viganò

In the mid-1980s, André Corboz highlighted a passage that is still fundamental today: *"Étant un projet, le territoire est sémantisé. Il est «discourable». Il porte un nom. Des projections de toute nature s'attachent à lui, qui le transforment en un sujet"* (Corboz, 1983). The territory has gone from being a "subjugated" object to a subject with a name or names – it is named; it is endowed with meaning and is not merely a dead and insignificant accumulation of objects; it participates in the construction of discourses as it is the recipient and inspiration behind projections and imaginaries (*"…il n'y a pas de territoire sans imaginaire du territoire"*). The territory *as a* subject is an agent territory and not a mere support.

Corboz's position contains many implications. The understanding (and consideration of these implications) is not yet complete; research and current territorial policies and design have not yet absorbed and reformulated them. The contemporary condition of ecological and socioeconomic transition instead calls for constructing a new outlook, a new epistemology of the territory as a subject. This is my hypothesis.

The analogical relationship Corboz established between the idea of the subject (therefore endowed with rights and duties), and that of the territory, likely contributes to the difficulty in understanding and assuming responsibility. To simplify things and support the start of a discussion, I will attempt to remove the "as", and for a moment leave the "territory-subject" to take shape in the open field of our imagination. More than subjectivity or identity, a process is here represented, namely "subjectification"; that is to say, the production, conscious or not, implicit or explicit, entirely internal or imposed from the outside, of a subject. In this case, the territory, in its multiple dimensions.

The territory-subject (or the subjectification of the territory) has some important epistemological and ontological implications, as well as very profound ethical and political ones. The subject is a single, unique individual, but it is also,

DOI: 10.4324/9781003145905-34

and in any case, the result, expression, and product of relations and logics of power (Foucault, 1994; Revel, 2014).

In research and teaching experiences – reading, interpretation and design – conducted in recent years, the questions raised by the idea of the territory-subject and their controversies have emerged regularly and with severity, without ever retreating. It is as if it were no longer possible to go backwards but only to advance, even with difficulty, within the radical recontextualisation of the territory-subject and territorial subjectivation. In this approach, I include all the indeterminacies and problematic issues contained within the process, and that emerge with the ambiguity of the very term "subject" (*sujectum-subjectus*, as Étienne Balibar recalls) (Balibar, 2012). At the same time, the subject is individual, anthropological-political, but also subjected and subjugated, placed underneath. It is relational and not abstract (Pelluchon, 2020), a material, textured and bodily relationship, which exchanges with the other subjects.

I will focus on just some methodological and design consequences that the idea of territory-subject has generated, and among them the assumption of the "reasons for the territory" as an object of investigation. Territorial rationalities – the long-term construction processes that have transformed natures, economies and landscapes by developing new ones – have entered in a clearer and more fundamental way into revising the idea of the city, its form, structure and design (for example, by enhancing the widespread and isotropic networks that have an ecological, economic and political rationale) (Viganò, 2008; Viganò, Secchi and Fabian, 2016). After seasons with almost no spatial memory, attention is once again placed on the material construction of the territory which is necessary to decipher the logics that form and transform it, the adaptations, abandonments and recoveries. Through these readings and interpretations, the idea of *capital space* and *agent territory*, not just a support, have fostered the revision of the territorial project, restoring to the territory the position of a subject around the table of the choices that concern it. However, the subject-territory, with its characteristics and qualities, states and dynamics, the agent territory that can perform actions, can also suffer them: it is weak, manipulable and violable. It is indeed "subject to", subjugatable and subjugated. Here too, an understanding of territorial rationalities, and among them the complex, ancient and recent sphere of rationalities related to the transformations of agricultural production and settlement patterns, accessibly reveals the countless paradoxes that constitute our late and tired modernity. Even if each of the recognised rationalities has at some time in human history had a collective, political and social meaning, and performed essential functions for our survival and development in constructing territorial habitability, this does not prevent us, today, from measuring with ever greater precision and knowledge the distance that separates us from them and their inherent contradictions. This is especially so for those that respond to strong, destructive and hegemonic rationalities, expressing resistance logics that are the most difficult to adapt and rethink (Figure 28.1).

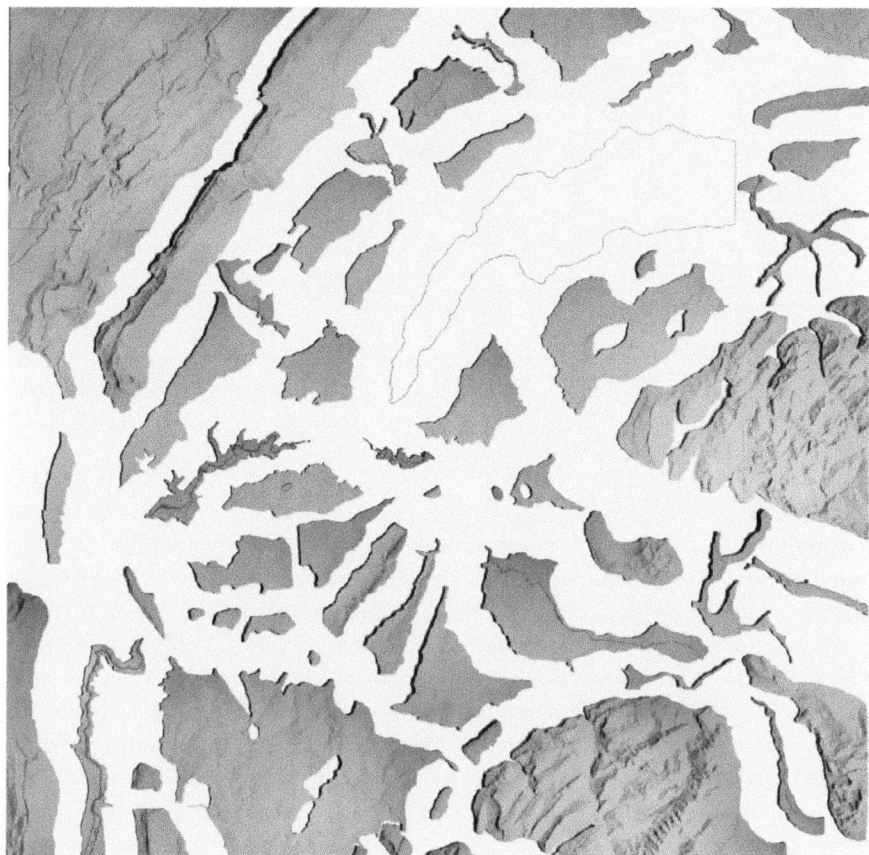

FIGURE 28.1   Emerging territorial figures. EPFL Masters project, 2020. Noélie Lecoanet,
teaching staff: P Viganò and T Pietropolli. Source: Noélie Lecoanet.

The consideration of territorial rationalisations is part and parcel of the criticism of modernisation, the process of transformation that we have critically
observed for some time and which poses fundamental problems in light of the
ecological transition, climate change and scarcity of resources. The transition
project implies a widespread shift, at territorial scale, to a different eco-socio-
economic and spatial model.

Weak Structures

To discuss the paradox of the subject-territory, simultaneously endowed with
rights and subjugated, I am assuming the structuring capacity of some of its elements, and of the territory itself, is not only pertinent or entrusted to strong structures (for example, large infrastructures) representing development ambitions of

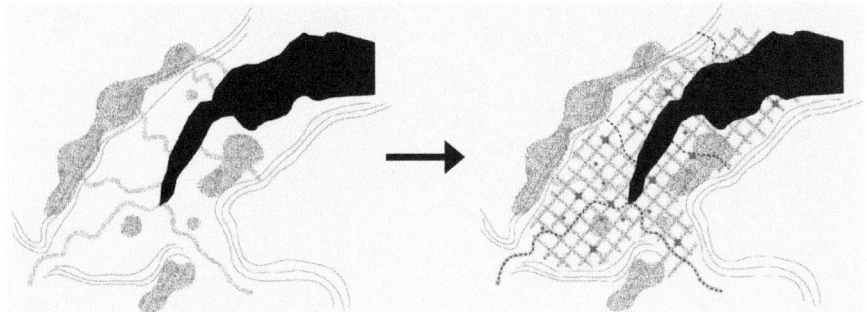

FIGURE 28.2 Weak structures for the transition. 38% of greater Geneva (excluding the Jura, Saleve and Chablais massifs) of which: 55% woods, 9% wetlands and flood areas, 11% built, 25% agriculture. EPFL Habitat Research Center, 2020. Vision for Greater Geneva.

a region, town or city. There are structures that are less capable of imposing their point of view, "weak structures", terms I have taken, at least partially, from the specific context in which they arose – the postmodernity debate (Vattimo, 1985). Weak territorial structures, incapable of resisting, which are often violated and rejected: what interests me is their pervasiveness, their power to organise contemporary extended urbanity, to question the traditional drivers of development. They are fascinating for their ability to adapt. To fully appreciate them, however, we must return to the territory-subject and its rights (Figure 28.2).

Confirming that environments and pieces of territory are legal entities, as occurred for example in New Zealand with regard to rivers and natural areas (Iorns Magallanes, 2015), has been considered an important achievement. This is so even if it once again locks nature into a cage of precise social or community values, and precise hierarchies that are culturally and historically determined. Similarly, considering the value of the soil, a weak structure par excellence, it does not seem feasible to escape the mechanism of monetising its fundamental functions (an aspect that is not agreed on, even at European level). We are far from Aldo Leopold's *Land Ethics*.

A few questions arise. The first concerns the sense of extending rights to the entire territory, beyond the wilderness and sphere of nature protection and conservation. The second concerns the possibility of going beyond the utilitarian dimension, imprinted into the term "resource" (the territory as a resource, or as capital) which always recurs in these cases, especially when we speak about the "vocations" of a territory. It is a question of attributing value to the territory-subject for what it is, and not only for the reserves/resources/capital/vocations it contains, thus also to its less useful, impressive and trivial aspects.

Reflecting on the weak aspect of the structure calls into question the concept of order, and with it the foundations of modern thought and its totalising rationalities. Attention to weak structures reveals latent coherences and reconnects what has been separated and fragmented, often rediscovering conceptual trajectories that have subconsciously traversed victorious modernity, proposing alternative paths (Audier, 2017; Barcelloni Corte and Viganò, 2021). The aspect of weak territorial and spatial structures that interests me is that they continue to be devices of emancipation in a context of crisis and transition. To cross them, a high level of widespread infrastructure must be maintained. I am interested in the weak subject's ability to pose questions and upset the balance of traditional power relationships. In the field of urban and territorial design it means abandoning the omnicomprehensive claim, providing instead a common vanishing point, and highlighting possibilities demanded and supported by "gentle nudges" (Thaler and Sunstein, 2008). However, this attitude does not avoid facing up to the now inevitable need for extensive structural changes. This point of view also has consequences for "strong structures", indicating the possibility of new relationships between them and the various "subjugated" territorial strata.

Explorations, Variations

Territory-subject and weak structures are some of the many trajectories that have generated research and exploration in space, consolidating the hypothesis of design as a producer of knowledge, a powerful device for reading, deconstruction, recombination, conceptualisation and projection, within the universe of what is possible. By constructing "counter spaces" with students, within which to explore value systems that do not coincide with the current ones, the design studio allows them to systematically analyse and practice the set of "paradigm shifts" required by design for ecological and social transition. (This has been the case for the "Horizontal Metropolis" research developed together with students from different schools, including Harvard GSD) (Cavalieri and Viganò, 2019). At the same time, this hypothesis makes it possible to read the territory itself in a different way, through the dynamic lens that each design operation requires, namely changes to how we live, work and establish relations. It is a question of avoiding clichés when describing and imagining the future, pre-packaged programmes and univocal directions (Figure 28.3).

Often, the direction of our gaze in the present lies in imagining the future. The construction of idiographies (Viganò, 2012a) associates urban and territorial design with the idea that it is important to understand the forms of landscapes, economies and cultures; it develops critical readings which, alone, can highlight their paradoxical conditions of possibility. While description through the project brings us closer to the territory-subject and its uniqueness and difference, it does not rule out that the identification and reconceptualisation process typical of all design hypotheses may contain, in seed form and simply an intention to appropriate and manipulate the subject itself. It is here that the design studio emancipates itself from being a mere place of production and requires constructing a broader horizon of meaning that calls into question different worldviews and the

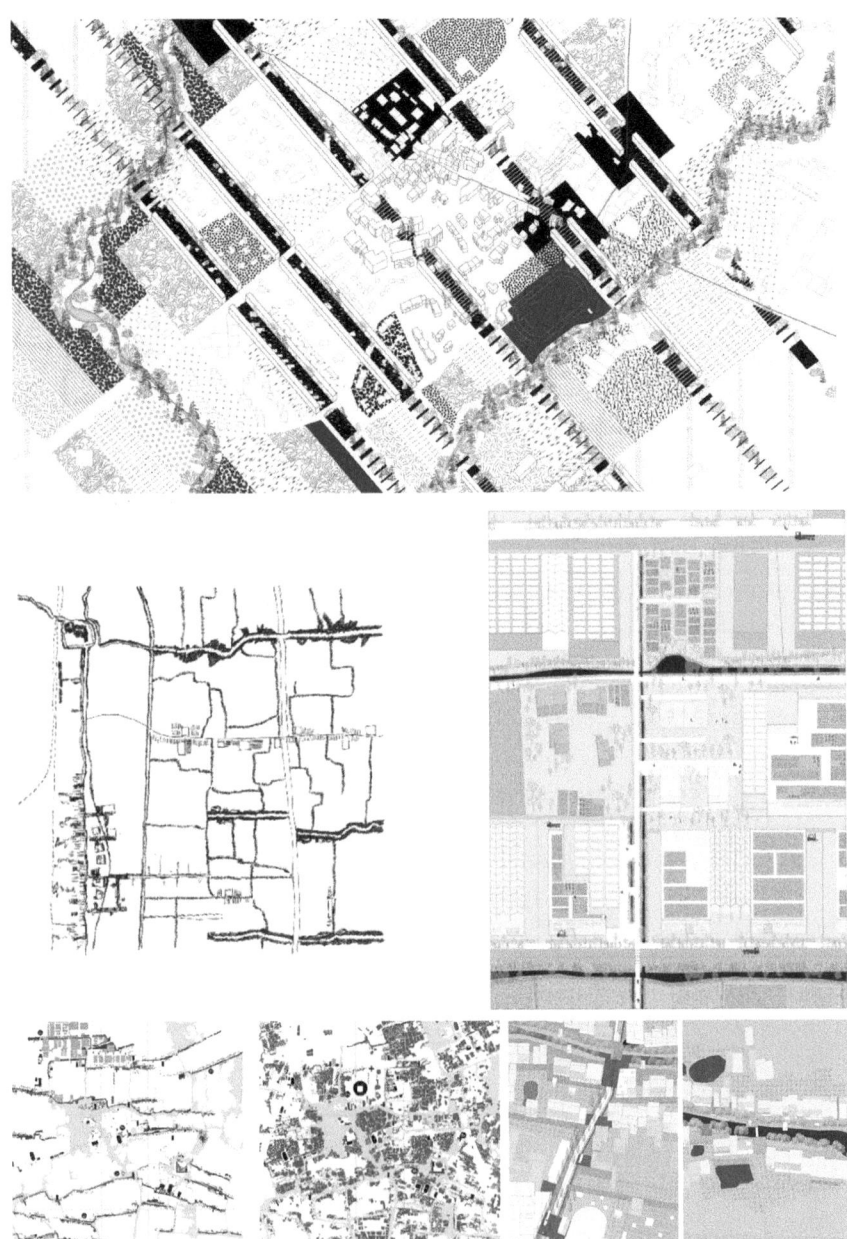

FIGURE 28.3 Top: Weak structures for the transition. EPFL Habitat Research Center, 2020 Middle: HM Project: through the transition, exploring the metamorphosis. HM Venice: T Pietropolli, Thesis IUAV, teaching staff: P Viganò and A Pagnacco. Bottom: HM Project: through the transition, exploring the metamorphosis. HM Hangzhou: Qinyi Zhang PhD thesis IUAV, supervisor P Viganò. Vision for Greater Geneva.

ethical-political positions that identify and support them. This is a delicate point that can never be resolved and which every new experience must address. Our own vision of the world is relativised through exchanges that transcend the protected world of the laboratory, bringing teachers and students face to face with a multitude of viewpoints, and with the hardness, malleability and ambiguity of "reality". Our vision is relativised through extended and repeated fieldwork, full immersion and workshops in different contexts. At the same time, these engagements enable us to fully grasp the expectation that each moment of "design", or imagination of the future, produces in communities that does not always find the strength and time to rethink itself and reconsider its practices in space.

Faced with the field of possible variations, uncertainty and the possibility that there is no solution, many of the traditional tools of urban and territorial design seem to lose their usefulness. But this difficulty drives us to further explore that very epistemological and ethical-political layer that helps us to develop a stance, however provisional and uncertain.

In my experience, some design studios have touched on fundamental aspects from this point of view. Such aspects include the relationship between the description of the present and the future projection: the present, read through latencies, emergencies and potentialities, approaches the future understood as an expression of possibilities. The key term that unites those different temporalities is "possibility" (Viganò, 2012b).

The territorial project directly confronts hypotheses of economic and social development that often challenge traditional models of territorial government. Many design studios have explored territories in which other trajectories have been defined outside institutional policies, revealing widespread strategies of resistance that foster design research and clarify the horizon of meaning. (This is the case for many design studios that have focused on the territories of diffuse urbanisation.) In this case, the key words are "resistance strategies".

On the other hand, there are territories that do not seem to have, and offer no, exit routes. We are faced with the inability of the territory and its communities to react to processes of exploitation and impoverishment, and of increased ecological, economic and social risk. In these cases, the project takes into account the hypothesis that no solution exists, definitively abandoning the safe path of "problem-solving". That path all too often leads to quick simplifications and superficial investigations. In such cases, the issue of structural redefining the problem arises, and the project must tackle the complicated terrain of issues apparently without a solution: in areas that have always been in crisis or marginalised, in areas subject to risks that can only increase and not decrease. In these cases, in addition to constructing scenarios, descriptions and reconceptualisations, operations come into play that reinforce the hypothesis of "design as knowledge producer" (Viganò, 2016): the narrative and design device of utopia. This structures criticism of the territory in the present – here and now – at the same time providing a projection similar to that returned by a mirror in which topics and issues are transferred and resolved under conditions different from the current ones. The sense of these latest explorations, outside the problem-solving field, is to provide radical and extreme keys to interpretation which are important for clarifying the gap between current conditions and what could or should be: these are not projects that lead to an immediate alteration of reality. They act as mediating elements between this and the action of future change; their ambition is to change the contours of the current discourse and establish new ones (Figure 28.4).

Territorial complexity cannot be simplified, especially when the territory becomes a subject. Design is simply part of it, contributing knowledge that is rooted and situated in its space, and imagination open to the future. Attempting to cross this complexity together with students, experts, institutional and political

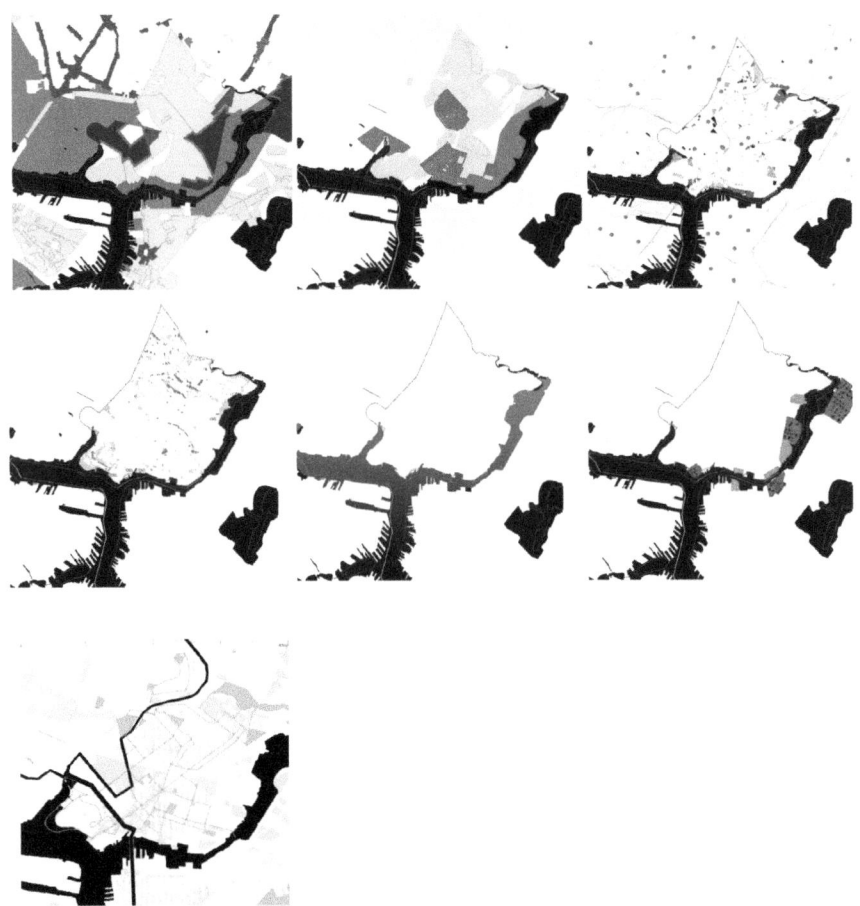

FIGURE 28.4 Chelsea, Boston: Servant as destiny. GSD autumn 2012, Dan Bier, Shuai Hao and Jana Vandergoot. Source: Shuai Hao and Jana Vandergoot, Now in: P Viganò, Territorialism, Harvard GSD, 2014.

decision-makers and the community is the pedagogical sense of what we call "teaching".

Note

1 The text expands the one I published with the same title in Marta De Marchi, Hessam Khorasani Zadeh (Eds), *Territori post-rurali/Territoires post-ruraux. Genealogie e prospettive / Généalogies et perspectives*, Rome, Italy: Officina Edizioni, 2020.

References

Audier, S. (2017). *La Société écologique et ses ennemis; pour une histoire alternative de l'émancipation*. La Découverte.

Balibar, É. (2012). *Citoyen-sujet et autres essais d'anthropologie philosophique*. PUF.

Barcelloni Corte, M., & Viganò, P. (Eds.). (2021). *The horizontal metropolis. The anthology*. Springer.

Cavalieri, C., & Viganò, P. (Eds.). (2019). *The horizontal metropolis. A radical project*. Park Books.

Corboz, A. (1983). *Le territoire comme palimpseste*. Diogène, 121.

Foucault, M. (1994). *Le sujet et le pouvoir*. In Michel Foucault (Ed.), *Dits et écrits, 1980–1988*, volume IV (p. 306). Gallimard.

Iorns Magallanes, C. J. (2015). Nature as an ancestor: Two examples of legal personality for nature in New Zealand. *Vertigo* 22 (hors série). Retrieved from https://journals .openedition.org/vertigo/16199.

Leopold, A. (1949). *A sand county Almanac*. Oxford University Press.

Pelluchon, C. (2020). *Réparons le monde. Humains, animaux, nature*. Payot et Rivage.

Revel, J. (2021). Tra Politica ed Etica. Italian version of the conference presented at Yale on 17 June as part of the international conference "Michel Foucault: after 1984" – Yale University, Whitney Center for Humanities, 17–18 October 2014", in *Euronomade* November 3, 2014. Retrieved from http://www.euronomade.info/?p=3572.

Stone, C. (1972). Should trees have standing? - Toward legal rights for natural objects. *Southern California Law Review 45*, 450–501.

Thaler, R. H., & Sunstein Cass, R. (2008). *Nudge: Improving decisions about health, wealth, and happiness*. Penguin Books.

Vattimo, G. (1985). *La fine della modernità*. Garzanti.

Viganò, P. (2008). Water and asphalt, the project of isotropy in the metropolitan region of Venice. *Architectural Design 78*, 34–39.

Viganò, P. (2012b). Situations, scenarios. In Giannotti Emanuel & Viganò Paola (Eds.), *Our common risk*. Edizioni.

Viganò, P. (2012a). Idiografia dell'agricoltura. *Territorio 60*.

Viganò, P. (2016). *The territories of urbanism: The project as knowledge producer*. Routledge-Epfl Press, 73–80.

Viganò, P., Bernardo, S., & Lorenzo, F. (Eds.). (2016). *Water and asphalt. The project of isotropy*. Park Books.

29

RELATIONAL URBANISM

Expanded Ecologies for a Capital Earth System

*Enriqueta Llabres-Valls, Zach Fluker
and Sheng-Yang Huang*

The symbiosis between ecology and digital design methods has led to a fertile but controversial design thinking and practice influence. Definitions and concepts about ecology and design thinking rapidly become obsolete. A necessary redefinition of these concepts constitutes an essential intellectual exercise to keep the discipline at the forefront of the environmental challenges of the time. The viability of the Earth System is the most relevant challenge at stake today. This chapter frames the discussion for an expanded ecology in the context of "planetary urbanisation" (Brenner, 2014, 2016) in which global ecology has become a capital-driven process (Harvey, 1996; Haraway, 2015). A capital-driven Earth System is one in which a large variety of agents interact in a non-linear way, returning different levels of organisation and hierarchies, each of them ruled by its law (Ellis, 2015). In such behaviour, resulting from a combination of top-down and bottom-up effects where small changes in the initial conditions produce large changes, extreme unexpected events occur more often, and interacting agents modify their strategies as experience accumulates. The designer will need to look at the main building blocks of such a complex system (Mitchell, 2009; Holland, 1995; Boyd and Folke, 2012) and identify the relevant scales and rules of law to put them at the forefront of the design question. This design approach embraces the non-linearity of the Earth System, leading to a much larger spectrum of disciplinary niches. Today, the discipline's main challenge is to incorporate these new niches and redefine humans' relation to other organisms and, ultimately, challenge contemporary trends in the global land use pattern and the urban fabric itself. In this context, design methods will need to move from flow modelling and time-based platforms representing change and evolution, to incorporate network technologies and data-mining strategies with the aim of encouraging positive behaviour and correcting negative environmental externalities caused by individuals, firms and institutions.

DOI: 10.4324/9781003145905-35

In the early 20th century, ecology started as an emergent natural science looking at the interaction between living organisms (Blaisdell, 1997; Amy, 2011). In the 1950s, complexity theory led to the development of ecosystem ecology, and complex physical systems models allowed to simulate interactions between populations (Odum, 1963). Ecosystem ecology was consolidated as a science in the 1960s with increased public awareness about pollution and biodiversity loss (Carson, 1963). At that time, The Great Acceleration started to leave its footprint in the physical world. Designers saw systems theory and cybernetics as a tool to introduce logical solutions to design problems. Similarly, the closed system idea also reflects early command and control attempts to manage the ecosystem by preserving its stability.

In the 1990s, cheap and ubiquitous computer-based communication technologies promised to remove social interaction and commerce from the physical space (Frazer, 1995, 2013). The development of pioneer design experiments of participatory authorship and interactivity in digitally augmented environments coincided with a new broad idea of ecology. The redefinition of ecology in the 1990s looked at how societies and culture shape humans' relation to their environment. They advocated for a vision in which humans are immersed in the ecosystem dynamics, challenging traditional ecosystem stability ideas (Botkin, 1990). Landscape architects started to look at how dynamic ecologies work, putting concepts of complex physical systems, emergence and morphogenesis at the core of the design discourse. This latest design approach coincided with the digital turn in architecture: digital technologies aided to bring forward a transdisciplinary framework looking at the territory by linking landscape, infrastructure and time-based modelling techniques, reflecting a tendency to increase the project's scale (Waldheim, 2006, 2016; Mostafavi and Najle, 2003).

Notwithstanding the well-established technology for ecosystem modelling, using remote sensing and geographic information systems to map and model complex data, the built environment disciplines still struggle to find design tools that can adequately respond to Earth System's challenges. Despite the increase in scale, the design methods still rely on the idea of a project as a conceptual framework which has limited its capacity for localised interventions. Currently, the discourse has moved away from solutions of stability and control towards developing organic, open-ended and resilient solutions that offer the environment the capacity to absorb unpredicted scenarios and adapt to a new reality. But localised interventions based on resilience and adaptation might be an insufficient and inadequate agency this time.

Over the last decade, scholars have begun bringing back epistemological questions challenging a city centrism approach to the urban phenomenon (Wachsmuth and Brenner, 2014). Instead, they propose planetary urbanisation to describe the extensive, uneven urban fabric shaped in a neoliberal globalisation context. This overarching condition results in a network of specialised territories that is, at the same time, highly connected and uneven. Planet Earth's habitable land accounts for 124 million square kilometres, where urban agglomerations occupy a small margin of 0.5 per cent of this area. Meanwhile, 70 per cent of this land constitutes intensive operational territories that sustain them. Ecosystem disruptions created by these intensive resource–extraction processes, and a widespread disregard for waste, are responsible for rampant biodiversity loss and the disturbance of established water and biogeochemical cycles that have kept the planet stable for the last half a million years. Currently, 55 per cent of the world population occupying a tiny 0.5 per cent of habitable land and are responsible for triggering a massive disturbance of the Earth's systems, without perceiving it. These urban agglomerations are far from what Edward Burtynsky has documented for the last 35 years: the Anthropocene's landscapes (Burtynsky and Adams, 2013). But the concept of the Anthropocene itself is a very elusive and slippery one. It misleads by the fact that not all humankind is causing environmental problems equally. Environmental challenges are deeply rooted in market economies' failures, contradictions between global-scale governance and international production systems, inadequate institutions, global inequalities and a general annihilation of a privileged share of humanity towards the material processes related to lifestyle and consumption. Environmental degradation intimately ties to social injustice (Ernstson and Swyngedouw, 2019).

At this time, a critical step in redefining ecology is to incorporate human behaviour as a complex adaptive system (CAS) looking at the agent's behaviour in ecosystem modelling. Agents learn and adapt to interactions with other agents, changing their performance through time by rating the usefulness of their behavioural rules. This performance is known as a signal-processing approach and is based on three principles: sensing, credit assignment and actuating. CASs are useful in a wide range of disciplines, from looking at how agents adapt their buying and selling conditions as markets change, to studying the performance of the immune system. The main question is how concepts from CASs can be useful design methods for the built environment.

Understanding Natural Capital in the Capitalist System

The basis of CAS behaviour is adaptive interactions resulting in a hierarchical structure. One way to intervene in CAS is by introducing lever points. A lever point in CAS is a small directed action causing large predictable changes in aggregate behaviour. One way to incorporate lever points in landscape architecture strategies is by introducing environmental policy instruments in the design process. Environmental policy instruments intervene in the context of both market failure and government failure to provide environmental protection (Helm and Hepburn, 2009; Hepburn et al., 2018). They look at appropriate environmental intervention models that lie between notions of corporate social responsibility, altruistic consumer or shareholder preferences and nationalising the delivery of environmental protection. The application of these tools by ecological agencies and policymakers is frequently through legislation, which has profound social and spatial consequences. However, both policies and instruments are too often detached from these places' spatial and material transformation; they lack vision and have a limited capacity for interaction with the local stakeholders. On the other hand, they encourage innovation that protects environmental systems, inducing and accelerating the development and diffusion of new clean products, processes or services. They incentivise positive behavioural patterns and correct negative environmental externalities.

We can identify two types of natural capital in the capitalist system: exhaustive and renewable natural capital. The risk of exhausting exhaustible natural capital assets is low because scarcity creates high prices, which leads to innovation and the development of substitutes. On the contrary, renewable natural capital tends to be scarce because it is often incorrectly priced or has no price. Examples of renewable natural capital include vital systems that support life on Earth, such as biodiversity. Market-based and flexible instruments such as emissions taxes or tradeable allowances provide individuals with flexibility to enhance innovation and foster organisational change. One type of market-based instrument is cap and trade. It is a tool used by governments to put an absolute limit – a hard cap – on pollution and allocate tradeable permits and some optional features. Intertidal Engine is a master research project developed in Bartlett Prospective (BPro) Research Cluster 18 as an attempt to introduce a cap-and-trade system as a lever point intended to prevent further environmental degradation and biodiversity loss in the context of coastal development in Abu Dhabi.

Abu Dhabi's rapid development, and the techniques employed for constructing artificial islands, have negatively impacted its coastal ecosystems. The project starts by developing the building blocks of the complex system on Abu Dhabi's coastline, identifying different coastal ecosystems of intertidal mangroves, salt marshes and subtidal seagrass meadows, algal mats and coastal sabkha (Schile and Megonigal, 2017). The project proposed a design methodology that departs from understanding the interaction between the local ecosystems and the geological processes in Abu Dhabi's intertidal coastline. The design methodology considers parameters related to the average depth and velocity of intertidal currents, and how these affect the coastal ecosystems. Based on the principle of relational construction of space-time and value (Llabres and Rico, 2016, 2012), the design team proposes an interactive dynamic interface as a governance tool to redirect agents' decisions. The proposed governance instrument, cap and trade, is an environmental policy instrument that allows flexibility and trade between local actors while maintaining their compromise with existing ecosystems and the ecological service. This case implies that the ecosystem's carbon sequestration capacity must always be maintained, leaving human actors the space to negotiate its location and the responsibility to guarantee its survival (Figure 29.1).

Current global ecology is a capital-driven process. Therefore, new design tools must deal with the primary agent of change, the capitalist system. In this case, the proposal introduced an environmental policy instrument as a lever point to try to change agents' behaviour. However, environmental policy instruments are

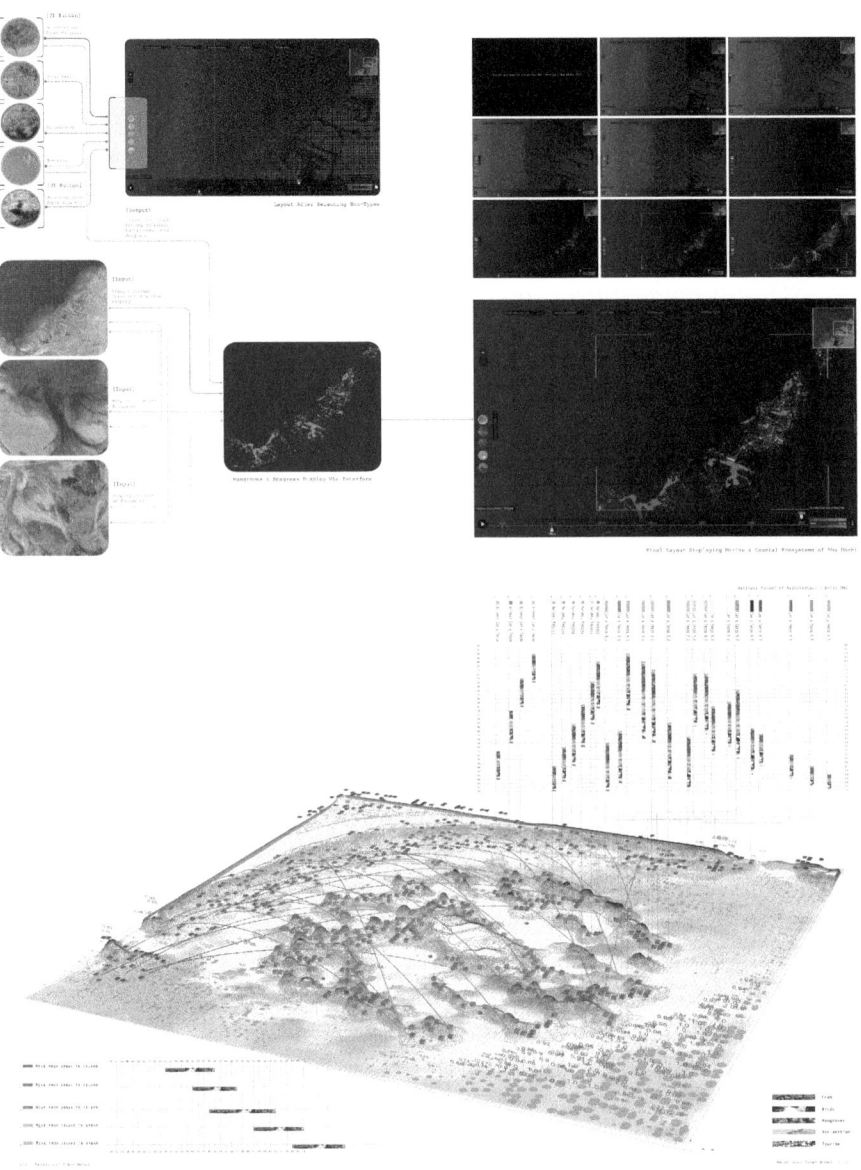

FIGURE 29.1 Top: Intertidal Engine. MArch Urban Design. Bartlett Prospective. (BPro) RC18. Tutors. Enriqueta Llabres-Valls, Maj Plemenitas, Zach Fluker and Claudia Pasquero. Skill Module Tutors. Immanuel Koh and Manos Zokohuras. Year (2014–15). Jiateng Sun, Xuyuan Yao, Yunyi Chen and Yiran Hu. Bottom: Intertidal Engine. MArch Urban Design. Bartlett Prospective. (BPro) RC18. Tutors. Enriqueta Llabres-Valls, Maj Plemenitas, Zach Fluker and Claudia Pasquero. Skill Module Tutors. Immanuel Koh and Manos Zokohuras. Year (2014–15). Jiateng Sun, Xuyuan Yao, Yunyi Chen and Yiran Hu.

more efficient if stakeholder compromise is substantial. A rapid transformation of the global ecology requires individuals and firms to assume environmental externalities and redefine their relationship with other non-human organisms. One of the challenges in redefining this relationship is the human population's concentration in highly anthropocentric settlements. In that sense, another aspect to consider is the use of immersive digital interfaces to develop empathy and environmental awareness. In addition, we can learn from the latest discoveries in neuroscience to progress agent engagement in the decision-making process.

Empathy and Heuristic Methods of Environmental Awareness

Advances in using immersive interactive environments to develop empathy and environmental awareness are related to the latest discoveries in neuroscience, the mirror neurons (Gallese, 1998, 2009). Their discovery has demonstrated that sensory-motor systems can serve as a mechanism to acquire new cognitive skills related to space and our relationship with other objects. How mirror neurons work shows that in embodied simulations, real actions and real things, or their digital representations, activate common parts of cerebral circuits (Gallese, 2013). These reciprocal links between action, perception and cognition, and how action frames perception, open new opportunities in which empathy becomes the central agency in developing these digital, artificial worlds.

Research cluster 18 in the Master of Architecture programme has explored the application of immersive interfaces in a wide range of applications. One example of this work is developing the physical and ecological simulation of landform patterns, and the implications in the decision-making process when these become a lived experience. It is the case of the work with scaled-down physical models, which has a long history within the realms of geology and engineering and offers a new value for designers through which different actors can experience the movement of materials, generally linked to fluid dynamics such as air and water (Figure 29.2).

Embodied simulation has proved a useful tool to arrive at compromises in complicated decision-making processes. Relational Urbanism used an interactive model as a consultation tool for design of Le Fanu Skate and Play Park, a people-centred design competition organised by the Irish Architectural Foundation. The design kit consisted of a sandbox model, a Kinect and a projector. The skater community used the tool to share their ideas about space and skating. The participants were able to see their models projected on the screen in real time. These data were analysed and used in the design proposal. The diversity of spaces in Le Fanu Skate and Playpark responded to the different interest group briefs. Using these interactive interfaces facilitated dialogue between different user groups and helped them arrive at compromises (Figures 29.3 and 29.4).

Further work in interactive, immersive interfaces has gravitated to developing empathy towards non-human organisms. That is the case with prosthetic

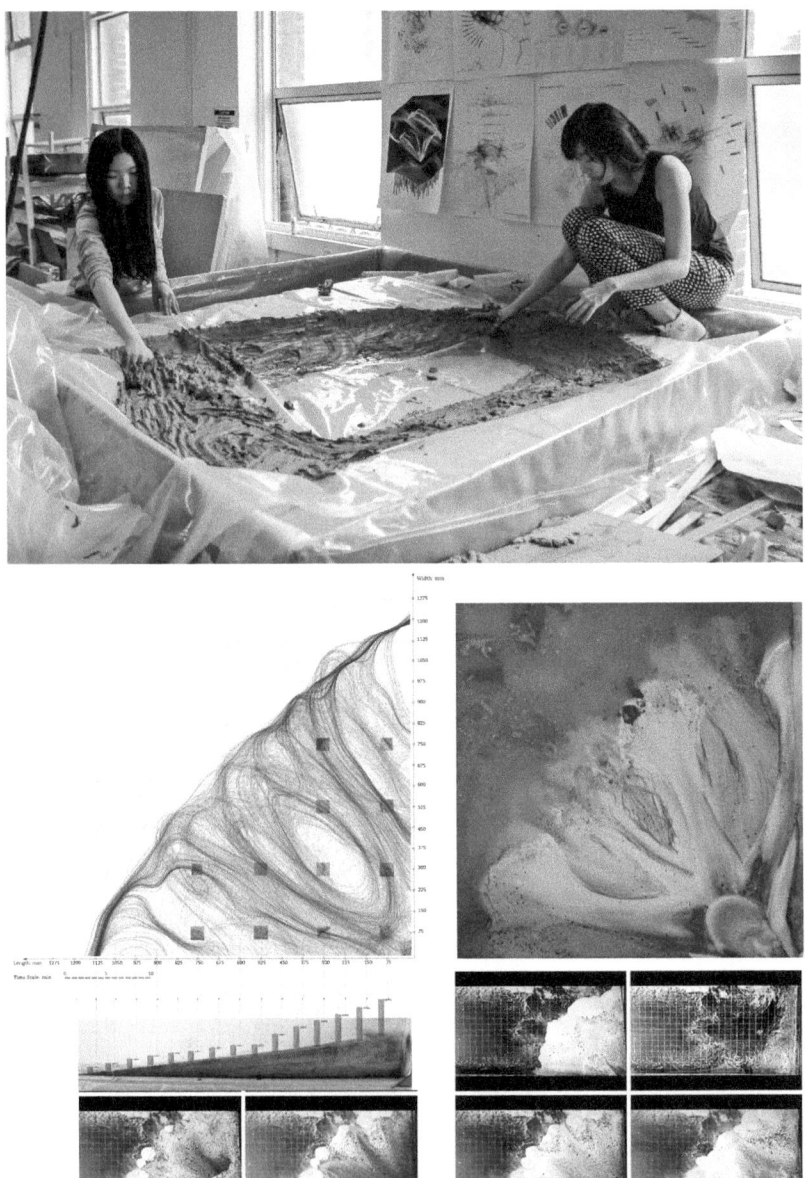

FIGURE 29.2 Top: Intertidal Engine. MArch Urban Design. Bartlett Prospective. (BPro) RC18. Tutors. Eduardo Rico, Enriqueta Llabres-Valls and Zach Fluker. Skill Module Tutors. Immanuel Koh. Year (2013–14). Bottom: Intertidal Engine. MArch Urban Design. Bartlett Prospective. (BPro) RC18. Tutors. Eduardo Rico, Enriqueta Llabres-Valls and Zach Fluker. Skill Module Tutors. Immanuel Koh. Year (2013–14). Waishan Qiu, Jia Zhang, Yunyi Chen and Yunke Zhang.

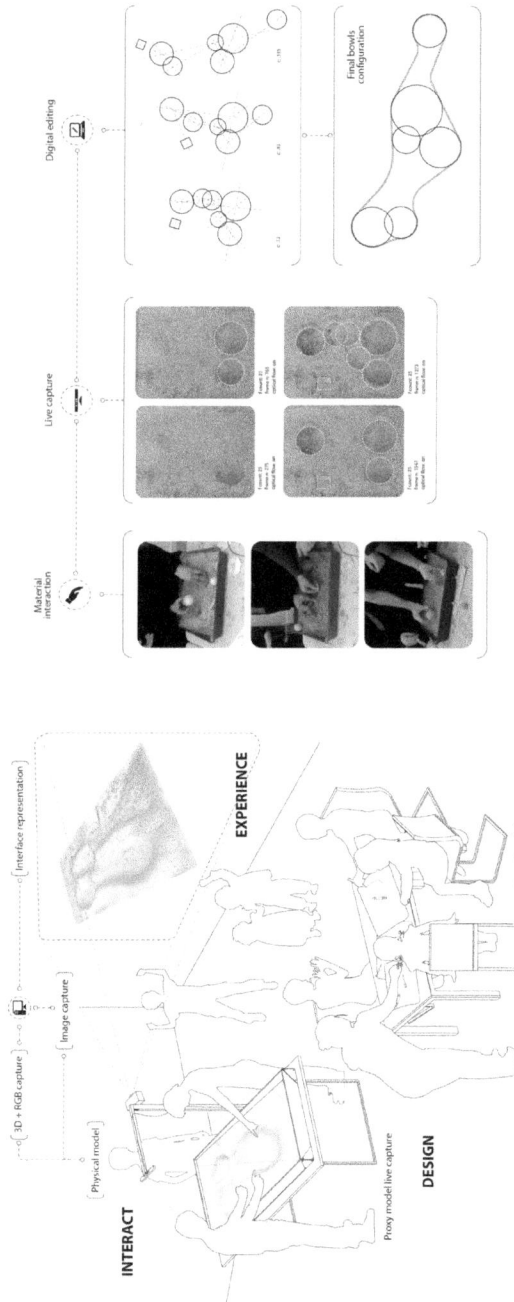

FIGURE 29.3 Le Fanu Skate and Play Park. Relational Urbanism. Eduardo Rico, Enriqueta Llabres-Valls, Giulio Dini, Dimitra Bra and Lida Driva.

FIGURE 29.4 Left: Prosthetic Ecologies. MArch Urban Design. Bartlett Prospective. (BPro) RC18. Tutors. Enriqueta Llabres-Valls, Zach Fluker. History and Theory Tutor: Nuria Alvarez Lombardero. Skill Module Tutors. Giulio Dini, Dimitra Bra and Martyn Carter. Year (2017–18). Farnoosh Fanaian, Haeam Jung and Luis Carlos Castillo. Right: E-Motion. Redefining London's Ecology of Intelligence. MArch Urban Design. Bartlett Prospective. (BPro) RC18. Tutors. Enriqueta Llabres-Valls and Zach Fluker. History and Theory Tutor: Ilaria Di Carlo. Skill Module Tutors. Dimitra Bra and Huang Sheng-Yang. Year (2019–20). Mochen Jiang, Yuankai Wang, Fei Chen and Haojun Cui.

ecologies, where the students use Normalised Difference Vegetation Index (NDVI) mapping to show photosynthetic organisms' performance in highly artificial environments. The project allowed for customising architectural mechanisms for a better symbiosis between non-human organisms and architecture (Figure 29.5). In other projects, such as E-Motion: Redefining London's Ecology of Intelligence, the research uses digital simulation technologies to understand migration patterns of critical species in London, helping to redefine the city's use of public space and its materiality.

Participatory authorship has been inherent in the development of digital design methods since the 1990s. The latest discoveries in neuroscience, and advances in immersive digital environments, open the opportunity to extend this participatory turn towards models that promote empathy and environmental awareness. However, a change in our global ecologies cannot exist without redefining how the global urban fabric operates – particularly tackling the uneven distribution of habitable land use. Notably, the design discipline must challenge cities' production and consumption habits, and the numerous environmental externalities they produce.

Redefining Urban Agglomeration Externalities

As the world continues to urbanise, cities have predominantly been preoccupied with successfully managing urban growth. By meeting the needs of their growing urban populations, the urban plan has historically been inward-looking, mainly focused on challenges such as housing, transportation, energy systems, public space and other infrastructure. This distinction between urban and non-urban environments, and the city's idea as a delimited entity, makes it difficult to understand contemporary trends in global habitable land use distribution, structure and specialisation patterns. One of these challenges is global food production and its effects on biodiversity due to crop specialisation. Currently, 60 per cent of global arable land cultivates four types of crops: rice, wheat, corn and soybeans. As the global population continues to rise, new food production patterns will be necessary, including increasing food production diversity and food production in artificial environments.

Agritecture, a new pattern for agrarian urbanism, is a master research project that explores a future scenario in which cities continuously reprogramme their land use. Food production is part of the urban fabric. The project explores new forms of intervention in the urban fabric using design methods based on CAS principles. An evolutionary learning algorithm is employed to calibrate the location criteria of vertical farms. In this way, the design method follows the principles of sensing, credit assignment and actuating. The research also looks at design methods of reprogramming an urban fabric share by humans and plants.

For this purpose, the research uses a Generative Adversarial Network (GAN) as a design tool to suggest possible spatial coexistence between vertical farming and humans. This class of machine learning framework employs two neural

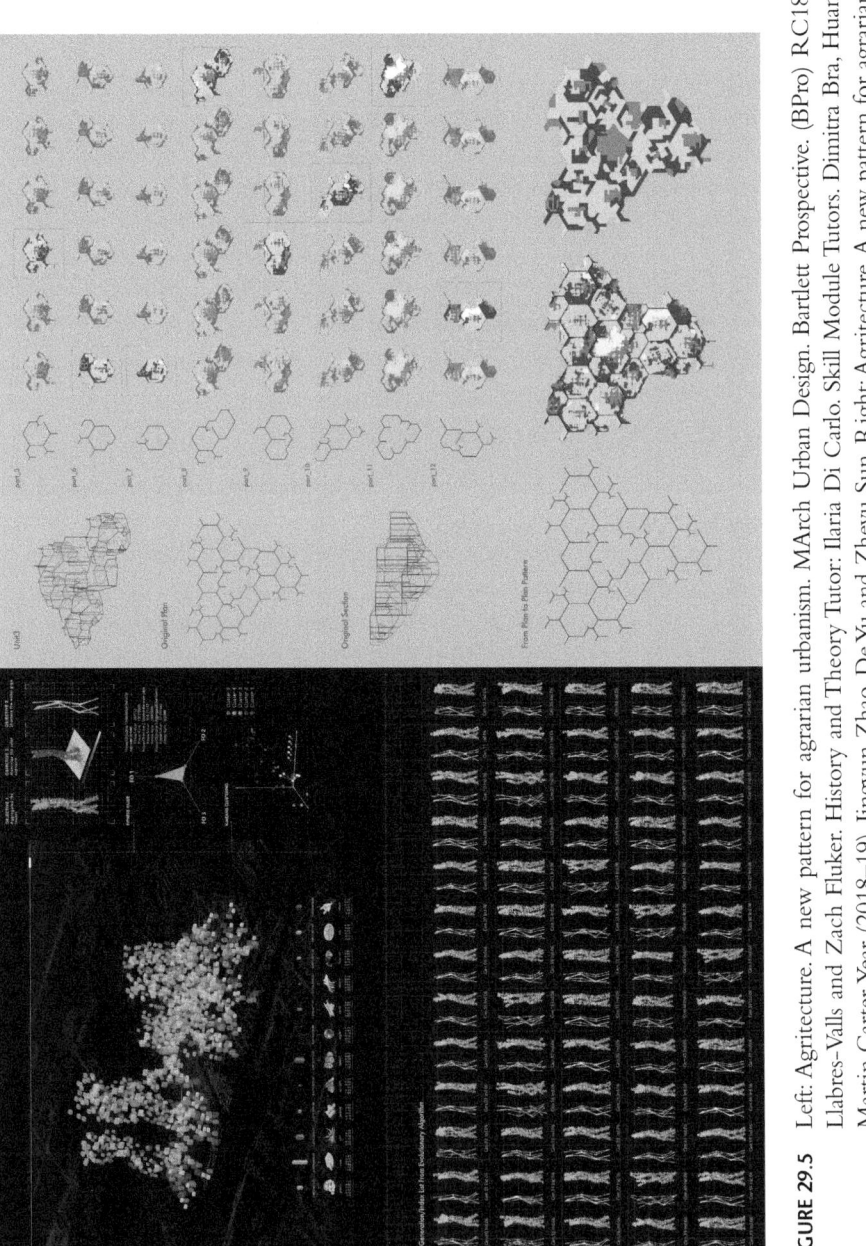

FIGURE 29.5 Left: Agritecture. A new pattern for agrarian urbanism. MArch Urban Design. Bartlett Prospective. (BPro) RC18. Tutors. Enriqueta Llabres-Valls and Zach Fluker. History and Theory Tutor: Ilaria Di Carlo. Skill Module Tutors. Dimitra Bra, Huang Sheng-Yang and Martin Carter. Year (2018–19). Jingyun Zhao, De Yu and Zheyu Sun. Right: Agritecture. A new pattern for agrarian urbanism. MArch Urban Design. Bartlett Prospective. (BPro) RC18. Tutors. Enriqueta Llabres-Valls and Zach Fluker. History and Theory Tutor: Ilaria Di Carlo. Skill Module Tutors. Dimitra Bra, Huang Sheng-Yang and Martin Carter. Year (2018–19). Jingyun Zhao, De Yu and Zheyu Sun.

networks. The generative network creates candidates based on a database, and the discriminator network evaluates them. The generator learns from a latent space of data distribution while the discriminator distinguishes candidates produced by the generator from the data distribution. In this case, the data consist of buildings historically dedicated to growing plants, from greenhouses to vertical farming.

Agritecture taps into design research questions such as the nature of creativity when technology challenges what it means to be human; specifically, how the latest advances in neural networks are accelerating the pace of change in our culture and society. Network-based technologies are currently entering the most tacit conceptual spaces that historically are built based on experience, embodiment and the accumulative employment of successive layers of knowledge and history. This unprecedented technological shift is expanding human forms of thinking.

Complex Adaptive Systems in Capital-Driven Planetary Urbanisation

The foremost challenge designers face today is how to intervene in a capital-driven Earth System. Many agents interact in a non-linear way, returning different levels of organisation and hierarchies, each of them ruled by its law. One of these design research questions aims to understand the hierarchical structure characteristic of CAS, and how particular combinations of agents at one level become agents at the next level up. Holland describes how this happens in the organisation of biological organisms. Similarly, the discipline should look at the hierarchical structure in the development of the urban fabric, considering forms of intervention that can help address negative environmental externalities currently challenging sustainability of Earth System cycles. This will lead to expanding the disciplinary niches in the study of ecosystem dynamics, and the incorporation of new design instruments based on neural networks and data mining.

References

Amy, A. (2011). Editorial: Celebrating the ecosystem's three-quarter century: Introduction to a "Virtual Special Issue" on Sir Arthur Tansley's ecosystem concept. *The New Phytologist, 192*(3), 561–563.

Blaisdell, M. L. (1997). Book review. A history of the ecosystem concept in ecology: More than the sum of the parts, by Frank Benjamin Golley. *Isis, 88*, 731.

Botkin, D. B. (1990). *Discordant harmonies: A new ecology for the twenty-first century.* Oxford University Press.

Boyd, E., & Folke, C. (Eds.). (2012). *Adapting institutions: Governance, complexity and social-ecological resilience.* Cambridge University Press.

Brenner, N. (Ed.). (2014). *Implosions/explosions: Towards a study of planetary urbanization.* JOVIS.

Brenner, N. (2016). *Critique of urbanization: Selected essays.* Birkhäuser.

Burtynsky, E., & Adams, A. (2013). Landscapes through the Lens. *Queen's Quarterly*, *120*, 348.

Carson, R. (1963). *Silent spring*. Hamilton.

Ellis, E. C. (2015). Ecology in an anthropogenic biosphere. *Ecological Monographs*, *85*(3), 287–331.

Ernston, H., & Swyngedouw, E. (Eds.). (2019). *Urban political ecology in the anthropo-obscene: Interruptions and possibilities*. Routledge.

Frazer, J. (1995). *An evolutionary architecture*. Architectural Association.

Frazer, J. (2013). *The architectural relevance of cyberspace*. John Wiley & Sons.

Gallese, V. (1998). Mirror neurons and the simulation theory of mind-reading. *Trends in Cognitive Sciences*, *2*(12), 493–501.

Gallese, V. (2009). Mirror neurons, embodied simulation, and the neural basis of social identification. *Psychoanalytic Dialogues*, *19*(5), 519–536.

Gallese, V. (2013). Mirror neurons, embodied simulation and a second-person approach to mindreading. *Cortex*, *49*(10), 2954–2956.

Haraway, D. (2015). Anthropocene, capitalocene, plantationocene, chthulucene: Making kin. *Environmental Humanities*, *6*(1), 159–165.

Harvey, D. (1996). *Justice, nature, and the geography of difference*. Blackwell.

Helm, D., & Hepburn, C. (Eds.). (2009). *The economics and politics of climate change*. Oxford University Press.

Hepburn, C., Pless, J., & Popp, D. (2018). Policy brief – Encouraging innovation that protects environmental systems: Five policy proposals. *Review of Environmental Economics and Policy*, *12*(1), 154–169.

Holland, J. H. (1995). *Hidden order: How adaptation builds complexity*. Addison-Wesley.

Llabres, E., & Rico, E. (2012). In progress: Relational urban models. *Urban Design International*, *17*(4), 319–335.

Llabres, E., & Rico, E. (2016). Relational urban models: Parameters, values and tacit forms of algorithms. *Architectural Design*, *86*(4), 84–91.

Mitchell, M. (2009). *Complexity a guided tour*. Oxford University Press.

Mostafavi, M., & Najle, C. (Eds.). (2003). *Landscape urbanism: A manual for the machinic landscape*. AA Publications.

Odum, E. P. (1963). *Fundamentals of ecology*. Saunders.

Schile, L. M., & Megonigal, J. P. (2017). *Abu Dhabi blue carbon demonstration project*. Smithsonian Environmental Research Center.

Wachsmuth, D., & Brenner, N. (2014). Introduction to Henri Lefebvre's "Dissolving city, planetary metamorphosis." *Environment and Planning D: Society and Space*, *32*(2), 199–202.

Waldheim, C. (Ed.). (2006). *The landscape urbanism reader*. Princeton Architectural Press.

Waldheim, C. (2016). *Landscape as urbanism: A general theory*. Princeton University Press.

30

CONVERSATION WITH JENNIFER DEGER

What is your definition of ecology in the context of your current research?

It depends where we're taking the definition from, but ecology in the broadest sense is the relationships between entities within an environment. That would be how I'd orientate myself, but of course, that can be the physical environment and also institutional environments, not to mention media ecologies and what we might call transdisciplinary ecologies. In my work, it's those varying and overlapping configurations of more-than-human worlds in transition and, indeed, crisis that focuses my energies. I'm interested in mediated forms of understanding and affective constellations within these contexts, which make it a nice broad and suggestive term.

Could you please describe your research practice?

I'm an anthropologist, and for more than 25 years I've been working up in north-east Arnhem Land (Northern Territory, Australia) with the Yolngu people. From the beginning, I've worked on collaborative forms of research, and I've been incredibly fortunate to have been invited to work in a context in where I've been invited to participate in creative projects from the outset. So to answer your question I would say that my research practice has been shaped by a commitment to creative, collaborative and contextual knowledge production – and a concern with the new forms of knowledge that can emerge through such practices. I'm part of two collectives: Miyarrka Media, an arts collective based in north-east Arnhem Land, and *Feral Atlas* which we can talk about later. Through those collaborations I work in what you might call multimodal ways of thinking. The practice is concerned with situated forms of thinking in ways that reach beyond text, that take up the relational and aesthetic potential of the digital to bring together different configurations of voices, perspectives and disciplines to render multi-sided and multi-sited ways of seeing and knowing the world, in new ways.

DOI: 10.4324/9781003145905-36

Could you please elaborate on what you meant by contextual issues in the context of the project you are undertaking?

Yes, let's answer the question in context! I'm working on a new project with my main Yolngu collaborator, Paul Gurrumuruwuy. We've been working together for about ten years now. It's a project about the beach as an important site in the Yolngu imaginary, formulated as an intervention into the broader Australian imaginary and its attachments to beach-based identities and performative possibilities. Gurrumuruwuy is motivated by wanting to tell his biography which relates to a specific site where sands have songs and stories, and where plastic is building up with the effect of literally choking the life out the ecology. Our aim is to record the stories of these places to create a website that will function as a ritual space based on Yolngu practices of ecological care and renewal, while reaching out to a broader public audience. It's a project that arises out of a specific context, in the sense that the urgency of the project emerges not just from kin-based relationships that authorise the project, but from the specific moment in time and culture, and the kinds of capacities we've developed together.

This project was conceived after ten years of making international exhibitions and films and books, and it's this prior experience that has given us the confidence to push our work with form and content further, especially in terms of this move to deliberate address and engaging the varying publics that the internet brings forth.

But context matters here too in terms of who will encounter this work – and where. The internet of course renders the question of context extremely complex. Miyarrka Media's projects are always made first and foremost to satisfy Yolngu social aesthetics and politics. Nothing that we record, and certainly nothing we make public, can go ahead without the authority of the relevant landowners and cultural managers who affirm the possibilities of that knowledge going public. But the internet brings a new kind of public interest and scrutiny – and so a kind of ritual participant.

But this is only one context that we will have to manage in this project. Miyarrka Media work with keen awareness of what you might call in other disciplines, site-specific knowledge. That kind of attentiveness – with an idea that we're also deliberately engaging audiences beyond those sites, reaching into a broader national and international arena with our work – means we have to satisfy local Yolngu cultural politics while navigating the epistemological challenges of meaningfully navigating between those worlds too. This is where the real work lies, in this attention to the play of sameness and difference through which meaning and relationships are found, made and strengthened. Somehow, we have to create work capable of working – communicating, moving, motivating, connecting, convincing even! – on multiple levels at the same time. That's the challenge and the motivation.

So it's context all the way down – and all the way across.

Could you please discuss your current project(s), and how your work has changed over the years?

This new project is called Rangingur which means "coming from the beach or belonging to the beach" in Yolngu Matha. What we are aiming at is what we're describing as "a digital art of renewal". We're just getting started. It's a collaborative research project that starts from, and uses digital methods to repeatedly return to, the threatened ancestral sand and saltwater worlds of East Arnhem Land. The idea is that Yolngu participants will record and re-enliven endangered songs, stories and associated ecological knowledge from three specific ancestral beach sites in East Arnhem Land. We will then take this archive of beach-based knowledge and connection and find ways for it to come alive on the internet, on a website especially designed for the renewal of indigenous history, knowledge and belonging.

So, at its core, this is an experiment with using digital media to care for the country in new ways, and in ways that can be relevant for indigenous and non-indigenous people elsewhere too. We got the funding in 2020 but we have yet to actually go and sit on a beach and talk and plan it out. On one level this is a legacy project for him Gurrumuruwuy, my friend and collaborator. He's in his mid-60s. He's interested in the ways we've been working to perform the possibilities of new kinds of social relations at multiple levels, in a creative way that energises him and has energised his family. We work intergenerationally. It comes off the back of all our experiences of exhibiting and speaking in places like New York, London, Sydney and Copenhagen, experiences that has helped him understand the kinds of things that worlds beyond the Yolngu world are interested in – and to identify new possibilities for connection and mutuality. It also rises out of his growing confidence that he and his family have a vision for anthropology as a discipline that brings worlds into relationship. The members of Miyarrka Media and their families value their own profound knowledge and set of values in relation to the broad anthropological questions of what it is to be human, or what it is to exist in a more-than-human world. It's a vision that has relevance and socialpolitical traction beyond the local, precisely because of its relational focus.

It's a very generous and outward-looking project. It's looking down through the generations to build not just digital expertise, not just training in media, but a vision of new forms of world-making that are both inclusive and generous and connective in the sense of creating a ritual space for non-Yolngu audiences. At the same time, it's insisting on the real particularities of knowledge that come out of the country, and authority that comes out of the country. That's the foundation from which we dreamed up this project, where we would go and sit in the outstation he's a custodian for and sit on the beach and yarn, first of all. From there we'll bring together a team of Yolngu and non-Yolngu filmmakers and artists to create the material that will go to the website. As I said, the website is envisaged to do the work of ritual through demonstrating Yolngu modes of connecting and caring for the country in the context of beaches, sending out an ethos and a performative demonstration of care that can engage audiences in other places.

So, this is a project conceived in the context of the shadowed futures now literally washing up on the beach, in these lands marred by the legacies of settler colonialism and industrial infrastructure projects. The beach, of course, is a powerful place of literal horizons and imaginative horizons, not to mention the various temporalities and connective vectors that time that sand itself materialises through ancestral songs, plastic residue and tidal forces.

The competence that Gurrumuruwuy and I and his family have developed over many years of doing these collaborations, and performative forms of public anthropology, means I think we're in a really good place to take this next step. Instead of travelling around the world and doing exhibitions, we are just settling down in-country and seeing what emerges from being in place.

We don't know quite how all this is going to work, but what sits at the foundation of the ideas and ambitions that Gurrumuruwuy and I are working on, is a shared, ongoing commitment to a generative research practice that cares for social relations. Our commitment acknowledges the precarity of the current moment, and the ways in which the social and the ecological are increasingly recognised to be bound together by non-indigenous people. This is not news to the Yolngu. What is new of course are the threats of climate change and all its ramifications that are already affecting the country and its landowners in extremely worrying ways.

You make reference to this notion of world-building as a creative methodology. Do you mind elaborating on what you mean by world-building? And its methods?

A good example of this would be Miyarrka Media's recently published book, *Phone & Spear*, which is a shared reflection on the ways mobile phones have intensified the push and pull of relationships in your life, for better and for worse. It features a range of collaged photographs where people will put together different family members who are in different places, often together with people who have passed away in order to reaffirm the vitality of an ancestrally ordered world in a highly contemporary context. The person who collages this family portrait will then choose elements from the internet in order to intensify the meaning and significance of the image in Yolngu terms. So for example, these two boys in a shark, reunite two young brothers living in different places by placing them side by side in the body of a shark which is their ancestor and the source of their Datiwuy clan power and identity. A generic image is made specific and deeply meaningful. When Yolngu look at the image, they see all kinds of details in what appears to be a generic image. For example, there is a special name for that splashing, shining and bubbling water the shark makes when it rises up. They call it *djarraran bunmirr*. Another favourite example depicts a group of sisters assembled at a funeral for a close family member, even though some of them had been unable to attend in person. The background is a vivid sunset chosen for the ancestrally significant stories associated with sunset for these women, and the deep affective register that the Yolngu associate with this particular and spectacular time of day. In the act of creating the image, and then circulating it amongst family members via their phones, connections between kin and country are affirmed and strengthened. The woman who made that image described that "sunsets tell us to go back to the land through our minds and hearts and to think about family who've passed away". So these images act as vectors of deep feeling, story and connection that bind people in place across time and space. In these ways the phone-made collages are acts of world-making and world-affirming using the materials at hand.

This cut-and-paste poetics informed the design of the book – and how we envisaged its performative purpose in terms of another register of relationship-making. When we designed it together, the idea was that we didn't want to just make an archive of these fantastic phone-made collages. The book itself had to enliven them again. Gurrumuruwuy has taught me that our work must always accord with a broader project of enlivenment-making, and so the design itself had to enact this enlivenment – and in doing so open up the potential to draw strangers into a form of mediated relationship. From the little collages we made a whole new field and world of pattern, bringing different ones together and putting them together, and then from a Yolngu point of view that did the energetic work of enlivenment and allure. Someone who is not familiar with the meanings here can sense it from the page, but works also from a Yolngu point of view

For us, it was a really successful way to take this form of social aesthetics that's about pattern, relationship, ethics, colour, and rework it into a form that would produce an effect on the reader. The book itself was understood to be a relational technology – designed to bring Balanda (European, or non-Yolngu) viewers into a feelingful relationship with the stories and images in the book – and in the process to connect them to their own families. It's quite amazing when I Zoom into classes and people say, "Oh, you know I was halfway through reading the book and I bought it for my grandmother". So there's this kind of stimulation for people to consider the relational dynamics that bind them within their own lives. It's amazing to us all that the book seems to have succeeded in activating a performative potential of relationship-making, in which Yolngu worlds and Balanda worlds (non-aboriginal worlds) can be brought into relationship without denying the constitutive differences that define where people come from and who they are. That's the foundational ethos with which we're going into the beach project again, enacting the potential of this relationship.

As *Phone & Spear* explores, we are all being drawn into new constellations of relationships made possible by the mobile phone itself. That was the metaphor of connectivity which the book unpacked and sought to activate anew (albeit on the page). And the beach one; it will draw precisely on an understanding that different people have their own relationships and attachments to beaches, and specific kinds of beaches. So the central question becomes: How do we render a Yolngu way of being attached in such a way that other audiences will be enlivened and motivated, and potentially reoriented at the same time to their beaches, to understand modes of connection that may not have occurred to them? Or maybe they might be brought into a new set of relationships and with that a deeper and broader sense of this country and the foundational stories that shape not only its past, but also potentially as my Yolngu colleagues would insist, its future possibilities too.

My next question potentially builds on that discussion. Could you please discuss what you see as the relationship between ecology and the Tipper Videos that have been developed for **Feral Atlas?**

Those films were made to demonstrate a core theoretical point of *Feral Atlas*, what we call mode of infrastructurally mediated state change. That's a real mouthful, which is why we've brought it down to *Tippers*. The idea is that infrastructures have non-designed effects. Paying attention to the everyday processes of infrastructures has an analytic value that we sought to mobilise within *Feral Atlas*. So, the films are made with a kind of a quiet observational frame, like a fixed frame in which to watch seemingly every day, maybe even banal actions like fruit being loaded on and off a truck, or a pile of plastic blowing in the wind, or a tractor ploughing a field. Through each of these videos, put together in a series, the underlying analytic concept becomes evident. Let's talk about the Tipper we called GRID: it's in that simplification of ecology that the plantation first developed, that certain things live and certain things die. There is a violence then to this infrastructural process of GRID that we aren't used to recognising (Figure 30.1).

The films associated with GRID show a Californian strawberry field where all the plants are covered with plastic, an automated milking system in Italy, a hydroponic farm in Bahrain and so on. Each of these short films ask you to slow down and pay attention. They invite you to start to recognise connections across a variety of infrastructural contexts and processes. They show you the processes by which land, sea and air scapes become transformed through infrastructural processes. They're designed to lead you into understanding the non-designed effects of that process. These are often banal, everyday moments of the industrial world in action. The films are made up of scenes shot in factories, markets, industrial farms, gas fields and nuclear plants, long distance transportation and garbage dumps. It took a long time to find the right footage because we wanted to show process. We wanted to show these infrastructures in action, to teach people to view footage that might in another context be taken as a celebration of the infrastructure but here it is offered up as a way to view infrastructural ecological violence. To this end the little films are quiet, they're meditative and they invite you, within the design of *Feral Atlas*, to also pause and to shift registers by reading about the concept. Hopefully through such a process users will eventually arrive at an "Aha" moment. "Oh, okay. That makes sense for me now", and then move onwards, perhaps to the field report. Like other aspects of the atlas, the effects of these films are intended as iterative. Each time you visit a Tipper page, you are offered another series of perspectives, and so another chance to reflect on the central argument of *Feral Atlas*, namely that these non-designed effects of imperial and industrial infrastructure are the Anthropocene.

FIGURE 30.1   GRID. Top: Andrés Camacho, Jennifer Deger and Duane Peterson, 2020. Reprinted from *Feral Atlas*: The More-Than-Human Anthropocene, Anna L Tsing and Jennifer Deger, Alder Keleman Saxena and Feifei Zhou, http://feralatlas.org/ published by Stanford University Press (c) 2020 by the Board of Trustees of the Leland Stanford Jr University. All rights reserved. Licenced under the Creative Commons License CC BY-NC-ND 4.0. Bottom: Still from video by Armin Linke, 2020. Reprinted from *Feral Atlas*: The More-Than-Human Anthropocene, Anna L Tsing and Jennifer Deger, Alder Keleman Saxena and Feifei Zhou, http://feralatlas.org/ published by Stanford University Press (c) 2020 by the Board of Trustees of the Leland Stanford Jr University. All rights reserved. Licensed under the Creative Commons License CC BY-NC-ND 4.0.

So there is work to do here, in piecing *Feral Atlas* together for yourself. And that too, is part of the point of the design and structure of the site. We wanted to not only to encourage people to take time to get lost in *Feral Atlas's* rendering of Anthropocene worlds, we wanted to counter the expectations of mastery that digital infrastructures tend to promote.

Could you discuss the significance of your work on the Feral Atlas *as a way or a means for conceptualising the entanglement between non-human entities and human-built infrastructures, as a shift in the ways we understand the built environment as territorial impacts?*

Feral Atlas offers more than human history over the last 500 years of imperial and industrial infrastructure projects. It was conceived and designed and realised to bring into play a range of disciplinary perspectives and situated empirical observations. The site invites you to choose your own adventure, if you like, and come to a cumulative understanding of the underlying argument through a process of exploring, looking, reading, pausing, reflecting and clicking onwards.

Certainly, that has been my own experience. About halfway through the process of curating *Feral Atlas* I started looking at the world around me differently. I started driving to work through the sugar cane fields of far north of Queensland, and instead of taking a kind of everyday joy in the way the green leaves caught the light, I started to see a long history of empire, slavery, plantations and industrial farming, and the ecological simplifications and forms of radical ecological change that are involved.

Feral Atlas offers these kinds of connections in the cumulative way I described so that seemingly unconnected processes and events start to be seen in relationship to each other through the lens of the more-than-human Anthropocene. So shipping pallets, for instance: these pallets move commodities around the globe, delivering not only commercial goods, but certain insects that go on to proliferate in new environments destroying forests with extremely worrying speed. The ways soil from the industrial nursery trade brings forms of disease: that is unchecked and also threatening ecologies in many parts of the world. Things are being moved around the world with a ferocious new speed and intensity that has consequences that we are still only coming to recognise. If we go back to the TAKE *Tipper*, it reflects on the presumption, and it's a very colonial presumption, that we can just pick up and take our worlds elsewhere and drop them down, and then upscale and spread on top of what was there before. There is no consideration of consequences. Or if the consequences have been considered in that ongoing arc of colonialism, in terms of the price paid by indigenous people and indigenous ecologies, they've proved horrifically secondary to the goals of capital.

So that's the aim of *Feral Atlas*: to offer a different way of seeing the Anthropocene that is all around us, in these multiple ways. I'm tempted to say if there's one takeaway from *Feral Atlas* it is that things are much worse than many of us yet perceive in terms of the proliferation of what we call feral entities that increasingly threaten the liveability of our environments for humans and non-humans alike.

Could you please discuss how **Feral Atlas** *shifts the way we think about sites of imperial industrial ruin?*

In curating *Feral Atlas*, we were really clear that we didn't want it to end up as just another pile of bad news. If this work was to indeed shift the way we see the Anthropocene, it's not enough to point to terrible case studies. It was really important that foundational understandings of the ways in which infrastructures work that the particular kinds of human projects, not all human infrastructures, but these ones that scale up from empire and industry, that art and anthropology both share an interest in that kind of gestalt, that moment when all of a sudden something clicks and you see the world differently (Figure 30.2).

FIGURE 30.2 Capital. Feifei Zhou. Reprinted from Feral Atlas: The More-Than-Human Anthropocene, Anna L Tsing and Jennifer Deger, Alder Keleman Saxena and Feifei Zhou, http://feralatlas.org/ published by Stanford University Press (c) 2020 by the Board of Trustees of the Leland Stanford Jr. University. All rights reserved. Licensed under the Creative Commons License CC BY-NC-ND 4.0, https://creativecommons.org/licenses/by-nc-nd/4.0/legalcode.

This is a really strange and super challenging moment, of course, in human history and as we approached the design of the site, we kept our focus on the challenge of telling terrible stories while offering an analytic framework and a narrative that helps us to understand the underlying processes that connect the atlas's 89 field reports. Here again lies the generative hope for the atlas: many different perspectives come together to tell a bigger story because examples are drawn from specific places in the world, and with a real emphasis on empirical knowledge and empirical understanding from a range of disciplinary perspectives – from a Yolngu artist through to biologists and oceanographers and artists and anthropologists and historians. By bringing these perspectives together, *Feral Atlas* demonstrates a new kind of potential for collective storytelling, as well as collective analysis. Hopefully the result in the viewer is that something is mobilised, that imagination, as well as wonder and horror, are mobilised in ways that will energise them to find their own way to take action.

My own take-away from working on *Feral Atlas* is that there is tremendous value in the work of holding open a transdisciplinary space that enables new kinds of political, intellectual, and even poetic, forms of collective thought in action.

What do you see are the future adaptations and extensions of your research? And how are you responding to emerging issues related to the climate crisis?

As you will gather by now, I am motivated by the conviction that there is so much more work to be done in terms of bringing digital and environment ecologies into relationship – in all the configurations I mentioned earlier – as we seek new ways to work together to understand, and critically and creatively respond to, the mounting challenges of our times.

I said at the beginning of our discussion that for me ecology was about situated environmental relationships, but it also about affective ecologies and media ecologies and institutional ecologies. My work with the Yolngu has really taught me to pay attention to all those things as constitutive of the ways worlds are made or might be remade. And so *Feral Atlas* has left me with a sense of possibility in the face of the horrors. It is precisely that we need new forms of knowledge, new ways of working, new forms of collectivity, to figure out the grounds of a new way forward, institutionally as well, of course. That's my commitment, and so I have ongoing projects with the Yolngu. I've mentioned the Rangingur project. I am also part of a team based at Charles Darwin University developing a broader mapping project, led by a team of accomplished Yolngu researchers which brings together 12 different groups along the Arnhem Land coast to produce forms of mapping in-country that don't primarily or necessarily at all rely on GIS. Rather the project refigures the question of what a map is in forms of sensuous, biographical, intergenerational orientations to being in-country.

The people leading this project, including Gurrumuruwuy, are the last generation to have walked and camped and paddled across that country. They're highly cognisant of the gap, the experiential gap, between themselves and the younger generation, and the consequent gap in confidence in authority for young people in speaking as emerging leaders, speaking for and from their country. The maps that they hope to generate through this project will at one level serve as an archive of endangered indigenous knowledge along an endangered coastline facing salt inundation and heat stress in places. At another, arguably more important level, these maps are intended as a new kind of community resource, providing the means by which a younger generation cannot just represent or demonstrate, but can actually speak with the authority of the country while making really directed interventions into ongoing colonial discussions.

As in *Feral Atlas*, we've developed a broad canvas of what potentially constitutes a map: video, drawing, painting, song, sound, drawing – all hold potential for a new genre of mapping informed by a Yolngu performative repertoire of sovereign knowledge transmission. The idea is to make maps that respond to particular issues around land management and governance. The Yolngu researchers have been very clear about the colonising power of maps that this project seeks to counter. They put it like this,

> "Yolngu have always had maps inside us. The trouble is we go to meetings and Balanda put their maps out on the table, and in the process they're covering up everything about our country. But they don't know that we're starting in the wrong place already in the meeting".

I'm really excited about this mapping project. I'm excited about the multiple dimensions of the kinds of mapping we've done in *Feral Atlas*, and the kinds of experiments in critical cartography and aesthetics about what actually constitutes a map. I'm also excited about what mapping itself can do to render shared worlds of relationship and mutuality, by bringing a shared attention to emergent points of connection and cleavage, flow and blockage, sameness and difference. Because as my Yolngu colleagues know better than most, maps don't just chart existing geographies, they determine the grounds of the possible. In doing so I reckon that they provide a crucial reference point from which to begin to answer the increasingly urgent question: where to from here?

31

ATTUNE AND ENTANGLE

Designing Multispecies Relations for the Sixth Extinction

Michael Ezban

Humans are relationally constituted; we are shaped through dynamic exchanges with other species, and we have been since our emergence (Marean, 2010). Paradigms of individualism and human exceptionalism that we have used to conceptualise ourselves as bounded entities set apart from our ecological surroundings are giving way in the 21st century to an understanding of all beings, from bacteria to blue whales, as comprised of complex, microscopic assemblages and wide-ranging, sometimes global, interdependencies (Hill, 2005). As physicist-philosopher Karan Barad describes it, "existence is not an individual affair" (2007: ix).

Beginning with the invention of agriculture in the Neolithic era, humans have played an outsized role in relations with other species and the ecosystems in which they are imbricated. There is overwhelming evidence that today most of the biodiversity and ecosystem processes of the terrestrial biosphere have been permanently reshaped by the actions of humans. The vast range of human land uses over millennia, and the legacies of that use, have unleashed planetary transformation that is now widely recognised as a geological epoch: the Anthropocene. The resultant novel ecosystems, which are the predominant form of terrestrial ecosystems on the planet (Ellis, 2014: 169), are the dynamic networks that contemporary landscape architects, and landscape architects-in-training, engage as they imagine, propose and direct socioecologies in an increasingly urban world.

Among the most salient effects of human engagement with other species and their habitat are the extinction of these non-human species and their forced relocation. Engagement here means destruction of habitat, consumption of non-humans, and deleterious agricultural practices, all of which have accelerated dramatically in capitalist economic systems. The work of landscape architects in the Anthropocene takes place in the context of both a massive contraction of the

DOI: 10.4324/9781003145905-37

diversity of life, commonly known as the sixth extinction, and also migrations and relocations of non-human populations, including mammals, birds, reptiles and amphibians (Leakey, 1996; Kolbert, 2014).

Faced with these challenges, many contemporary landscape architects create hybrid ecologies and multifunctional environments that are *designed* to be cohabited by humans and non-humans. Further, some designers theorise about, and put into practice, the assemblage that I call *multispecies coalitions*. These are gatherings of humans and non-humans into a constellation of relationships that enable symbiotic practices and increase animal agency in *coshaping* the physical and social production of landscape (Ezban, 2020: 4). "Life", wrote Lynn Margulis and Dorion Sagan, "is a network of cross-kingdom alliances" (2000: 191). Progressive theories and practices of cohabitation and coalition such as these counter backwards-looking restoration and conservation practices which aim to return ecosystems to some idealised previous state. They also eschew dated binary paradigms that place humans outside and apart from the ecologies and landscapes they inhabit.

While theories about landscapes of cohabitation and views of human–non-human relations are relatively limited in the discipline of landscape architecture, there is an increasing number of such theories in the field of human-animal studies. Geographers, anthropologists, philosophers, ethologists and environmental scientists contribute to the exploration of more-than-human perspectives and places. Their shared inquiry rests on the premise that relations between humans and non-humans are simultaneously biological, cultural, economic, ethical, geographical and political (Urbanik, 2012: 3). *Entanglement*, a term originating in quantum mechanics, has been widely adopted to describe complex processes of *becoming* and indeterminate possibilities that emerge from spatiotemporal relations between entities, including relations between humans and non-humans. According to this view, the self emerges from and only exists within these relationships, and it is "inside these multispecies entanglements that learning and development take place, that social practices and cultures are formed" (Van Dooren, 2014: 4).

To illustrate how an entangled world thrives on multivalent connections, theorists are using non-human formal patterns as metaphors. For example, Alan Rayner draws on the model of the fungal mycelium, Gilles Deleuze and Félix Guattari use the rhizome, and Tim Ingold appeals to the spider web (Anderson, 2016). Common to these metaphors is the challenge they pose to longstanding anthropocentric evolutionary analogues, such as trees and ladders, and the power and privilege they ascribe to humans in multispecies entanglements. By contrast the relational analogues being proposed emphasise non-hierarchical and lateral connections, intersecting branches and multidirectional lines (Hejnol, 2017).

To provide an entrée for students of design into the complexities of indeterminacy and change endemic to urban ecologies and globalisation in the Anthropocene, I have developed a design studio pedagogy centred on crafting *relations* between humans and non-humans. My pedagogy for teaching studios dealing with cohabited urban ecologies is organised around two phases of work: *embodied attunement* and *designing entanglements*.

Embodied Attunement and Designing Entanglements

Attunement is kinaesthetic rapport between two or more beings brought about through both sensing and reciprocal response. It is a form of knowledge derived through somatic experience and requiring attention, patience and willingness to understand the ways of others. Humans can be attuned to non-humans in a variety of ways, and in the process also may never really know them. They may gather static details about non-humans but fail to be attuned to them. In *Being a Beast: Adventures across the Species Divide*, naturalist Charles Foster (2016) set a high water mark for knowing non-humans by inhabiting their lives. He lived with Eurasian otters (*Lutra lutra*) and caught fish in his teeth. He honed his sense of smell by living with European badgers (*Meles meles*). In the company of red foxes (*Vulpes vulpes*), he dug through garbage cans. My pedagogy does not ask students to engage in such radical behaviour, but the embodied attunement phase of my studios with them does challenge their willingness and their ability to take on *imagined* perspectives in order to perceive the needs of others. I challenge students to build their awareness of and responsiveness toward another species through direct and embodied human-to-non-human experiences, as well as by means of human-to-human, team-based research on that species.

In a second phase of designing entanglements, my students are asked to craft encounters and interdependencies between species. Lori Gruen observes in her book *Entangled Empathy*, that "rather than trying to accomplish the impossible by pretending we can disentangle [from other species], we would do better to think about how to be more perceptive and more responsive to the deeply entangled relationships we are in" (2015: 30). Following Gruen's insights, my studios work through cycles of first perceiving and then respond-ing to complex entanglements. I select non-humans endemic to an urban ecology and challenge students to bring to light existing interdependencies between species (at local, regional and global scales), and then propose mutu-alistic forms of interrelating. Of crucial importance is that the landscapes, structures and ecologies designed to entangle are evident and palpable to humans, which is a necessary component to them being influential and col-lectively valued (Meyer, 2008).

Pedagogy based on attunement and entanglement aims to build students' capacities to join and support multispecies coalitions beyond the studio. It shapes their development as effective partners in cross-species and cross-disciplinary teams that shape urban ecologies in the Anthropocene. It also develops their sensibilities as advocates for non-human agency and symbiotic relations between humans and the non-human inhabitants of socioecologies.

Anacostia Aviary Studio

In the discussion below, I describe the application of these pedagogical phases and their outcomes in a design studio I wrote and directed. The studio, titled "Anacostia Aviary: Bird Blinds as Apertures for the Sixth Extinction", was offered in 2019 and 2020, and was offered again in 2021 at the University of Maryland School of Architecture, Planning & Preservation. In the Anacostia Aviary studio, students are asked to develop attunement through awareness of and responsiveness to birds found at Kingman Island, an island with a unique urban ecology in the riverine corridor of the Anacostia River in Washington, DC. They are then challenged to design productive entanglements, using landscape and architecture as mediums, that shape encounters and interdependencies between humans and birds.

Kingman Island is a 40-acre artificial island built of dredge spoils in the Anacostia River. It is a wooded island in an urban setting. The land itself is under the jurisdiction of the National Park Service and is renowned as one of the most vibrant communities of non-humans and plants in the city. It is home to more than 100 species of resident and migratory birds. Students participating in the Anacostia Aviary studio are asked to think of the island as a landscape that has been co-created by birds and humans. Following the mechanical deposition of dredged sediment more than a century ago, birds have played a major role in developing a variety of ecologies and habitats through seed dispersal, pollination, predation and insect consumption. Today the island hosts beech-oak-tulip upland forests where red-headed woodpeckers (*Melanerpes erythrocephalus*) thrive; it has floodplain forests where bald eagles (*Haliaeetus leucocephalus*) and orchard orioles (*Icterus spurius*) are found; forest-meadow ecotones are occupied by barn swallow (*Hirundo rustica*) and indigo bunting (*Passerina cyanea*); and tidal freshwater wetlands are home to great egrets (*Ardea alba*), blue heron (*Ardea herodias*) and ruddy ducks (*Oxyura jamaicensis*).

Paradoxically, the urban river corridor is experiencing the growth of both human and non-human populations simultaneously. Commercial and residential development along the river has decreased viable non-human habitat, and this has coincided with ongoing riverine rehabilitation and clean-up efforts. Conservationists have observed rising populations of fish and mammals, and thanks to these programmes the area is supporting a growing population of predator avian species, such as eastern osprey (*Pandion cristatus*) and Cooper's hawks (*Accipiter cooperii*). In addition to increases already observed in the populations of humans, birds and other non-humans, even higher numbers are expected as the river becomes a major migration and relocation route for non-human species due to climate shifts (Schlyer, 2018: 200). Kingman Island, embedded in this changing riverine landscape, is a populous urban ecology in flux. It is a dynamic, open-air aviary – jointly constructed and under multispecies management – that is ripe for intensification of mutualistic human-avian entanglements through design.

In the embodied attunement phase of this studio, students spend time with birds, engaging in close visual and auditory observation, and pursuing a variety of creative practices intended to help them inhabit avian worlds. Students are asked to create a map that collects all manner of avian activity that they can perceive with their senses, most especially the various movement patterns of birds: flight paths, nesting, foraging, squabbling, evading and flocking. This mapping activity is a recurrent graphic recording process, undertaken 20 minutes once a week throughout the semester, and results in visually compelling maps that are dense with lines, marks and notes. Mapping is intended as an active practice undertaken in situ, in the presence of birds, a means of appreciating what science writer Jennifer Ackerman calls the avian "mapping mind" (2016: 195). Birds, she writes, rely on sensing, learning and extraordinary spatial and temporal ingenuity to support their mental cartographic practices (2016: 196) (Figure 31.1).

Alongside this *parallel practice* of mapping, students are also asked to explore avian perspectives, and decentre themselves from their own lived experiences, through narrative writing. In this medium they develop descriptions of the world from imagined avian vantages. Students are shown examples of this form of writing, where narrative conventions are often flouted in an effort to invoke non-human lived experiences, such as in Karen Macdonald's description of a northern goshawk (*Accipiter gentilis*) scanning and "seeing serially" in her acclaimed book *H Is for Hawk*: "Everything the hawk saw was raw and real and drawn hair-fine … her attention [was] catching on everything serially. *Field-fence-fieldfare-win g-flick-pheasant-feather-on-path-sun-on-wire-twelve-woodpigeons-half-a-mile-distant-t ick-tick-tick* …" (2014: 186–7). The writing of ecological philosopher Martin Lee Mueller, who narrates from the ichthyocentric perspective throughout his book *Being Salmon, Being Human* (2017), is another instructive exemplar of this practice.

With a foothold in multiple attunement activities, students then enter the designing entanglements phase. This phase is centred on developing speculative designs for bird blinds to be sited on Kingman Island. In this studio, blinds are envisioned as apertures, windows in and of the landscape that foster multidirectional views – outward and inward. They simultaneously enable us to witness and record the effects of avian–human entanglements out in the world, and they sponsor insight into our individual roles and collective responsibilities in the midst of the sixth extinction. "Paying attention to avian entanglements", writes extinction philosopher Thom Van Dooren in his book *Flight Ways*, "unsettles human exceptionalist frameworks, prompting new questions about what extinction teaches us, how it remakes us, and what it requires of us" (2014: 5) (Figure 31.2).

The value of this exercise is more than speculative. Each year people around the world inhabit blinds to contribute bird observation data to research hubs like the Cornell Lab of Ornithology. The project incorporates massive crowdsourcing of data gathered by citizen scientists, and this aggregation of individual observations helps to paint the larger picture of avian extinction in the

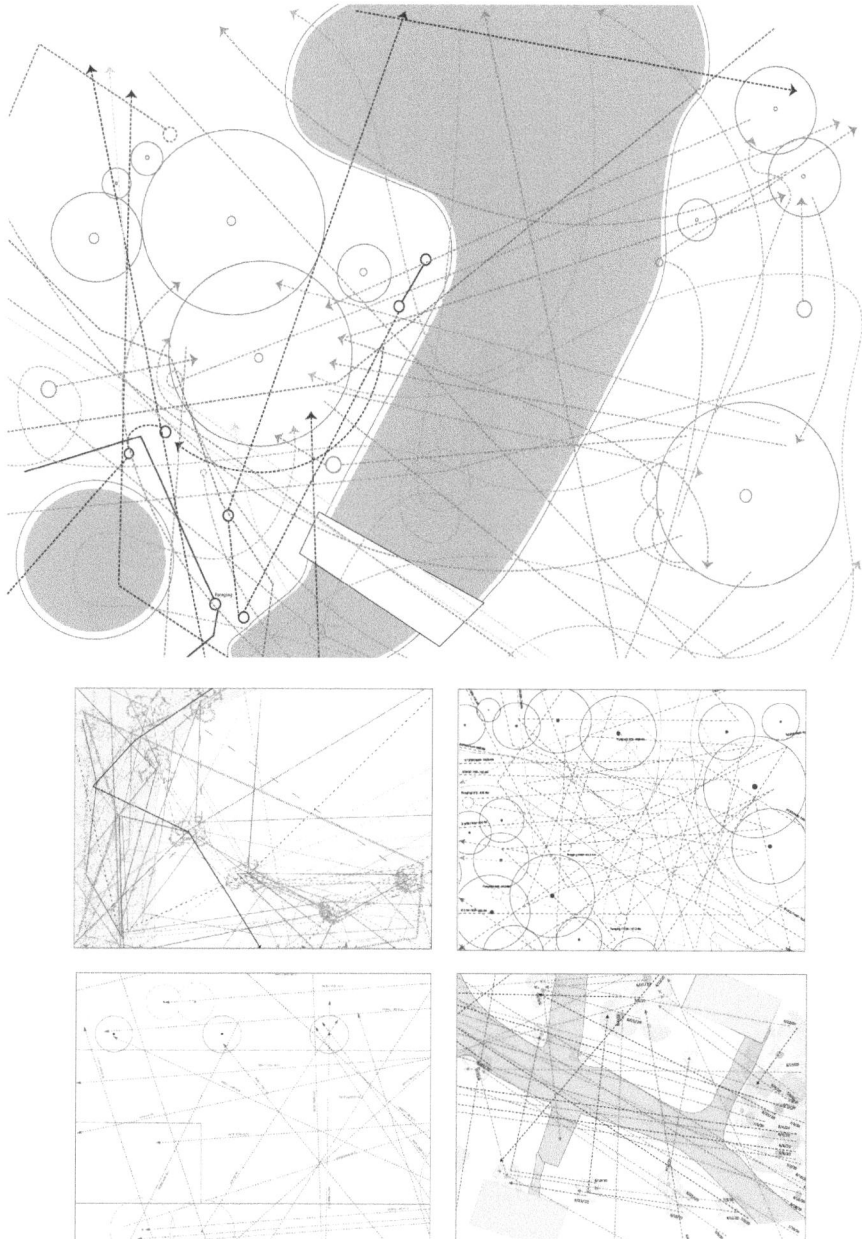

FIGURE 31.1 Avian attunement mappings that collect all manner of avian activity, including flight paths, nesting, foraging, squabbling, evading and flocking. By Maura Beste, Wyatt Rhal, Leticia Lacerda, Juleesa Jolley and Kat Schwab.

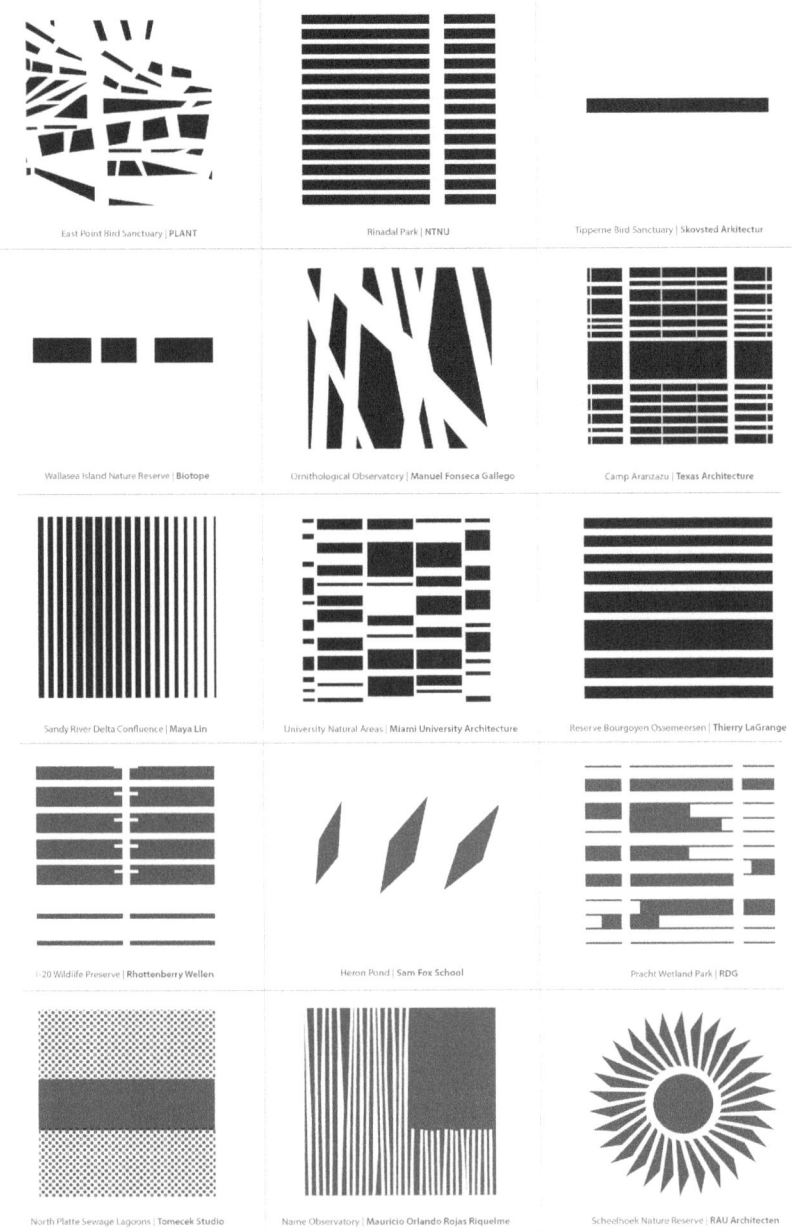

FIGURE 31.2 Catalogue of bird blind apertures designed by architects, adapted from precedent research produced by students. In the Anacostia Aviary studio, blinds are seen as apertures for the sixth extinction. They form a broad, global network of windows from which millions of birders observe beauty and collect evidence of alarming change. Michael Ezban.

Anthropocene. The blinds at Kingman Island would facilitate such observation, providing designers a critical role in constructing a collective understanding of how birds of the Mid-Atlantic region are adapting behaviours and migration cycles as temperatures change, insect populations drop and non-native vegetative species spread (Narango et al., 2018).

At the same time that blinds serve as apertures to facilitate data collection by citizen scientists, they also foster intensely intimate and introspective experiences for observers who venture into them. In her essay "Windows into Another World", writer Terry Tempest Williams (2017) describes her experience in a blind, scanning for sandhill cranes (*Grus canadensis*) in marshes on the Platte River. Williams casts herself as "the pupil of the bird blind, an eye squinting for insight beyond [her] own kind", and from this vantage she witnesses cranes engaged in "the amorous play of lovers, leaping and bowing to the embodied knowledge that the future depends on each gesture granted to the other" (2017). For Williams, the blind is a place to "practice patience", and to "rise to an awakened state of being" (2017). This leavening, born of staying still enough to notice, engenders empathy, helping us to "make sense of living" (Gruen, 2015: 30).

The designing entanglements phase of the Anacostia Aviary studio begins as a series of team-based research experiences that reveal the local, regional and global systems, economies and ecologies in which Kingman Island and its avian and human inhabitants are imbricated. To build knowledge collectively, we gather together to outline the research ahead, and through a multi-hour charrette, the students determine the range of topics that will be explored. All topics speak to the open-ended processes being inaugurated by climate change: non-native species spreads, flooding forecasts, ongoing dredging and sediment deposition, and shifts in the annual cycles of various species. As research proceeds over two weeks, students are asked to check in frequently with each other to report their findings, open new topics or close existing ones, and negotiate a graphic strategy for the output of work. Skilful navigation surfaces, as do frustrating logjams. Through both types of experience, students who practised noticing and responding to birds in the embodied attunement phase may find themselves more aware of their actions and those of others as they interact, a practice of honing their "response-ability" with their colleagues. The result of this two-week, human-to-human collaboration is a rich database of information that also will inform their design of multispecies entanglements.

Following this period of research and remote sensing, students return again to the island itself for guided and individual visits as they begin the speculative development of their blinds. Student responses to this design challenge vary. Many students approach the blind as a discrete aperture nestled within a single habitat zone, sized for single occupants, and directed toward observation of a single species. Others, however, challenge traditional relationships between blinds and landscapes. One student proposed replanting the abrupt transition between forest and meadow on the island to create a broad, sinuous ecotone, and then sited her blind in this area of enriched species diversity and interaction. Another student studied flooding projections, and then sited her blind at the upland edge of a wetland expected to be inundated in the coming decades; her blind is designed to adapt its form relative to rising waters by rotating, skewing and extending toward changing species and, as needed, to remain accessible despite radical contextual changes. A third student eschewed typological conventions and conceptualised the blind not as a single structure, but as a trail system that spans floodplain forest and wetlands, transforming the landscape itself into an instrument for observing, measuring and tracking invasive species growth and its effect on bird populations over time. No matter their approach, students are encouraged to graphically depict their designs from both human and imagined avian perspectives, and to challenge privileging anthropocentric ways of seeing and ways of being (Figures 31.3–31.5).

The blinds proposed for Kingman Island are apertures for the sixth extinction. They join a broad, global network of blinds from which millions of birders observe beauty and collect evidence of alarming change. It is from these apertures in the western hemisphere that we have witnessed the loss of more than one in four birds in the last 50 years (Rosenberg et al., 2019), and from which we will watch as two-thirds of North American bird species grow increasingly endangered as global temperatures rise (National Audubon Society, 2019). Van Dooren writes that "[w]hen species are understood as vast intergenerational lineages, interwoven in rich patterns of co-becoming with others, then their departure from the world cannot help but be felt" and "possibilities of life and death for everyone—not just the 'endangered'—are made" (2014: 12–13). Designing blinds at Kingman Island encourages attunement to the perennially entangled and increasingly imperilled world of birds and humans, and might provide opportunities to "relearn our place in a shared world" (Van Dooren, 2014: 12)

Co-creating Ecologies

The Anacostia Aviary studio involves observation and projection, during which students gather information through somatic experiences and cognitive processes, and then in turn imagine and represent methods of mediating relations with birds in an urban ecology. A studio like this could also be written to position non-humans as partners in the design process, thus fostering interdependencies with non-human actors in acts of real-time *co-creation*. Roxi Thoren, Associate Professor of Landscape Architecture at the University of Oregon, illustrates such an approach in her studio "Co-creating with Animals" (2018). This multispecies studio foregrounds encounter and engagement with non-humans as formative to the pedagogy. In her studio, collaborative practices, such as artful monitoring and co-created artworks, are developed by students and earthworms, ash borers, newts and deer. The iterated prototypes that emerge from the studio expand

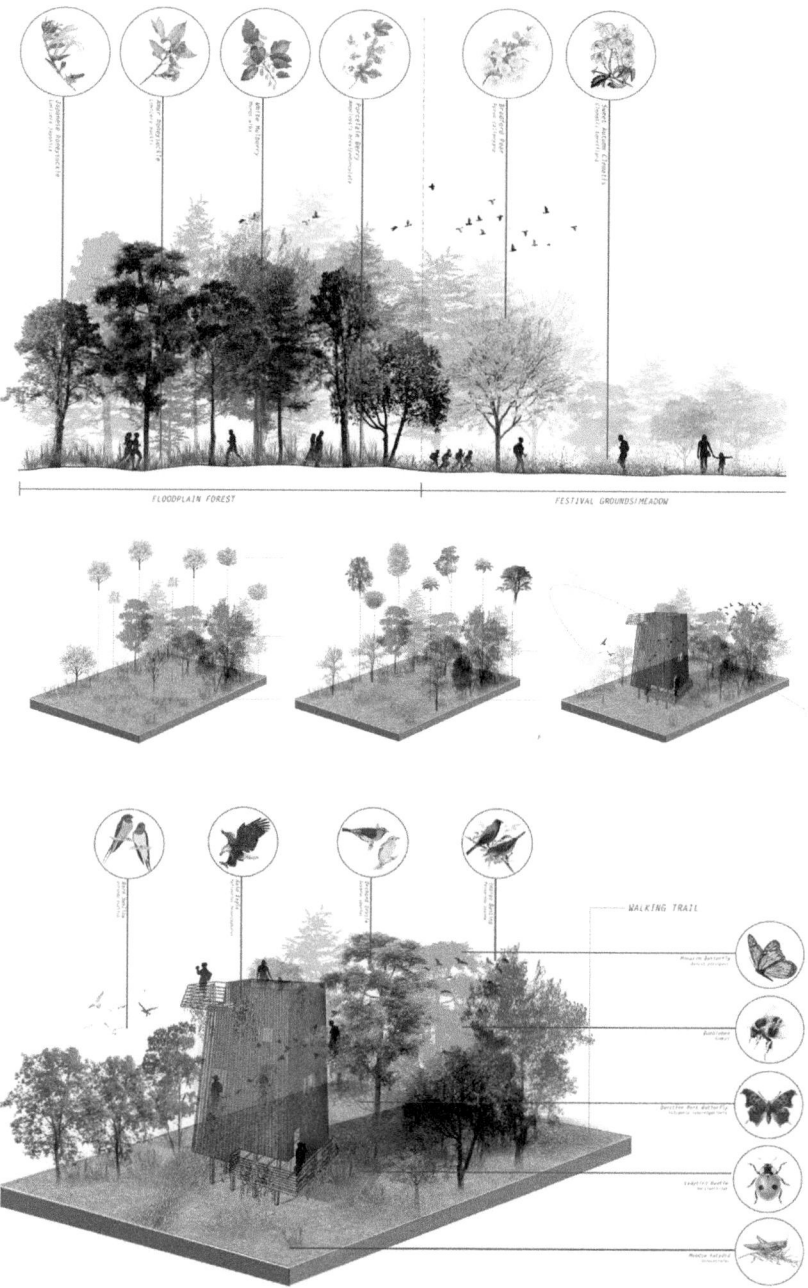

FIGURE 31.3 Proposal for a blind sited within the redesigned transition between forest and meadow; the goal was to develop a blind and a broad, sinuous ecotone of enriched species diversity and interaction simultaneously. Abisayo Omisore.

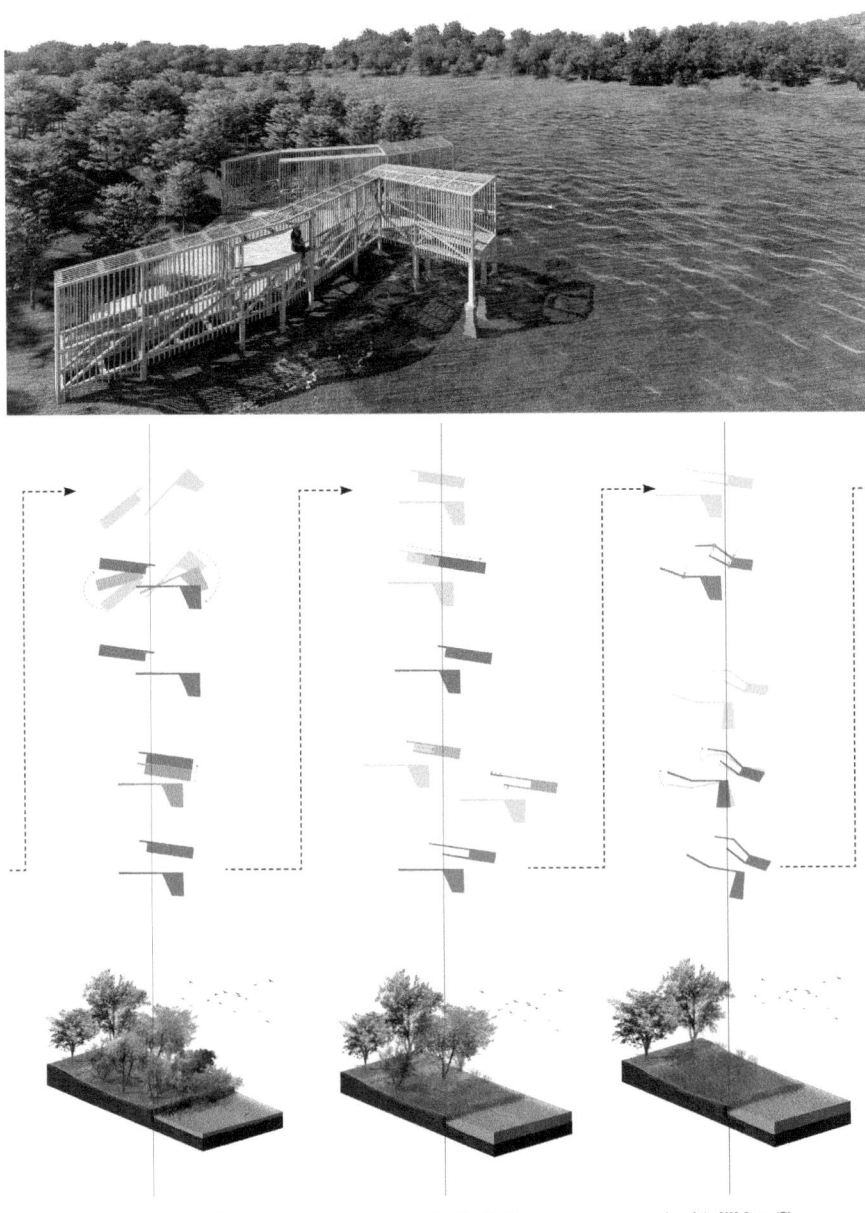

Innundation 2020: Rotate and Realign   Innundation 2050: Shift and Extend   Innundation 2100: Cant and Tilt

FIGURE 31.4 Blind proposed for the upland edge of a wetland expected to be inundated in the coming decades; the blind is designed to adapt its form relative to rising waters by rotating, skewing and extending toward changing species and as needed to remain accessible despite radical contextual changes. Laura Mardesich.

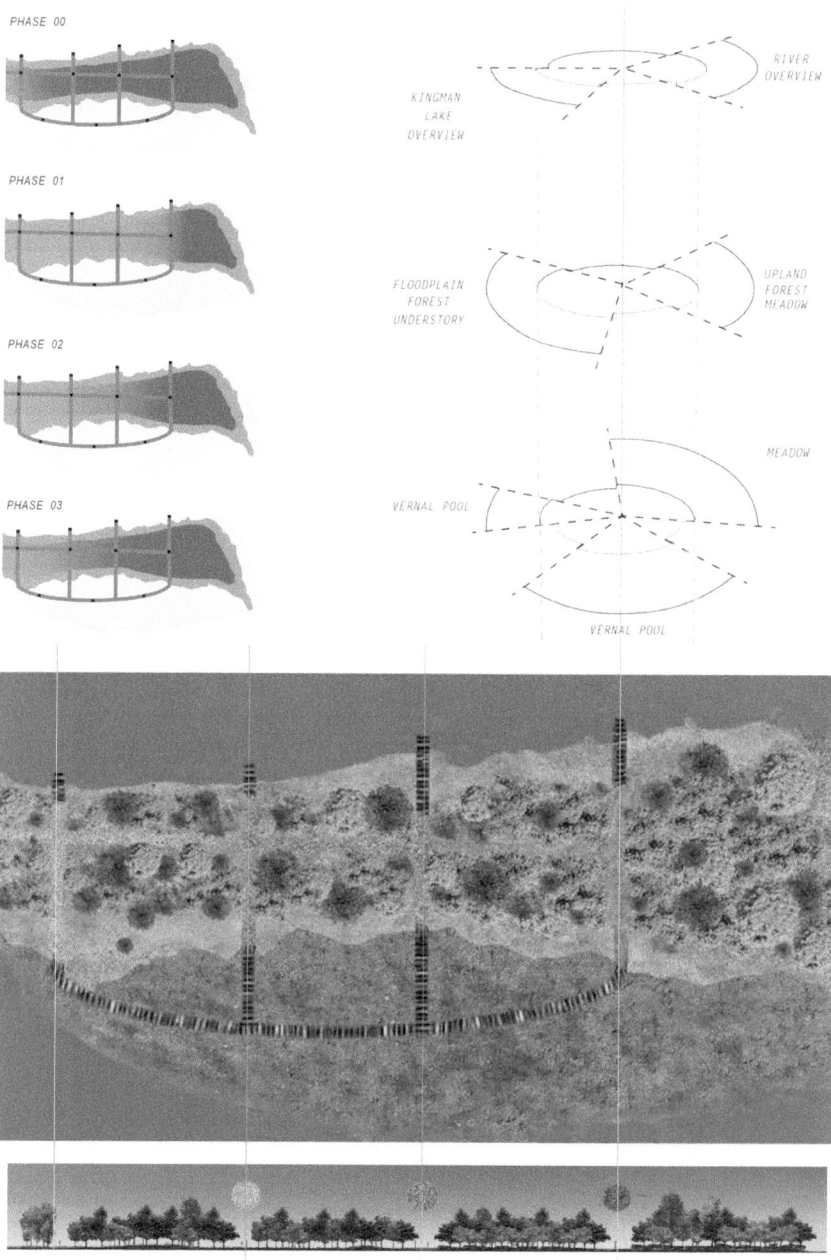

PHASE 00

PHASE 01

PHASE 02

PHASE 03

RIVER
OVERVIEW

KINGMAN
LAKE
OVERVIEW

FLOODPLAIN
FOREST
UNDERSTORY

UPLAND
FOREST
MEADOW

MEADOW

VERNAL POOL

VERNAL POOL

FIGURE 31.5 Blind conceptualised as a trail system that spans floodplain forest and wetlands, transforming the landscape itself into an instrument for observing, measuring and tracking invasive species growth and its effect on bird populations over time. Cody Gardner.

the ecological functions of a landscape, as well as "forge a way of designing *with* animals rather than *for* animals" (italics in original) (2018: 24).

Processes of designing both with and for non-humans are increasingly necessary in the multispecies metropolis. We should expect our entanglements with non-humans of all kinds to continue to proliferate and intensify as urban areas expand further into habitat, as soft and hybridised infrastructure undergirds cities, and as landscape-based solutions for urban waterfront resiliency are inaugurated. The abilities to assemble and steer multispecies coalitions through thorny problems and toward shared goals will be an essential skill for designers working within urban ecologies in the coming century. Ecological systems designed through and with symbiotic coalitions and creative practices that entangle humans, birds and other non-human species underpin what Jennifer Wolch calls *zoöpolis*, a co-created urbanism that is characterised by biodiversity, resiliency and vitality (Wolch, 1998).

Ironically, in an epoch we have named for ourselves, designers of urban ecologies in the Anthropocene must work to discard anthropocentric paradigms that separate "us" from "them" and seek much flatter hierarchies. In place of these old paradigms, we do better to recognise that "life and death do not take place in isolation from others; they are thoroughly relational affairs for fleshy, mortal creatures" (Van Dooren, 2014: 4). With training, a new cohort of designers may intentionally craft dynamic relationships by imagining and then designing embodied, unexpected and vivid interactions between species that cohabit *zoöpolis*. The possibilities include fleeting encounters, day-to-day living, recurrent working partnerships, artistic collaborations and more. These engagements create the "local, situated, everyday knowledge of animal life required to grasp animal standpoints or ways of being in the world" (Wolch, 1998: 124), and help us realise that "our sense of who we are as humans is mirrored in our lived relationships with other creatures" (Mueller, 2017: xvi).

In bringing to the foreground the design of multispecies relations in the studio and beyond, designers become aware of and directly participate in the ancient and open-ended processes through which humans and non-humans become, in Donna Haraway's evocative phrasing, "unpredictable kinds of 'we'" (Haraway, 2007: 5). Practices of embodied attunement to these vital and generative bonds, rhythms and exchanges between beings, assembling cross-species coalitions, and co-designing the entanglements that are enacted within urban ecologies and landscapes can inform a pedagogical approach for explorations of 21st-century multispecies urbanism.

References

Ackerman, J. (2016). *The genius of birds*. Penguin.

Anderson, L. (2016). "Entanglement," in making worlds: Art, materiality and early modern globalization. Retrieved from https://www.makingworlds.net/entanglement.

Barad, K. (2007). *Meeting the universe halfway: Quantum physics and the entanglement of matter and meaning.* Duke University Press.

Ellis, E. C. (2014). Anthropogenic taxonomies: A taxonomy of the human biosphere. In C. Reed & N-M. Lister (Eds.), *Projective ecologies* (pp. 168–183). Harvard University Graduate School of Design.

Ezban, M. (2020). *Aquaculture landscapes: Fish farms and the public realm.* Routledge.

Foster, C. (2016). *Being a beast: Adventures across the species divide.* Henry Holt and Co.

Gruen, L. (2015). *Entangled empathy: An alternative ethic for our relationships with animals.* Lantern Books.

Haraway, D. (2007). *When species meet.* University of Minnesota Press.

Hejnol, A. (2017). Ladders, trees, complexity, and other metaphors in evolutionary thinking. In A. Tsing et al. (Eds.), *Arts of living on a damaged planet* (pp. 87–102). University of Minnesota Press.

Hill, K. (2005). Shifting sites. In C. Burns et al. (Eds.), *Site matters: Design concepts, histories, and strategies* (pp. 131–155). Routledge.

Kolbert, E. (2014). *The sixth extinction: An unnatural history.* Henry Holt and Co.

Leakey, R. (1996). *The sixth extinction: Patterns of life and the future of human kind.* Anchor.

Macdonald, K. (2014). *H is for hawk.* Grove Press.

Marean, C. (2010). When the sea saved humanity. *Scientific American, 303*(2), 54–61.

Margulis, L., & Sagan, D. (2000). *What is life?* University of California Press.

Meyer, E. (2008). Sustaining beauty: The performance of appearance. *Journal of Landscape Architecture, 3*(1), 6–23.

Mueller, M. L. L. (2017). *Being salmon, being human: Encountering the wild in us and us in the wild.* Chelsea Green Publishing.

Narango, D. L., Tallamy, D. W., & Marra, P. P. (2018). Nonnative plants reduce population growth of an insectivorous bird. *PNAS.* www.pnas.org/cgi/doi/10.1073/pnas.1809259115.

National Audubon Society. (2019). *Survival by degrees: 389 bird species on the brink.*

Rosenberg, K., et al. (2019). Decline of the North American Avifauna. *Science.* https://science.sciencemag.org/content/366/6461/120/tab-pdf.

Schlyer, K. (2018). *River of redemption: Almanac of life on the anacostia.* Texas A&M University Press.

Thoren, R. (2018). Co-creating with animals: Crossing the "narrow abyss of non-comprehension." *Landscape Review, 18*(1), 22–36.

Urbanik, J. (2012). *Placing animals: An introduction to the geography of human-animal relations.* Rowman & Littlefield.

Van Dooren, T. (2014). *Flight ways: Life and loss at the edge of extinction.* Columbia University Press.

Williams, T. T. (2017). Windows into another world: Take a tour of bird blinds across the country. Retrieved from https://www.audubon.org/magazine/summer-2017/windows-another-world-take-tour-bird-blinds.

Wolch, J. (1998). Zoöpolis. In J. Wolch & J. Emel (Eds.), *Animal geographies: Place, politics, and identity in the nature-culture borderlands* (pp. 119–138). Verso.

32

ECOLOGY AND TWO THESIS LAB CASES

Roberto Pasini

Premises

The stability of the planetary metabolism is severely compromised, according to the *Global Assessment Report on Biodiversity and Ecosystem Services* issued in 2019 by the Intergovernmental Science-Policy Platform on Biodiversity and Ecosystem Services.[1] By assessing biodiversity and ecosystem services, the report assesses the current state of life on Earth. The document describes environmental degradation as extensive, with three-quarters of terrestrial environments severely altered by human actions. Natural ecosystems have declined by half relative to their earliest estimated states, with an ongoing trend of decline. Water stock quality is progressively degrading due to the untreated discharge of four-fifths of wastewater. At the same time, the global consumption of materials is increasing, as is the manufacture of disposable products: for example, plastic pollution is increasing at a rate of 250 per cent per decade (IPBES, 2019).

On one hand, these data describe expanding environmental destruction and degradation from human occupation coupled with incremental depredation of resources for human utility. On the other hand, there is increasing production of unsustainable goods for short-term consumption. International bodies' efforts to assess and counteract the environmental crisis have been intensified since the beginning of the 2000s, but concerted policies struggle to achieve practical implementation.[2] In the meantime, the land is being mined for the parasitical exploitation of natural resources. The fields are being expansively converted to intensive and polluting agricultural productions. The oceans are being swept for fish beyond renewability, and their natural habitats pillaged or devastated by collateral effects.

Authoritative voices of technocratic thinking support non-catastrophic prospects in which the increasing impact of human activities on nature is specular

DOI: 10.4324/9781003145905-38

to the increasing ability of technology to govern the disruptions produced (Ellis, 2018). In such a perspective, the productivity of the exploitation of nature can be reconducted to a sustainable proportionality with the proliferation of the human species (Ellis, 2014) and a novel equilibrium can be artificially governed (Baccini and Brunner, 2012).

Beyond speculations on the plausibility of miraculous technological progress in the short-term, broad scientific literature interprets the manifestations of climate change that we are experiencing as prodromes of a catastrophic biological crisis (Steffen et al., 2004). Following five similar events in the history of planet Earth (Raup and Sepkoski, 1982), a sixth mass extinction is about to end life on the planet as we know it (Mason, 2015; IPBES, 2019). The human species is likely to be one victim among thousands of others. However, in the fullness of time, the Earth will probably renew its biosphere, evolving into a novel regime as it has happened before.

The Role of Landscape Architecture

Under a looming planetary crisis, we are urged to reconsider our current approaches to natural resources and the environment in all aspects of our lives and academic fields. In the design disciplines, at the architectural, urban, landscape and planning scales, several short-sighted tactics have been tackling localised manifestations and symptoms of a much broader phenomenon: a crisis threatening the present regime of life.

Scenarios of protective coastline parks to prevent the flooding of settled regions by rising seas propose to act on the manifestations of the issues rather than on their causes. The aestheticisation of informality dissimulates the urban misery of the unprotected and commodifies it into an aesthetic paradigm for the remote consumption of the privileged.[3] A hegemonic delegitimisation of overall planning has dismissed the intrinsic value of a general coherence of development as an unnecessary burden that undermines the vitality of money flows. In several metropolitan contexts, the result is the expansion of heterogeneous environments of exclusiveness that occupy the centres where cities previously lay. Simultaneously, open territories worldwide have been uncontrollably covered with expanses of undifferentiated suburban developments and encumbered with bundles of dirty infrastructure, resulting in an unsustainable degree of fragmentation of the remnants of nature (Collinge, 2009).

Despite the centrality that landscape architecture has acquired, both in academia and practice under present circumstances, the discipline has often remained ancillary to the logics of the construction industry. The celebrated High Line linear park in Manhattan has come to instantiate an exemplary model for landscape urbanism, a movement that has advocated for the emancipation of landscape design from architecture. Alongside its spontaneous look, its high-demand maintenance and its aesthetically sophisticated flora, the elevated walkway crossing Chelsea was born out of a grassroots initiative of residents, has propelled and legitimised the creation of one of the most exclusive city corridors in the world with a proliferation of luxurious towers on both sides.

In sharp contrast, the "Atlas of the End of the World" project attempts to audit humankind's present status in the frame of the Earth system, with specific reference to spatial production and metabolic management. Forty-four maps render the state of global biodiversity vis-à-vis the pressure from global anthropic activities, and specifically the production of dwelling space.[4] Stemming from the McHargian legacy, and soundly founded on Formanian ecology, the atlas culminates in a planetary-scale proposal for creating two world parks that cross the Americas south–north, and Eurasia and Africa east–west. Reconnecting the planet's biological hotspot regions,[5] containing the world's paramount and most endangered stocks of genetic heritage and vitality, might grant a future to humanity. The world park is a visionary project representative of a radical change of approach in the disciplines of planning, designing and space-making: it acknowledges that a threshold has been reached, beyond which any acts of new construction are wounds against nature.

Rather than producing an immediate profit by adding new construction bulks in a choking world, the challenge confronting new generations of planners, architects and designers worldwide is that of finding new ways of generating long-term wealth (Magnaghi, 2005). The challenge is met by reducing the number of existing structures, dismounting obsolete accretions from the past (both material and intangible) or adapting and retrofitting useful parts. The idea of wealth generated by reducing human-made structures in the world resonates with the positions of the degrowth movement, with appeals for replacing GDP data with happiness indices as indicators of prosperity and calculating the hidden environmental damages associated with the production processes (Latouche, 2009).

Studio Work

In the last decade, design studios have demonstrated the urgency of a response to the ongoing ecological crisis. In parallel, a demise of the aura surrounding the practice of architecture and urban design has occurred, while landscape architecture has been represented as a naive sculptural play or as a disguise to perpetuate grey investments. The result of this demise is a novel interest in reassessing an equitable relationship to nature. A significant number of students exhibit a formed position about such issues. They are emancipating from a conventional approach to the idea of metropolitan dwelling. Rather than exploring the potentials of the combinatory play, or the eases of stylistic emulation, their interests are reoriented towards reassessing the natural components that come close to our communities' life. This combines a rediscovery of the dynamics of the pristine natural platform on which settlements have thrived and the analysis of the hybridising of natural and artificial systems characteristic of contemporary environments.

The interest in incremental building patterns and linear growth is diffusely fading in favour of the exploration of novel forms of balanced correspondences between human communities and nature. Now that the logic of infinite growth that has propounded urbanisation so far is proving obsolescent, an overall redesign of the planet is to occur by downsizing the human presence and making it more amiable to the environment. The challenge today lies in dismounting the built landscape, along with the expansive mindset that has produced it, based on the drives of colonisation and exploitation.

Two Thesis Lab Cases

Among the thesis labs of the last decade at the University of Monterrey, Mexico, two rather different projects best represent this momentum towards a novel equilibrium with nature through reduction and this reach towards biological dynamics. One tries to dissolve the human presence within the landscape. The other mitigates the metropolis by incorporating in it the airy logics of ecology.

The "Anthropic Matorral"[6] is a thesis lab generated by the fascination with the matorral, an ecosystem characterised by a high degree of biodiversity that covers the lower topographies of the Sierra Madre Oriental in north-eastern Mexico. Its low vegetation cover, typical of shrubland, has been bulldozed without hesitation in the last half-century to clear the surface for the rapid expansion of Monterrey's metropolitan area. The poetic nucleus propelling the project is the idea of a receding human community that dismantles its metropolitan infrastructures to withdraw into the fractures produced in the landscape by the human impacts of a demised urban age. At the pace of progressing demographic awareness, large portions of the cities previously devoted to production are dismantled. Groups of a reduced human population can even accommodate into landscape interstices deploying minimal installations. These light interventions mainly consist of artificial bio-tissues which mend a severed landscape and reconnect interrupted ecosystem dynamics. The designed spaces of a human abode are plugged into an abandoned trail, deeply excavated in the rocky hillside, once used to reach an old water tank.

The exploration of the matorral recollected through objective observations and subjective perceptions over multiple individual and group site visits. Various activities, including flora and fauna surveys, art performances, ritual dances and overnight camping, were conducted onsite by lab members and various guests to establish physical contact with the locus and its spirit. The project is based on a thorough analysis of the ecosystem of the matorral, starting from the visual capture of the superficial aspects of soils, flora and fauna, and ending with the acknowledgment of deeper mechanics of the ecologic cycles. The reading emphasises the connectedness of physiological cycles of air, water, energy, nutrients and essential functions such as phytodepuration, pollination and seed propagation, all involving interacting flora and fauna species. The integration of a small human group into the matorral starts with identifying a set of fundamental acts of human life. The actual design focuses on the sustainable incorporation of those acts into the ecosystem cycles. The spaces of human/nature interaction are ordained according to a scale on seven levels. Through the scale, domestic spaces, protected from the perils of wildlife, gradually merge into natural environments, protected from the impacts of human presence. Intermediate spaces accommodate varying levels of risk from external agents for the residents and, conversely, varying levels of interference from the human presence on the ecosystem.

The project envisions an epochal change in the present anthropocentric mindset, through which humans should conceive themselves as just one part of a larger unity. Intellectual and emotional faculties structure self-awareness through scientific knowledge and perceptual capture (Figure 32.1).

The "Dynamics of the Void" thesis lab[7] starts from the same imperative of reducing human-made structures, their submission to broader ecological principles, and the idea that equilibrium generates forms of wellbeing more desirable than metropolitan intensification and congestion per se. The project starts with the observation of the fundamental ecological dynamics alimenting the vitality of natural ecosystems. A stochastic event that hits the periphery of an ecotope provokes a perturbation penetrating for a certain depth. That perturbation triggers responses of the ecosystem ranging from resilience to renovation. For instance, a tornado might hit the woods and carve a void by felling the trees on its way. Over time, the glade will be colonised by old and new plant communities proceeding from inside and outside the ecotope. The further articulation of the perimeter of the ecotope, due to the perturbation, extends the area of ecological opportunity represented by the ecotone mediating between the ecosystem and the outer context. At the periphery of an ecosystem, any autochthonous species' opportunity to retrieve alien resources is higher as is, though, the risk of encountering alien predators (Farina, 2009). However, the magnitude of the stochastic event must be commensurate to the size of the ecosystem and its dynamics for it to maintain a beneficial character and not become destructive.

The project deploys the logics of ecological perturbations over the metropolitan platform of Monterrey. External stimuli are intended to revitalise the asphyxiating urban dynamics of the city. In the first phase of the work, a central ensemble of the city is identified, and its ongoing urban dynamics are analysed. A virtual model is generated through an algorithm combining categories such as structure, land cover, land use, productive units and capital gain. Iterations are run on a cellular-automata system as for an ecosystem, casting scenarios of urban evolution under the dynamics of the present urban

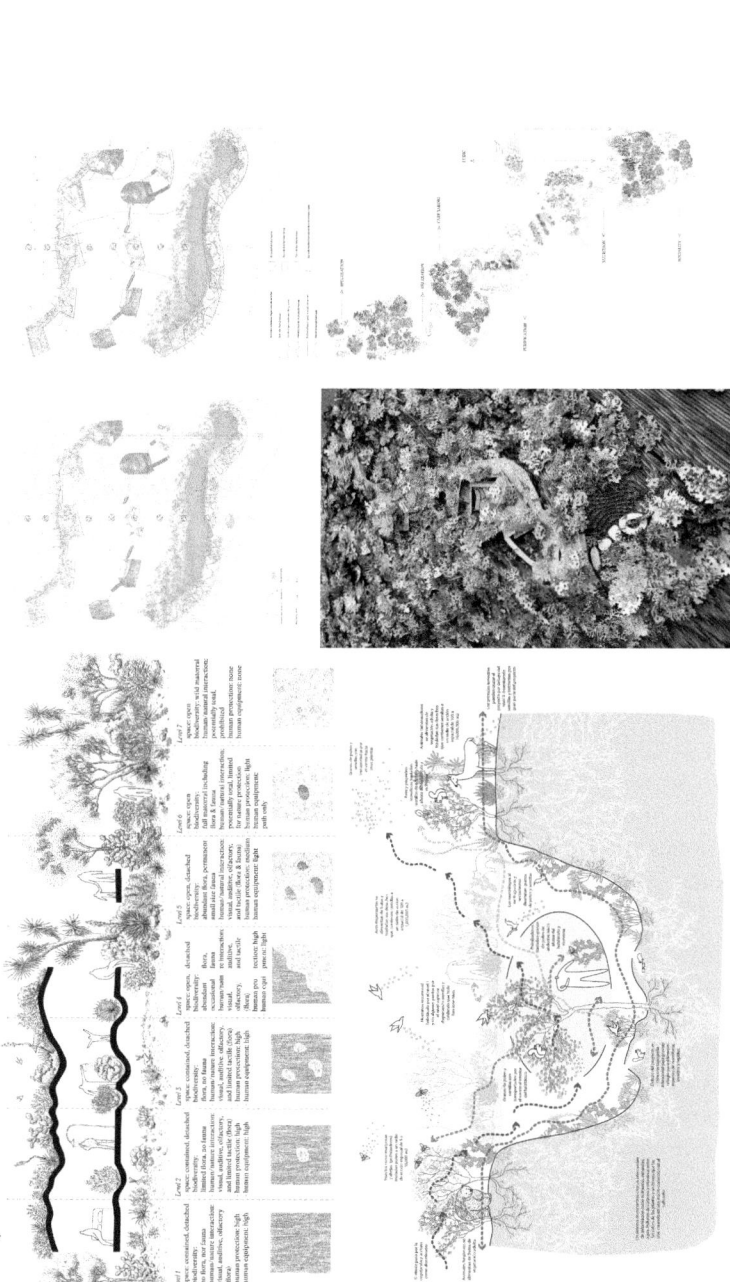

FIGURE 32.1 Top left: Seven levels of human/nature interaction: conceptual scheme. P Garza, A Ramos and F Rosas; digital/hand-drawings A. Ramos. Text: R Pasini. Bottom left: Metabolic diagram of human/flora/fauna interaction cycles. P Garza, A Ramos and F Rosas; digital/hand-drawings: A Ramos. Top right: Metabolic schemes of human dwelling in the matorral: water cycle and floral-faunistic cycle. P Garza, A Ramos and F. Rosas; digital/hand-drawings: A Ramos. Bottom Left – model of the human dwelling plugged into a fracture of the matorral. Bottom Right – Fundamental acts of human life. Model: P Garza, A Ramos and F Rosas; watercolours: A Ramos.

regime. A set of intentional perturbations is then introduced as modifiers of the characteristic rules of the algorithm. The perturbations introduced into the virtual model of the urban regime, in particular, favour the generation, expansion or further articulation of urban voids, the way perturbations puncture an ecotope. The emergence of urban voids initiates cascade effects of a successional sort over the surrounding built and social fabric. The abstract notion of urban perturbation in the model is conjugated according to real-life agencies such as direct public interventions, public policy enactments and public-private consortia of transformation. The iterations are eventually repeated, including perturbations and alternative scenarios of urban evolution cast under the dynamics of an altered urban regime. A set of parameters provides a virtual twin interface of the metropolitan ensemble to toggle across the interval between the projected urban scenarios relative to present and altered regimes (Figure 32.2).

Conclusions

Regarding pristine nature on Earth as a system in a permanent balance is deceptive. Nature has been characterised by constant change through smooth transitions of millennial evolutionary phases or abrupt caesurae following catastrophic events. It is indeed not only in the interest of the unique species that compose the irreplaceable beauty of the biosphere but also of the human society to maintain the evolution of nature along a harmonic trajectory. The ecologic crisis that is now threatening the planet's biodiversity and our civilisation is associated with the enormous proliferation of the human apparatus within the planetary machinery. From an anthropocentric perspective, the ecologic crisis is also coupled with a humanitarian emergency. Some assert that the world has become one city (Sarkis, 2011), but the divide between the privileged and the unprotected has grown incrementally crude under geopolitical schemes expanded to the planetary scale. At the root of both the ecologic crisis and the humanitarian emergency, there lies the logic of a globalised economy of unlimited growth. Our response today is bound to an ecology of the planet and an ecology of the community.

In brief synthesis, the fundamental aim of the work in our labs is that of reimagining the figure of the designer of spaces, be they architect, landscape architect or planner. We regard the landscape continuum that is expanding over the entire planet simultaneously as form and process, spatial pattern and functioning metabolism, *paysage* and environment. However, at a point of discontinuity of our civilisation, we endeavour to project our focus beyond the aesthetic performance and the utilitarian efficiency of a spatial organisation to scrutinise their interactions. By focusing primarily on the reciprocal correspondences of these terms, the labs favour a broader vision of the surrounding reality, based on notions of harmonic dynamics and mutuality among the natural and human agencies.

FIGURE 32.2 Top left: Selected metropolitan ensemble: classification of urban conditions exhibiting potential to accommodate new urban voids. Bottom right: Cellular automata model: algorithm governing the iterations over the selected metropolitan ensemble. Top right: Superimposition of iterations through input perturbations: emergence of virtuous voids in blue. Bottom right: Detail of urban assemblage: successional effects of void emergence over surrounding fabric. C Huerta, A Garrido and D Franco.

Notes

1 The IPBES is an independent intergovernmental body established by states to coordinate efforts on the assessment of the state of nature in relationship to humankind.
2 The Millennium Ecosystem Assessment of 2005 is a foundational characterisation of the state of natural ecosystems and biodiversity on Earth, consequently determining urgent targets to recalibrate humankind/nature equilibrium. Unaccomplished ambitions have been further articulated in the Strategic Plan for Biodiversity 2011–2020 and the Aichi Biodiversity Targets of the Convention on Biological Diversity of 2010 and following sessions until 2018. The 2015 UN Sustainable Development Summit has adopted an overall reprogramming of the targets as the 2030 Agenda for Sustainable Development. From the overall picture resulting from the report, the Aichi Biodiversity Targets have mainly been missed. In consideration of the poor progresses achieved, even the reprogrammed 2030 Agenda appears today highly implausible, unless ground-breaking changes transform the economic, social, political and technological status quo.
3 The lack of appropriate public and private space, material deprivation of elemental facilities, social malaise, contextual insecurity, civic marginalisation in which individuals are struggling to survive with inventiveness, have been forged into the promise of a colourful, exotic life.
4 An initiative by Richard Weller, the atlas unfolds over the notions of biome, ecoregion, anthrome, measures of land, water and landcover degradation, biological diversity, urbanisation dynamics, climate change effects, conservation targets and environmental performances.
5 The biological hotspots, first defined by Norman Myers and Russell Mittermeier and focus of the activities of the Critical Ecosystem Partnership Fund, play a relevant role in the narrative of the atlas.
6 The project by Patricio Garza Zambrano, Andrea Ramos Gándara and Fernanda Rosas Velasco was awarded the mention of excellency 2016 at the University of Monterrey. The work on the matorral was inspirational for a research project conducted during the following two years under the title of *Symbiotic Matorral* involving schoolchildren from primary schools and supported by the University of Monterrey and the Parque Ecológico Chipinque (Pasini, 2019).
7 The project by Carlos Huerta Fernández, Ana Garrido Chávez and Daniel Franco Pérez was awarded the mention of excellency 2020 at the University of Monterrey.

References

Baccini, P., & Brunner, P. (2012). *The metabolism of the anthroposphere*. The MIT Press.
Collinge, S. K. (2009). *Ecology of fragmented landscapes*. Johns Hopkins University Press.
Ellis, E. (2014). Ecologies of the anthropocene. Global upscaling of the socio-ecological infrastructures. In Ibañez, D., & Katsikis, N. (Eds.), *Grounding metabolism: New geographies no. 6* (pp. 20–27). Harvard Graduate School of Design.
Ellis, E. (2018). *Anthropocene: A very short introduction*. Oxford University Press.
Farina, A. (2009). *Ecology, cognition and landscape*. Springer.
IPBES. (2019). *The global assessment report on biodiversity and ecosystem services. Summary for policymakers*. IPBES Secretariat.
Latouche, S. (2009). *Farewell to growth*. D. Macey (trans.). Polity Press.
Magnaghi, A. (2005). *The urban village: A charter for democracy and local self-sustainable development*. Zed Books.

Mason, R. (2015). The sixth mass extinction and chemicals in the environment: Our environmental deficit is now beyond nature's ability to regenerate. *Journal of Biological Physics and Chemistry, 15*(3), 160–176.

Raup, D. M., & Sepkoski, J. J. (1982). Mass extinctions in the marine fossil record. *Science, 215*(4539), 1501–1503.

Sarkis, H. (2011). The world according to architecture: Beyond cosmopolis. In Jazairy, E. H. (Ed.), *New geographies n. 4: Scales of the earth* (pp. 104–108). Harvard Graduate School of Design.

Steffen, W., et al. (2004). *Global change and the Earth's system: A planet under pressure*. Springer, 81–141.

33

FROM "GUTTER TO GULF" TO THE 'GLADES

A Decade of Urban Landscape Climate Resilience Studios at the University of Toronto 2008–2018

Fadi Masoud, Elise Shelley and Jane Wolff

Introduction

Addressing the climate crisis lies at the core of today's landscape architectural studio pedagogy. This agenda is central – and relatively new. As recently as the mid-2000s, questions of adaptation and resilience (and the vocabulary, discourse and definitions that underpin current studio agendas) were marginal at institutional, professional and disciplinary levels. Hurricanes Katrina (2005) and Sandy (2012) transformed the ways North American designers understood and dealt with the climate crisis. They showed many design educators that a region's adaptability and resilience were fundamentally linked to its landscape, its broader environmental systems, and its urban development and governance structures. Since then, these issues have evolved and expanded to encompass questions of land policy, civic engagement, social justice and racial equity. Increasing awareness of the climate emergency gave rise to (and was informed by) high-profile interdisciplinary competitions and exhibitions that aimed to address climate change through design, including Waterfront Toronto's Lower Don Lands competition (2006), MoMA's Rising Currents exhibition (2010), Rebuild by Design (2014) and Resilient by Design (2018). At the same time, climate-related frameworks came to the fore in professional design education. Climate change has become a principal topic in graduate programmes in landscape architecture around the world; the participation of more than 70 universities in the Landscape Architecture Foundation's (LAF) global Green New Deal Super-Studio demonstrates the ubiquity of climate concerns.

In this chapter, we look back at a decade of teaching about landscape-based urban climate adaptation and resilience at the University of Toronto. Bracketed by Hurricane Katrina (2005) and Hurricane Irma (2017), this time period has involved two generations of teachers and more than 200 graduate students of

DOI: 10.4324/9781003145905-39

landscape architecture and planning. It included two core studios, one focused on New Orleans, Louisiana, and the other on Southeast Florida, and in particular, Broward County. Each of these studio initiatives brought student work to the attention of local communities and influenced public discussion: studio research and design on New Orleans was shared through the website *Gutter to Gulf*,[1] and studio research and design on Broward County was published in *Coding Flux*.[2] Both studios also produced award-winning student work[3] and prompted academic conference papers on teaching methods.[4]

Despite increasingly visible threats, coastal communities around the world continue to grow faster than inland zones.[5] Estimates indicate that sea level rise may force 13 million Americans, many of them in the southern states of the United States, to move by the end of this century.[6] Nevertheless, urban growth and development in and beyond coastal zones carry forward because of interrelated, outdated tendencies. One is the heavy reliance on fixed structures for engineering landscapes, and the other is the adherence to static regulatory planning and zoning protocols. Both exacerbate climate vulnerability and social inequity. Our University of Toronto studios on New Orleans and Broward County questioned this status quo. The sequence that produced *Gutter to Gulf* contended that rigid infrastructure would never solve the dilemmas of the dynamic Mississippi River Delta, and the sequence that produced *Coding Flux* advocated for a flexible land use code to adapt to the changing landscapes of southern Florida.

This chapter has been an invitation to revisit course materials and publications that emerged from the studios. We have re-examined our work with four topics in mind: *ecology*, which we define through the design of landscapes that consider human systems in active relationship to more-than-human and abiotic systems; *dynamism and flux*, defined through the understanding that change is a given; *coding*, meaning the development of prescriptions for action under specified circumstances, or the choreography of design tactics and strategies; and *materials composition*, meaning the movement and recombination of materials to instigate, shape and take advantage of landscape processes. Over ten years, student work provided evolving operational definitions of each of these subjects. As we reflected on their meaning, we asked ourselves about changes in our teaching – and changes in the context of our teaching. The chapter begins with a timeline that situates the studios in a decade when catastrophe brought the climate emergency into widespread public consciousness. It concludes with an examination of student projects that put forward *in situ* studies of what ecology, flux, coding and materials composition mean in landscapes affected by climate change.

Resilience Studios at the University of Toronto: A Timeline

This timeline situates our work in the context of climate-related catastrophes (noted in **bold type**) and high-profile resilience policy and design work (noted in *italic type*).

2005: Hurricane Katrina devastated New Orleans and the Gulf Coast of Louisiana. Weeks later, Hurricane Rita hit the same areas, plus the Florida coast.

2006: Waterfront Toronto, a public agency charged with revitalising the city's Lake Ontario shoreline, sponsored an international design competition for the Lower Don Lands and brought attention to flooding hazards at the mouth of the Don River. Michael Van Valkenburgh Associates was awarded the commission, as landscape architecture-based strategies for flood adaptation and urban design were central to all the finalists' entries.

2008: Hurricane Gustav wrought significant damage on Louisiana's Gulf Coast.

2008–2010: Jane Wolff and Elise Shelley began to use New Orleans as the site of a required fourth-term studio about urban landscape systems. The goal was "to understand how landscape processes can be a means to examine and reshape cities, rhetorically and practically; to relate the historical, cultural and ecological evolution of urban landscapes".[7] The studio treated New Orleans as a case study, a vivid demonstration of dilemmas that existed in Toronto but were harder to register. Fadi Masoud was one of their students.

From the start, the endeavour offered a critique of current teaching practices, which often focused on end-state solutions for predictable conditions. Instead, it "emphasised the need to confront and address problems of ecological integrity, public space and aesthetic quality under contemporary conditions of extreme uncertainty".[8] Its methods emphasised the building of meaningful relationships to place and community. Wolff had been working with grassroots organisations on rehabilitation strategies for the post-Katrina landscape since early 2007; Shelley had deep family ties to southern Louisiana.

Working collaboratively with water activists and experts in New Orleans, and with a studio group led by Derek Hoeferlin at Washington University in Saint Louis, students carried out research and fieldwork to investigate a deceptively simple question: how did water travel from any given gutter to the Gulf of Mexico? Their designs began in critiques of the status quo and proposed strategies that could simultaneously mitigate risk and create beautiful public landscapes. The studio's first two iterations produced a taxonomy of New Orleans's water infrastructure and drainage systems. Because it filled large gaps in public information, the studio's work supported (and was supported by) the *Dutch Dialogues* initiative, which brought together water managers from the Netherlands and North America to consider the future in New Orleans (Figures 33.1 and 33.2).

2010: MoMA's Rising Currents *exhibition sought "to jump-start a dialogue on the urgency of climate change and rising sea levels among public officials, policy-makers and the general public".*[9]

FIGURE 33.1   Aquaculture Canal – Many industrial waterways in New Orleans were abandoned as active port activity moved out of the city. This resulted in hazardous and contaminated sites, especially vulnerable to storm surges due to their size and location. This project uses aquaculture as an attraction and amenity to return economic value to an obsolete industrial channel. By closing the Industrial Canal with productive infrastructure, the threat of storm surge is eliminated, and the dimensions of the new intervention invite recreation. Catfish, crawfish and rice are grown and harvested with downstream wetland vegetation absorbing nutrients and releasing naturally treated water back into the Mississippi River (2009). Fadi Masoud.

2010–2011: Building on the two previous years' investigation of the New Orleans water system, and proposals for how water might transform the everyday landscapes of the city, the studio's third iteration concentrated on transforming water infrastructure from urban obstacle to urban catalyst. The studio's work took on an expanded public presence when Wolff, Shelley and Hoeferlin launched the website *Gutter to Gulf,* which provided technically rigorous information about the city's hydrology and hydraulics in broadly accessible terms to "address community-based watershed awareness, protection and management questions in the city of New Orleans".[10] A demonstration of the power of the academic labour pool to address questions beyond the conventional scope of design commissions, it became an important resource for designers, planners, advocates and citizens.

2011–2012: The studio redefined the scale and context of urban landscape design work. It considered the city as part of a region and examined the gradient of metropolitan landscape systems from New Orleans's Lower Ninth Ward to the Gulf of Mexico.[11] Work on *Gutter to Gulf* continued. The website became an important resource for developing the Water Management Strategy for Metropolitan New Orleans, and Jane Wolff was asked to serve as a technical adviser to the project.

2012: Hurricane Sandy wrought havoc in New York and along the eastern seaboard of the United States and brought national attention to the climate emergency. Hurricane Isaac pounded Louisiana's coast again.

FIGURE 33.2 The Drainage Filter: This project mobilised two critical issues in South Florida – remediation of the sensitive Everglades ecosystem and the development pressure that threatens it – to create a land management strategy that is mutually beneficial and transformative. This project uses suburban and exurban community development to help restore the Everglades. By rethinking the grading and vegetation on open and available land the students envision a network of channels that integrate water into developed areas to lengthen stormwater flow and allow for natural filtration, engaging topography and phytoremediation to make functional and aesthetic landscapes. Unused and seemingly abandoned lands are connected as a living, breathing landscape filter, improving water quality, ecosystem value and the standard of living for the community (2017). Right: Qiwei Song, Meikang Li and Chaoyi Cui.

2012–2013: Sandy had demonstrated that sea level rise, climate change and ageing infrastructure present design challenges for cities all over North America. Using New Orleans as a datum and point of departure, the studio used research and design methods developed in previous iterations to consider a wide range of places subject to sea level rise, from Prince Edward Island and New Brunswick to

Philadelphia and Charleston. Students investigated the hydrological infrastructure of different cities at risk, developed a shared taxonomy of common dilemmas, likely sites for design and useful precedents, and brought those examples to bear on a design project in New Orleans.

2013: Rebuild by Design – Rebuild by Design began as a design competition, launched by the US Department of Housing and Urban Development (HUD) in partnership with non-profits and the philanthropic sector, in response to Hurricane Sandy's devastating impact on the eastern US. Driven by innovation and collaboration, Rebuild by Design's Hurricane Sandy Design Competition became a model to help governments create research-based, collaborative processes that prepare communities and regions for future challenges. Rebuild by Design invited a mix of sectors – including government, business, non-profit and community organisations – to gain a better understanding of how overlapping environmental and human-made vulnerabilities leave cities and regions at risk.[12]

2013: Catastrophic flooding on the Don River floods Toronto's downtown. (Figures 33.3 and 33.4)

2013–2014: Joined by Nina-Marie Lister, associate professor at Ryerson University's School of Planning, Wolff and Shelley shifted gear to investigate Toronto's hydrological infrastructure, and to determine how lessons learned in New Orleans could help inform resilient transformations of Toronto's urban systems. As in its previous iterations, the studio examined urban and metropolitan landscapes as a series of integrated systems at multiple scales of space and time. Students were required to ground design proposals in substantial, rigorous research on environmental challenges, new technologies and policy frameworks. After six years, the studio had produced a clear set of methods for design in conditions of uncertainty. It emphasised the close relationship between rigorous documentary work and defensible proposals. Proposals had to meet criteria for hydrological performance, community value, political feasibility and aesthetic quality: targets for resilience given reasonable scenarios for everyday and extreme environmental demands at the moment and according to predictions for climate change in the coming decades".[13] The design of specific proposals for particular sites had to be able to adapt to change and reveal environmental dynamics. The studio also put forward the idea of design as the scripting of tactics, strategies and scenarios to guide interaction around the more open development of sites and situations.[14]

A decade before it became mainstream, the University of Toronto studio considered the increasingly complex urban problems presented by climate change as an urgent task for landscape architectural education. It played a pivotal role in the core studio sequence, and its disciplinary questions and methods lay at the foundation of a new generation of studios Fadi Masoud taught at the University of Toronto and MIT.

2013: Masoud's University of Toronto option studio, Nissological Codes, taught with Fionn Byrne,[15] became the first of a studio sequence called Coding Flux. Nissology is the study of islands, and the course explored the Florida Keys

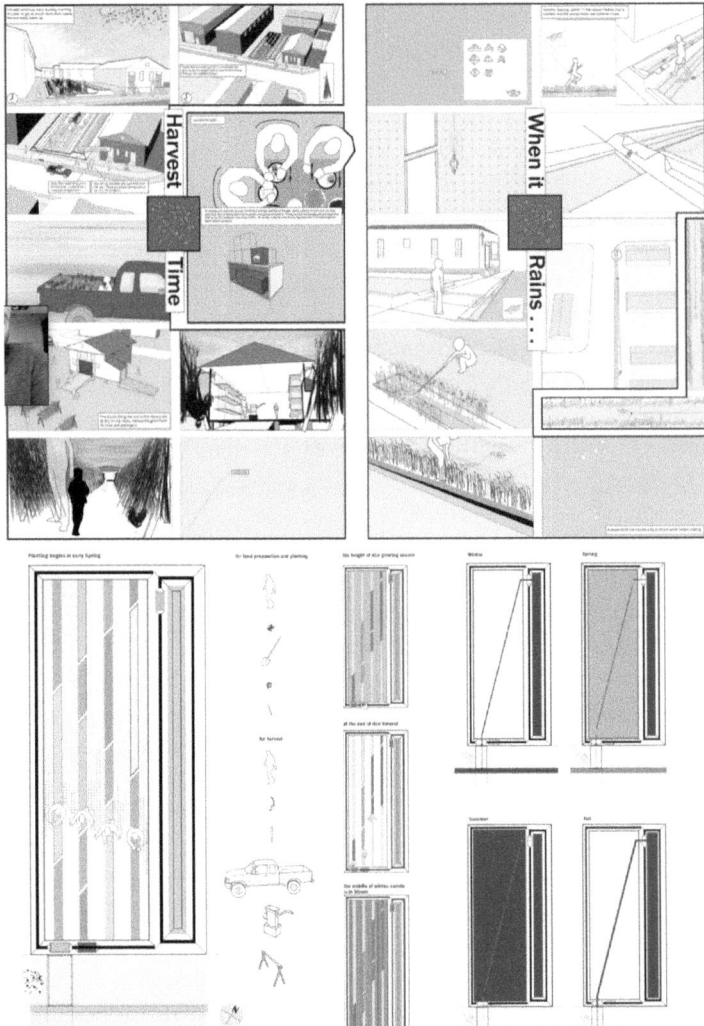

FIGURE 33.3   Left: Rice Farm – The New Orleans neighbourhoods that occupy the lowest ground suffer from both routine flooding and food insecurity. The long-standing history of inundation prevents economic investment. This proposal introduces small-scale cooperative agriculture at the community scale as a means to store and convey seasonal water levels while also creating economic opportunity. Rainfall is collected and conveyed within streetscapes to rice paddies constructed on city-owned properties. The scale of production is easily managed locally and makes the cycles of planting and harvesting a source of pride and community celebration. The productive landscape engages residents in the process, creates a seasonal register and produces enough rice to generate profit for the neighbourhood (2010). Left: Adam Bobbette and Karen May.

FIGURE 33.4 The Via Network: Saltwater intrusion and tide-related sunny-day floods have corroded the aging sub-surface sewer infrastructure of south-east Florida. This project sees plans for the municipal replacement of this system as an opportunity for a more resilient and flexible strategy in the face of unpredictable hydrological events. This proposal replaces roads with an alternative network of vehicular and hydrological rights of way throughout the city – the "Via" system, a hybrid between a street and a canal – providing space for water during times of inundation, increasing hydrologic capacity while maintaining circulation routes and creating green amenity space for social and recreational activity (2017). Right: Marianne Lafontaine-Chicha and Niloufar Makaremi.

to test the limits of urban planning codes. The studio promoted landscape strategies for addressing climate threats and expanded the agenda through a critical examination of planning policy tools. It used landscape architecture as a framework to reconsider the legislated spatial planning tools that shape climate change responses.[16]

Like New Orleans, South Florida exists in a state of constant hydrological flux. Its low-lying cities have water on either side – the Everglades to the west and the Atlantic Ocean to the east – and the Biscayne Aquifer lies below its porous limestone surface. At present, maintaining the urbanised area (6,137 square miles or 15,890 square kilometres) in its dynamic estuarine environment requires 35 miles (56 kilometres) of levees and 2,000 hydraulic pumping stations. One of North America's fastest-growing regions, South Florida faces extreme threat from climate change. Projections indicate that property worth US$152 billion will be below sea level by 2050.[17]

2015: Working with Alan Berger and Adele Naudé Santos, Masoud coordinated the first of two Coding Flux urban design studios at MIT's School of Architecture and Planning. With Broward and Palm Beach counties as case studies, the studios asked planning and design students to recognise that ground conditions shift constantly along the continuum from wet to dry. This flux was a prompt to develop innovative design proposals, typologies and urban codes based on hydrological shifts and gradients. The goal was to plan for environmental indeterminacy (Figures 33.5 and 33.6).

2016: Hurricane Matthew, a Category 5 hurricane, threatened the Floridan coastline. The region's first major hurricane in 11 years, the storm cost billions of dollars in damages.

2016: Masoud taught the second MIT Coding Flux studio in collaboration with Miho Mazereeuw. On a field trip following Hurricane Matthew, students witnessed a king tide and its associated storm and sunny-day floods.

2017: Hurricane Harvey, a Category 4 hurricane, struck Louisiana and Texas and produced record rainfall and massive flooding in Houston. The damage made it one of the costliest natural disasters ever recorded.

2017: Hurricane Irma, a Category 4 hurricane, struck Florida. The third strongest Atlantic hurricane ever recorded at landfall, it produced a 10-foot (3-metre) surge in the south-eastern part of the state, and caused 21 deaths in Broward County.

2017: Masoud and Elise Shelley taught the final instalment of the Coding Flux studio sequence at the University of Toronto. In continued partnership with Broward County's Environmental Protection and Growth Management

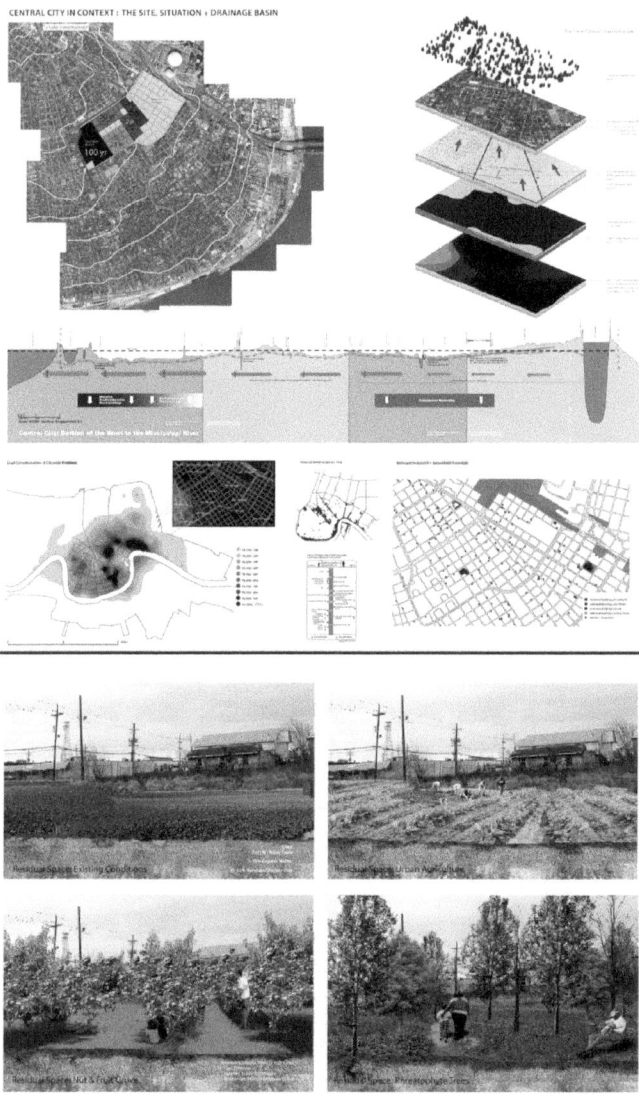

FIGURE 33.5 Left: Decontamination – Many neighbourhoods in New Orleans and throughout the United States suffer from toxic levels of lead contamination in the soil. Soil contamination was made worse through the flooding caused by Hurricanes Katrina and Rita. This poses a great threat for children who live in these neighbourhoods. This project introduces phytoremediation to ameliorate lead and other contaminants found in the soil. A phased strategy is presented for remediating soil in this central city neighbourhood so that clean soil can then be redeployed for urban agriculture, redevelopment sites and public space (2009). Left: Laurel Christie and Anson Main. Right: Carlos Portillio and Aidan Loweth.

FIGURE 33.6 The Other Coast: This project challenges the existing condition of raised land within the Everglades being dedicated to singular infrastructural functions of levees and arterial roadways. This proposal posits that this raised land becomes the primary site for future urban development, rather than the sensitive and challenging low ground on either side. A Kit-of-Parts assembly choreographs a future condition of settlement and density, through tactics that transforms the inherently safer higher circulation routes into linear, mixed-use urban communities (2017). Right: Carlos Portillio and Aidan Loweth.

Division, the studio included an animated mapping exercise, research on urban zoning codes and precedents, and an urban-scale design problem. Students projected animated analytical maps of dynamic processes (water table fluctuations, tides, wind patterns, ocean currents and solar exposure) onto physical models of the region. Students prepared design proposals that took these variable systems into account. Through analysis of their proposals, using the lenses of existing local planning polices and land use zoning regulations, they discovered their projects did not conform to legal standards. The exercise revealed that current codes perpetuated risk and vulnerability.[18]

In 2018, the Coding Flux studio was awarded the Studio Prize and Sloan Award by *Architect Magazine*. The jury selected six exemplary studios from across the US and Canada. The studio was also awarded the Sloan Award which recognises sustainable water conservation and management. In 2019, Drainage Filter, a project by Meikang Li, Qiwei Song and Chaoyi Cui, was awarded a World Landscape Architecture Award from a shortlist of eight projects in the student category. The same project was also awarded the 2020 National Urban Design Award Certificate of Merit, jointly administered by the Royal Architectural Institute of Canada, the Canadian Institute of Planners and the Canadian Society of Landscape Architects.

2018: Resilient by Design was launched as a design challenge, modelled after Rebuild by Design, to address the need for greater regional climate adaptation collaboration and innovation in the San Francisco Bay Area.

2019: Launch of Flux.Land, an interactive digital platform designed to increase awareness about climate risk and resilience in Broward County. The platform is a collaboration between the MIT Urban Risk Lab and the University of Toronto's Centre for Landscape Research. It collects and combines eco-morphological, spatial and climate data for cross-departmental decision-making, participatory planning, education and civic engagement. This tool helps the county identify Climate Action Zones – areas of higher risk and adaptation opportunities. Flux.Land also provides recommendations and actionable insights for its users in the form of adaptation design strategies. The first phase of Flux.Land was presented in 2017 at the Southeast Florida Climate Leadership Summit. In the autumn of 2020, the platform was used as part of a climate educational module in the Broward County District School Board.

2020: The 2020 Atlantic hurricane season was the most active, and the fifth costliest Atlantic hurricane season on record. Exhausting the Latin alphabet, 2020's hurricane season used the Greek alphabet for only the second time ever. (Figures 33.7 and 33.8)

Cross-Cutting Project Themes

Over the decade of our resilience studio teaching, student projects made use of consistent strategies to serve similar goals. In their work, students addressed ecological connections across local and regional scales. They drew on the specific, nuanced characteristics of their sites to develop programmes. They developed flexible spaces to allow for uncertain conditions in the future. They integrated infrastructure across boundaries of land and water. They worked synthetically to create spaces that had both utilitarian and civic value. They collaborated with local experts, agencies and citizens. The goals across the board were to create

FIGURE 33.7 Left: Crevasse – Levees that have been built along the Mississippi River to protect from flooding also prevent the natural deposition of sediment on adjacent land. This results in the degradation of wetlands and exacerbates subsidence. This proposal creates a crevasse – or a controlled diversion – that brings water from the Industrial Canal into Bayou Bienvenue through a meandering channel. The speed the water travels encourages the deposition of sediment, that can then be harvested and used to replenish soil in low-lying areas elsewhere in the city. The crevasse is productive and also protective – the introduction of fresh water into the bayou helps ameliorate saltwater intrusion and wetland degradation (2010). Left: Justin Miron.

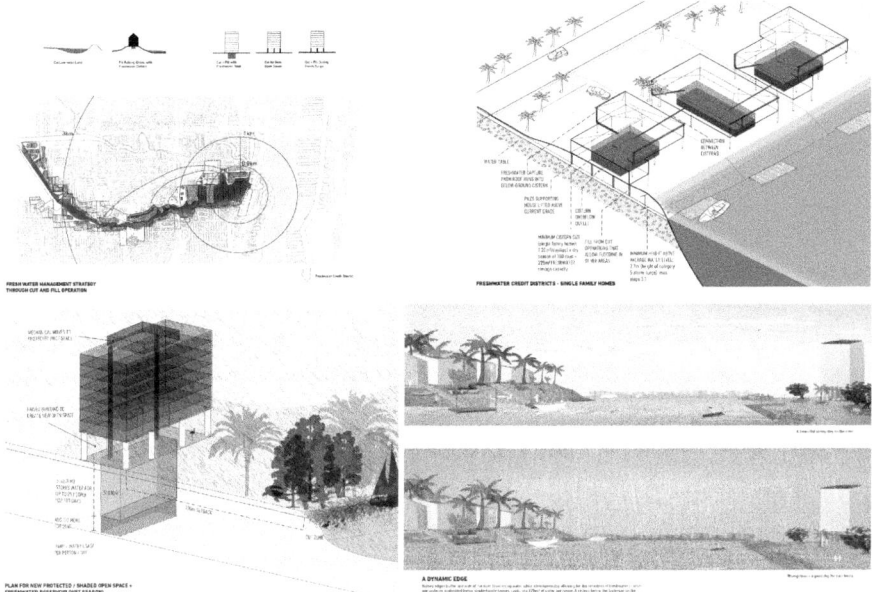

FIGURE 33.8 Buffer + Fill – A Freshwater Credit District: South Florida freshwater aquifers are threatened by saltwater intrusion and this concern increases with seasonal freshwater scarcity. This project redirects water from high-risk residential areas that flood into open spaces that can act as flood plains. Cut-and-fill techniques are employed to protect private property from flooding while creating a network of residential-scaled freshwater storage basins. This neighbourhood infrastructure manages, dispenses, stores and uses the fresh water as a resource – turning the hazard of too much rain from increasingly frequent storms into a benefit for the community (2017). Right: Leslie Norris and Natalie Schiabel.

landscapes that could fail without catastrophe, and to support and create places grounded in community. In selecting projects for this chapter, we curated pairs of projects from 2009–10 and 2017–18, under the following categories: *ecology, dynamism and flux, coding,* and *composite materials.*

In the category of *ecology,* defined by the relationship of living and non-living systems in the landscape, we see the designs of two places that allow human systems to engage and coexist with abiotic systems. In *Aquaculture Canal,* a project completed by Fadi Masoud (2009), New Orleans's defunct Industrial Canal was turned into a fish farm, water storage basins and recreational trails, transforming a single-purpose infrastructure into an economic generator as well as a recreational amenity. In *Drainage Filter,* by Qiwei Song, Meikang Li and Chaoyi Cui (2017), the proposal strategises how suburban and exurban neighbourhoods near the Everglades can restore water quality and habitat function using residential and agrarian swales, ditches and yards.

The category of *dynamism and flux* assumes change as a given condition and embraces cyclical variability. *Rice Farm*, by Adam Bobbette and Karen May (2010), was a project that calibrated seasonal and annual cycles of flooding, farming and socialising. The *Via Network*, by Marianne Lafontaine-Chicha and Niloufar Makaremi (2017), transformed roads into hybrid networks of park-road-canals for storage of high water.

The category of *coding* is understood as the development of design tactics and strategies, choreographed and prescribed to address a given set of circumstances. In *Decontamination*, by Laurel Christie and Anson Main (2009), soil contamination was addressed with a prescriptive neighbourhood strategy for phytoremediation, urban agriculture and public space. In *The Other Coast*, by Carlos Portillio and Aidan Loweth (2017), a kit-of-parts assembly is established to facilitate urban development on elevated levee infrastructure.

The last category, *materials composition*, features projects where the movement and redistribution of materials instigates, shapes and positively exploits landscape processes. Both these projects use strategies that move earth to manage water. *Crevasse*, by Justin Miron (2010), initiated tactical flooding to build land in the Mississippi River Delta. *Buffer and Fill*, by Leslie Norris and Natalie Schiabel (2017), used cut and fill to protect inhabitable land and store fresh water within a residential neighbourhood.

The project pairings demonstrate parallels despite different locations, scales and time periods. Their differences represent an evolution of thinking over the years.

Conclusions

Over 15 years post-Katrina, it has become clear that New Orleans was the harbinger of change for climate-based design education. The ten studios of *Gutter to Gulf* and *Coding Flux* demonstrate that working in extreme environmental circumstances provides an operational window to reflect on local issues. The opportunities for knowledge mobilisation and transfer are immense. Students gained skills, tools, resources and perspectives that would have been impossible in their institution's hometowns. In both studio courses, site visits were a necessary form of site documentation and relationship building. Each year, students travelled to New Orleans and South Florida to experience the complexities and richness of those regions. Field trips included visits to numerous neighbourhoods, pumping stations, levees, bayous, marshes, mangroves and degraded wetlands. Students also witnessed first-hand the impacts of climate change on the urban landscape – from tidal sunny-day floods, to saltwater intrusion, to the limits and failures of drainage systems. The visceral and personal connections students made on the trips amplified the projects' successes.

The documentation and critique of existing conditions became a way of defining design problems and proposals in which the technical problem of "flooding" was always situated within a sociocultural context. Both studios argued for the importance of resilience: the "rate at which a system returns to a reference state or dynamic after a perturbation"[19] in the design of landscapes. Both studios asserted that living in coastal regions required embracing the inherent fluctuations of that landscape. Both studios analysed existing practices and policies and concluded they were not adequate in addressing the challenges posed by the climate crisis.

Unlike professional consultants, the academic studio platform enabled design responses that were varied and experimental. Landscape architecture students envisioned climate-resilient projects at the intersection of urban design, public open space, environmental restoration, public health, social wellbeing and landscape infrastructure provision. They considered the ecological, topographic or geomorphological conditions that impact an area's adaptive capacity, instead of its political or administrative boundaries.

A decade of landscape-based climate resilience studios at the University of Toronto point to a trajectory of pedagogical ambitions that paralleled the climate adaptation and resilience discourse. Looking towards the next decade of studios, educators will need to foreground design pedagogy that addresses climate resilience's intersection with issues of social equity and power. These questions are especially pertinent given the conundrum of climate-induced gentrification that could displace established low-income populations into areas of unseen future risk. Design educators will also need to anticipate the subordination of equity to investments in climate-resilient green infrastructure at the expense of other social programmes. This would require us to conceive of design studios at the nexus of social and environmental needs that operate in an era of continued and heightened uncertainty.

Notes

1 Wolff, Jane, Elise Shelley, and Derek Hoeferlin, *Gutter to Gulf: Legible Water Infrastructure for New Orleans*, 2021, John H. Daniels Faculty of Architecture, Landscape, and Design, University of Toronto and Sam Fox School of Design & Visual Arts, Washington University in St. Louis, 2021, http://www.guttertogulf .com.

2 Masoud, Fadi and Elise Shelly (eds.), *Coding Flux: Seeking Resilient Urbanism in South Florida*, Daniels Faculty Studio Publication, San Francisco, https://www.amazon .com/Coding-Flux-Elise-Shelley-Masoud/dp/1388035677.

3 Risen, Clay, "Coding Flux: In Pursuit of Resilient Urbanism in South Florida," *Architect Magazine*, September 13, 2018, https://www.architectmagazine.com/ awards/studio-prize/sloan-award-coding-flux-in-pursuit-of-resilient-urbanism-in -south-florida_o. Bobbette, Adam and Karen May, "I Am a Second Responder," American Collegiate Schools of Architecture (ACSA) Competition, Spring 2011, https://guttertogulf.wordpress.com/2011/05/16/acsa-i-am-a-second-responder -competition/.

4 Shelley, Elise and Jane Wolff, "Landscape Studio Teaching for Cities in Flux," in the European Council of Landscape Architecture Schools (ECLAS) "Landscapes

in flux," September 2015. Published in ECLAS conference proceedings. Masoud, Fadi, "Coding Flux," in *2019 Annual Conference of the Council of Educators in Landscape Architecture Conference Proceedings*, Sacramento, CA, March 6 – 9, 2019, p. 26.

5 Seto, Karen C., Michail Fragkias, Burak Güneralp, and Michael K. Reilly, "A Meta-Analysis of Global Urban Land Expansion," *PLOS ONE* 6, no. 8 (August 18, 2011): e23777, https://doi.org/10.1371/journal.pone.0023777.

6 Hauer, Matthew, "Migration Induced by Sea-Level Rise Could Reshape the US Population Landscape," *Nature Climate Change* 7, no. 5 (April 2017): 321–325, https://doi.org/10.1038/nclimate3271.

7 "Course Objectives," LAN1014 course outline, 2009.

8 Shelley, Elise and Jane Wolff, *op. cit.*, p. 3.

9 https://www.moma.org/explore/inside_out/2010/11/01/rising-currents-looking-back-and-next-steps/.

10 Shelley, Elise and Jane Wolff, *op. cit.*, p. 1.

11 Ibid., p. 5.

12 http://www.rebuildbydesign.org/about.

13 Shelley, Elise and Jane Wolff, *op. cit.*, p. 1.

14 Ibid., p. 5.

15 Byrne, now an assistant professor of landscape architecture at the University of British Columbia, was a student in the first *Gutter to Gulf* studio in 2009.

16 Masoud, Fadi and Elise Shelley, *Coding Flux: In Pursuit of Resilient Urbanism in South Florida*, John H. Daniels Faculty of Architecture, Landscape, and Design, University of Toronto. Toronto, ON, 2017, p. 4.

17 Keys, Benjamin J., and Philip Mulder, *Neglected No More: Housing Markets, Mortgage Lending, and Sea Level Rise*, National Bureau of Economic Research, October 12, 2020, https://doi.org/10.3386/w27930.

18 Masoud, Fadi, "Coding Flux," in *2019 Annual Conference of the Council of Educators in Landscape Architecture Conference Proceedings*, Sacramento, CA, March 6 – 9, 2019, p. 26.

19 Pimm, Stuart L., quoted by the Stanford Encyclopedia of Philosophy in "Ecology," accessed May 17, 2013, http://plato.stanford.edu/entries/ecology/.

Tending towards a Matter of (Ethics of Ground)

We require no convincing that the climate crisis is affecting the planet's ecosystems and species, with the most vulnerable being susceptible to a fluctuating series of extreme effects. The crisis, the extreme effects, arise from rapid transformation of the earth's climate caused by increased concentrations of greenhouse gases in the atmosphere.

This is a profound calamity for humankind and the planet. It requires a critical shift in landscape architectural education. We must turn to new ways of developing pedagogical ambitions and designing approaches for engaging students in the discipline.

With a focus on studio teaching, this requires a repositioning of the design project, the designer's role within applied knowledge systems, and their relationships to the landscape. Acknowledgement of the role that Western thinking and design of the landscape has played in causing the crisis is fundamental. This shift means the designer is explicitly enmeshed within, and affects, multiple sets of relationships across a range of operational scales. This shift asks for the assumption of responsibility to the land, to humans and to non-humans.

Similarly, within the studio frame, projects within a teaching and learning framework need to be reconceived. They need to consider multiple stakeholders and communities that move from speculative projects that proceed from a conventionally assumed paradigm of endless growth, development and resource supply. Instead, they must consider a collective act of repair. To achieve this, it is important to move towards models of teaching that emphasises collective forms of contribution.

Contributors to *Studio Ecologies* argue that a business-as-usual approach in design education will be inadequate to face the challenges of the climate crisis; therefore, a deep reconsideration of the role of design and the designer is a required starting point for adapting the curriculum. The design studio work

DOI: 10.4324/9781003145905-40

described in this book's five sections make clear that global and planetary crises have provided the incentive to rethink studio teaching. That rethinking moves away from reinforcing goal-driven problem-solving techniques. Moving to design approaches that formulate novel ideas and concepts to pose alternatives to near future spaces, landscapes, territories, cities and regions, ones that embrace heterogeneity, accept indeterminable change, and redress existing inequalities.

This conclusion takes the form of a series of principles for "*Tending to*" and "*Matters of*" that enable a different approach to those that primarily facilitate the acquisition of skills. These actions create the conditions for the emergence and refinement of *qualities* that are ways to approach design, that allow for engagement with the complexity of the climate crisis. The emphasis on ecological thinking across the book's five sections gives rise to a number of thematic matters of concern, each described as calls to action, that formulate principles for *Tending to*.

The act of *Tending to* describes ways in which these actions may be engendered. As a form of making, *Tending to* implies care, empathy and accountability towards other human and beyond human-centric endeavours. It implies an ongoing, reflective pedagogical practice that signals a shift from modes of production to actions that prioritise cultivation for design studio teaching.

Principles for Tending to …

The Challenge

Tending to Matters of Mutability

The matter of concern is with current governance systems, institutional structures and infrastructural determinism that emphasise rigidity and resistance to change due to the persistence of archaic legal systems and codes that hark back to periods of white settlement when land was considered "terra nullius" – nobody's land. These impediments continually repeat and reinforce injustice. They underscore the need for radically different alternatives to the way we design, for example, the coastal and urban interface, and the way we consciously use design to dismantle social disparities. Stubborn adherence to injustice is demonstrated through heedless plans to reinforce coastal flood walls as a means to combat sea-level rise, and boorish conformity to rigid bylaws that safeguard monocultures and singularity, both ecological and social.

Tending to matters of mutability presupposes that indeterminable change is an ambition, an understanding and an act to be imbued in the facets of teaching and learning.

This is achieved through actions that:

- *Acknowledge* "multi-sited" ways of seeing and knowing the world.
- *Develop* design approaches for seeing, representing and working with material change of the landscape, both through landscape and planetary

dynamics and through the movement of materials, their provenances and associated labours.

- *Generate* material perspectives, narratives and performance as means to evaluate and design for multidimensional understandings of material ecologies.

Tending to Matters of Reciprocity

The matter of concern is imperial knowledge systems that include geology, ecology and biology. Historically, these systems have objectified species, and constricted understanding through flawed measures, inherited and acquired, of traits, fitness and reproduction. The objectification of these knowledge systems does not allow for broader, performative understandings of species and/or systems. A desire to limit ideological differences and knowledge becomes the dominant paradigm, rather than considering cross-pollination and cross-sections across historical lineages. Wrongdoings are repeated when we act on the desire to restore lost ecologies, when we constantly renew landscapes under the guise of maintenance, when we invoke a misguided "quick fix" agenda comprised of technological solutions for transitioning to green energy or for greening urban environments.

Tending to matters of reciprocity asks for models of learning that seek to work across, within and between areas of knowledge. These models are animated through actions that:

- *Evaluate* historical figures and projects in the context of current social norms.[1]
- *Position* history, and theoretical components of studios and the broader curriculum, within contemporary contexts of labour, justice and inequality.
- *Assemble* complexity as a collective endeavour that embeds diverse knowledge systems in the teaching and learning process.

Tending to Matters of Exactitude

The matter of concern is with projected systems of measures – cartographic systems such as geographic information systems – that have constructed a particular view of the world through coordinates and geometric projections. These systems of measure are often driven by prospecting for capital accumulation, extraction or surveillance. They emphasise the demarcation of land and water, resource and tailings, more and less. They exacerbate the desire for infinite growth and development, and consequently intensify waste production, resource extraction and deforestation.

Tending to matters of exactitude requires a reframing of systems and techniques of measure, expression and idea-forming. This is achieved through actions that:

- *Question* the physical reality of the designer in the field as an agent who affects the environment and question the foundations of methods of measurement.

- *Employ* modes of attunement to living and non-living systems.
- *Create* tools of measure and forms of representation that enact worldbuilding. Such tools imbue a deep palimpsest of material performativity through considering the relationship between things that are not scalar but require an ecological shift in ordinance systems.

Tending to Matters of Novel Agents

The matter of concern is the current value systems we impose on non-living systems, such as carbon offsetting that gives licence to continue current modes of living, that avoids accountability for polluting the environment and that induces large-scale ecosystem loss.

Tending to matters of novel agents establishes ways of rendering visible forms of entanglement with, attachment to, and detachment from known, and yet unknown, ecological actors. Such rendering frames a view of the world. Not through networks and connections, but through ecologies and relations. It reveals the productive planetary value of large-scale ecosystems. This is achieved through actions that:

- *Identify* knowledge systems of landscape that are deeply embedded in the processes of being in the field.
- *Utilise* the performative nature of ecology as a means to evaluate and design, and as an alternative to performance metrics which neglect life as an indicator.
- *Devise* alternative value systems that debunk the doctrines of capital and reinforce the *Rights of Nature — that all life on earth has a right to exist, thrive and evolve.*[2]

The Studio Environment

Tending to Matters of Multiplicity

The matter of concern is the studio environment that historically emulates and serves corporate and private practice models that consequently enforce singular design approaches and models of learning.

Tending to matters of multiplicity frames the design act and its forms of engagement as a generative process of variation. This embrace encompasses multiple positions and domains of knowledge. This is achieved through actions that:

- *Construct* collective frameworks that embed diverse knowledge systems and domains in the teaching and learning process.
- *Prescribe* methods that embrace multiple actors, agents and their tendencies. Prescribe models of teaching that emphasise ideas for the "near future", that

project landscapes, cities, territories and regions to embrace complexity, that prioritise rewiring current material flows and processes.

- *Generate* models of curation and learning that encourage diverse approaches, that question what, for whom and to what extent design can provide service.

The Institution

Tending to Matters of Collectivity

The matter of concern is current institutional models that support segregated modes of knowledge production where the design studio isn't the primary focus.

Tending to matters of collectivity argues for a shift from singular domains of knowledge. It cultivates formation of territorial and knowledge commons, cooperative structures and assembling incongruous fields charged with the purpose of change through co-production within the design studio. These goals are achieved through actions that:

- *Frame* practice and academia as distinct undertakings with modes of learning, research and contributions specific to each.
- *Advocate* for forms of assessment that are collective testimonies of the design work being evaluated.
- *Create* plurality of approaches, along with their lineages and subsequent emerging specialisations.

Notes

1 Refer to Jesse Keenan, *Climate Core: A Roadmap for Climate Education in the Built Environment*.
2 Refer to "Rights of Nature" by the Australian Law Alliance.

INDEX